Data Analysis
in Astronomy IV

ETTORE MAJORANA
INTERNATIONAL SCIENCE SERIES
Series Editor:
Antonino Zichichi
European Physical Society
Geneva, Switzerland

(PHYSICAL SCIENCES)

Recent volumes in the series:

Data Analysis in Astronomy IV

Edited by

V. Di Gesù
University of Palermo
Palermo, Italy

L. Scarsi
University of Palermo and
Institute of Cosmic Physics and Informatics/CNR
Palermo, Italy

R. Buccheri
Institute of Cosmic Physics and Informatics/CNR
Palermo, Italy

P. Crane
European Southern Observatory
Garching/Munich, Germany

M. C. Maccarone
Institute of Cosmic Physics and Informatics/CNR
Palermo, Italy

and

H. U. Zimmermann
Max Planck Institute for Extraterrestrial Physics
Garching/Munich, Germany

Springer Science+Business Media, LLC

Library of Congress Cataloging-in-Publication Data

Data analysis in astronomy IV / edited by V. Di Gesù ... [et al.].
        p.   cm. -- (Ettore Majorana international science series.
    Physical sciences ; v. 59)
        "Proceedings of the Fourth International Workshop on Data Analysis
    in Astronomy, held April 12-19, 1991, in Erice, Sicily, Italy"--T.p.
    verso.
        Includes bibliographical references and index.
        ISBN 978-1-4613-6496-2   ISBN 978-1-4615-3388-7 (eBook)
        DOI 10.1007/978-1-4615-3388-7
        1. Astronomy--Data processing--Congresses. 2. Astrophysics--Data
    processing--Congresses.   I. Di Gesù, V.   II. International Workshop
    on Data Analysis in Astronomy (4th : 1991 : Erice, Italy)
    III. Title: Data analysis in astronomy 4.   IV. Series.
    QB51.3.E43D28   1992
    520'.285--dc20   92-10775
                                                               CIP

Proceedings of the Fourth International Workshop on Data Analysis in Astronomy,
held April 12–19, 1991, in Erice, Sicily, Italy

© 1992 Springer Science+Business Media New York
Originally published by Plenum Press, New York in 1992
Softcover reprint of the hardcover 1st edition 1992

PREFACE

In this book are reported the main results presented at the "Fourth International Workshop on Data Analysis in Astronomy", held at the Ettore Majorana Center for Scientific Culture, Erice, Sicily, Italy, on April 12-19, 1991. The Workshop was preceded by three workshops on the same subject held in Erice in 1984, 1986 and 1988.

The first workshop (Erice 1984) was dominated by presentations of "Systems for Data Analysis"; the main systems proposed were MIDAS, AIPS, RIAIP, and SAIA. Methodologies and image analysis topics were also presented with the emphasis on cluster analysis, multivariate analysis, bootstrap methods, time analysis, periodicity, 2D photometry, spectrometry, and data compression. A general presentation on "Parallel Processing" was made which encompassed new architectures, data structures and languages.

The second workshop (Erice 1986) reviewed the "Data Handling Systems" planned for large major satellites and ground experiments (VLA, HST, ROSAT, COMPASS-COMPTEL). Data analysis methods applied to physical interpretation were mainly considered (cluster photometry, astronomical optical data compression, cluster analysis for pulsar light curves, coded aperture imaging). New parallel and vectorial machines were presented (cellular machines, PAPIA-machine, MPP-machine, vector computers in astronomy). Contributions in the field of artificial intelligence and planned applications to astronomy were also considered (expert systems, artificial intelligence in computer vision).

The third workshop (Erice 1988) was dedicated mainly to data analysis methods (chaotic processes in astronomy, search for galaxy chains via cluster analysis, search of bursts with adaptive cluster analysis) for solutions in the new frontiers of astrophysics (γ-astronomy, neutrino-astronomy, gravitational waves, background radiation, and extreme cosmic ray energy spectrum).

The aim of these workshops was to create a cultural bridge between astronomers and computer scientists, and to follow the evolution of data analysis in astronomy during the last decade.

The "Fourth International Workshop on Data Analysis in Astronomy" represents the natural evolution of those held previously. In fact, most of the methods and systems developed in the past are now used on working large scale experiments in astrophysics.

During the workshop 60 talks were presented. In the five sections of this book appear 36 selected papers. The first section is dedicated to the presentation of general lectures in data analysis, archives, and image processing. The other sections provide an overview of large operating experiments in astronomy (Hubble Space Telescope, ROSAT, GRO-COMPTEL, GRO-OSSE, GRO-EGRET) and future space mission (SAX, SPECTRUM-X-JET-X, UEI).

Demonstrations of working data analysis systems on specialized computer machines were presented, compared, and discussed. The organization of these demonstrations was successful due to the computer support available at the Ettore Majorana Center and to the hardware provided by DELPHI-Italy.

The success of the Workshop was the result of the coordinated efforts of a number of people, from the Workshop organizers to those who presented a contribution and/or took part in the discussions. We wish to thank the entire staff of the Ettore Majorana Center for Scientific Culture for their support and invaluable help in arranging the Workshop.

V. Di Gesù

L. Scarsi

R. Buccheri

P. Crane

M.C. Maccarone

H.U. Zimmermann

CONTENTS

SYSTEMS AND ARCHIVES IN ASTRONOMY

THE HST MISSION

THE ROSAT MISSION

THE GRO MISSION

FUTURE MISSIONS

SYSTEMS AND ARCHIVES IN ASTRONOMY

ASTROPHYSICS DATA SYSTEM

Frank Giovane

NASA Headquarters
1600 Independence Ave. SW
Washington, D.C. 20546

INTRODUCTION

The present decade will be an exciting time for astronomers, and one that will be filled with changes in the way they work and do research. The introduction of large and highly sensitive detector arrays in the visible and infrared, the development of several 8-m ground based optical telescopes, and the completion of the Very Large Baseline Array (VLBA) radio telescopes will start this process. A large amount of astrophysical data has already been collected from very productive space missions in the 80's (IRAS, the HEAO series, and IUE). The 1990s, with the National Aeronautics and Space Administration (NASA) Great Observatories, promises to add an immensely greater quantity of data to NASA's holdings. This data will have to be reduced, analyzed, and stored, and will significantly tax the procedures and systems used previously.

Recognizing the need to better organize the scientific data generated by the present and future astrophysics mission, the Science Operations Branch was created as part of the Astrophysics Division of NASA, Headquarters. Some of the charges of this Branch are assuring the proper operations of the space missions and the appropriate distribution and storage of space generated data. Principals in this process are the astronomical operations and data centers whose task is the operation of the spacecraft and assisting the community in observation, reduction, and analysis. At these centers scientists and technical experts work closely with the community to provide a reference point for the space missions. However, since each center is organized about a discipline-specific mission, they tend to operate independently of each other. They provide the research community with independent data services which utilize mission-unique hardware and software, and generally are only accessible from the physical location of the center or by direct communication with the center.

In 1988 it was realized that improved methods of handling and of access to the rapidly increasing amounts of astrophysical data was required. A committee of scientists and systems experts was organized by the Science Operations Branch to evaluate how best to achieve these goals. The results, a "Final Report of the Astrophysics Data System Study", recommended that for future operations both a data set and the human expertise which supports it should be maintained at the same physical location. This is the specific strength of the centers and one that the committee felt should not be disrupted. The committee also recommended that these centers, and their databases, should be physically linked through high speed communications networks, and the various data sets should be accessible through a common set of tools. In order to fulfill this recommendation NASA set up a program called the Astrophysics Data System (ADS).

The Astrophysics Data System is intended to make NASA's current and future data holdings more broadly and efficiently accessible, and make the data itself more interpretable. It is to provide a common information system infrastructure for science analysis which reduces duplication of effort while increasing the scientific return from the various astrophysical missions. To achieve these goals the ADS has been developed as a truly distributed system in which the data and processing tools are physically distributed. The ADS utilizes a server/client architecture which allows services and data to be added or replaced without having to change the basic architecture or interface of the system. It can thus accommodate both anticipated and unanticipated growth, and changes in technology. Further, the ADS is modular and layered, to enable smooth evolution of the hardware and software without interruption of service to the user community.

OPERATIONS

The ADS is presently being developed in a way to initially provide basic essential services, with the development of more sophisticated processes and services undertaken later. The basic ADS consists of a directory of holdings; distributed databases (or host nodes) providing astrophysical data, catalogues, and other services; and a distributed set of users. The directory of holdings provides descriptive material about all the nodes, services, software, and data that the system supports. It holds general information from all the participating nodes. In addition each of the nodes maintains its own copy of this information and other information of specific interest to users of the node. Each node has a specific set of services that it offers. These services are primarily specific to the databases and observation sets, but also may include documentation, data, software and processing functions. The host nodes maintain control of their databases and place on-line those data and services that they choose. The ADS provides the user, with an interface to this distributed set of data and services, and a consistent way of interacting with them.

The system architecture is illustrated in Figure 1. It involves a Character User Interface (CUI), a non-graphic display which provides a menu system form of user connection. It allows efficient access to the ADS with as rudimentary an interface as a dumb terminal. CUI gives the look and feel of the system to the user and thus provides access to all the other services within the ADS. It is linked with the Knowledge Dictionary System (KDS), developed by Ellery Systems Inc, Boulder, CO, with which it exchanges information.

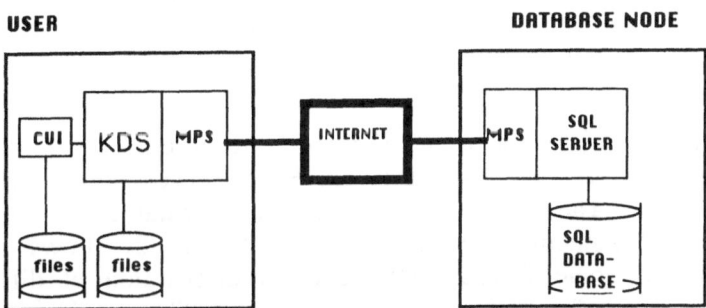

Figure 1. The Astrophysics Data System Components.

This KDS is a commercially supported processing environment from which the rest of the ADS can be accessed. Part of this environment includes the tools for managing the user's own collection of information, local data sets, notes describing current projects, sets of observations, and other private user material. By means of Query By Example (QBE), KDS can help the user develop Structured Query Language (SQL) statements which is the standard language used by ADS to direct queries to the remote databases or observation catalogues. The KDS also controls the users' ADS operations and manages the various transactions required by the user and system. It is linked to the Message Passing Service (MPS).

The MPS utilizes the Advanced Network System Architecture (ANSA), which was developed at Cambridge University and is now distributed by APM of Cambridge, UK. Ellery Systems Inc. provides a modified superset version of ANSA, which includes additional functions for transaction processing and robustness. This modified ANSA allows the MPS to enable remote inter-operability and data transfer/translation among the ADS services via the Internet science network. It provides homogeneity across networks and operating systems for process requests and responses. Thus, the MPS implements the functionality of a Remote Procedure Call (a mechanism by which the subroutines of a program can call each other); Remote Process Invocation (a process by which a program on one server can spawn programs on other servers); and Remote Inter-Process Communication (a procedure whereby two or more programs running concurrently on different servers can communicate among themselves).

In this way the user's ADS interface allows specialized table display and interactive manipulation; a complete relational database; menu bar/pulldown menu and multiple windows (split-screen); context-sensitive help and a dynamic tutorial facility; and full-featured text management facility that supports browsing, plain-text inquiry, and cut-paste editing of selected files.

At the host node, the user's inquiry is received by the MPS from the Internet network and passed to an SQL Server which accesses the SQL database. The actual processing is somewhat more complicated by routing and authentication operations. Figure 2. shows several data nodes and some future ADS services distributed at different locations. These future services are shaded. The routing and authentication is show in the figure to be at the Administrative Node, however, the actual placement of these two functions are not specific to one location but can be distributed through the system. In practice, the user inquiry is interpreted and the correct databases are identified by an administrative operation function located somewhere in the system. This is of course transparent to the user.

The administrative function uses the Kerberos security system, developed at the Massachusetts Institute of Technology, in order to assure the authentication of the ADS user and the databases that can be access. It should be noted that in the ADS it is not necessary for a user to establish an account at each and every node intended for use, but it is sufficient to register and be accepted as an ADS user in order to have access to all the ADS Databases. This does not exclude some host nodes from restricting part of their database to qualified users. An example of this kind of restriction is the handling of proprietary mission data reserved for the Principal Investigator and his research

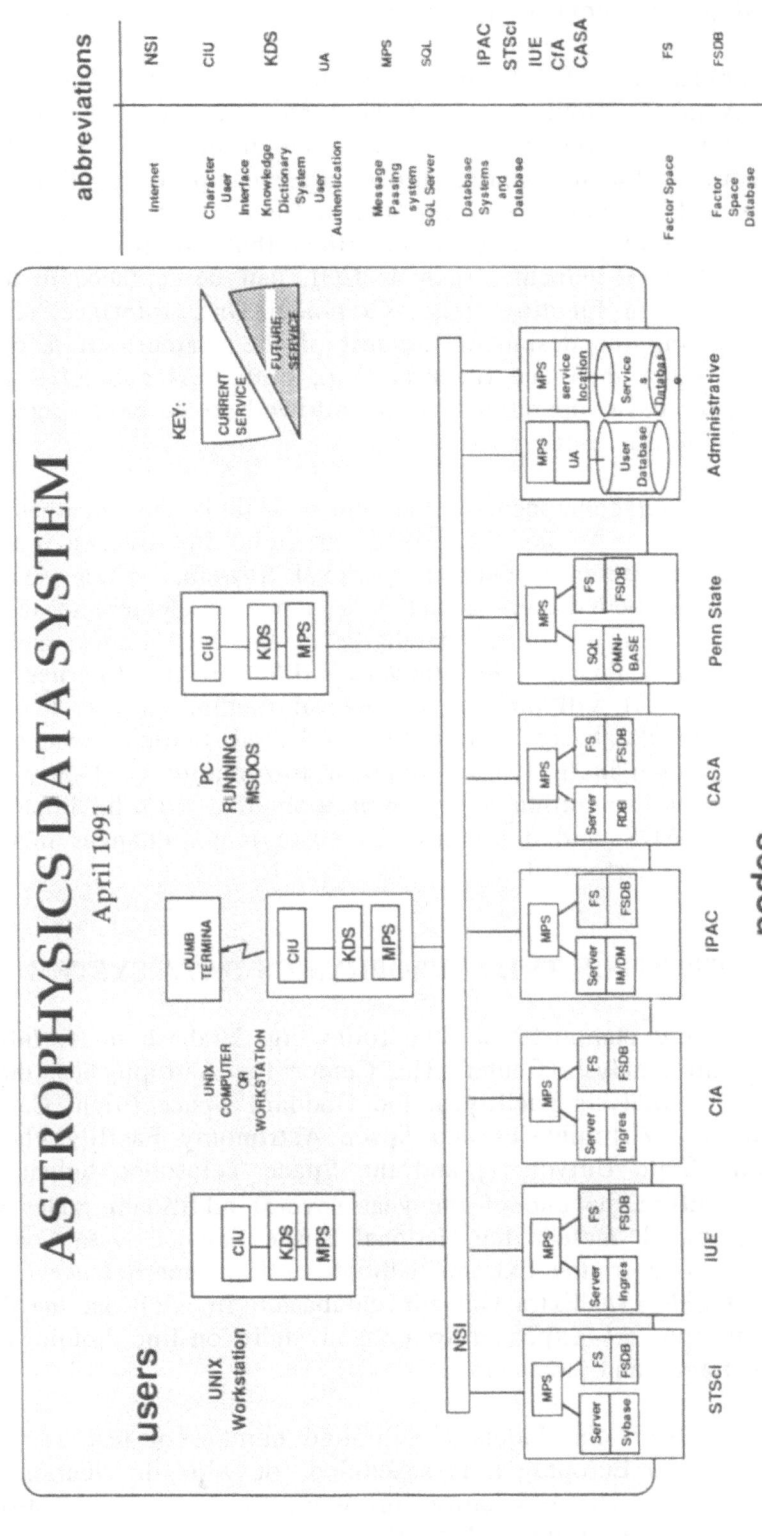

Figure 2. Active Nodes of the Astrophysics Data System Illistrating the Node-User Relationship.

7

team for a period of time. The Kerberos system can provide this service and general control of the management and use of the ADS.

The envisioned ADS Functional Architecture is illustrated in Figure 3. Here, as in the previous figure, the shaded regions are yet to be implemented by the ADS project. The modular nature of the ADS is apparent here. This modularity allows the system to function at a basic level and provide useful services before all of the contemplated functions are developed. System functions that are now handled by any single software element, such as CUI, can be replaced in the future by improved functions (e.g., Graphical User Interface, XUI) without impacting the design of the rest of the system. It is even possible that more than one function (e.g., both CUI and XUI) performing the same basic operation can be allowed to co-exist, selection being based on the user's choice.

An important component of the future ADS is the document search capability. Among these is an artificial intelligence tool called the Factor Space Based Document Retrieval System. This system, which appears in both Figure 2 and 3, as a future element of the ADS, will interface to the document databases at the host nodes, and will be accessed itself through the standard MPS and the Internet scientific network. It will provide a powerful method of searching very large document databases, where keyword and boolean search methods bog down in the vast volume of information. Factor Space is one of a new generation of research tools that will be implemented as part of the ADS, and it promises to make major changes in the way research will be performed.

CURRENT STATE OF THE ASTROPHYSICS DATA SYSTEM

The ADS is now composed of the following database node: Infrared Processing and Analysis Center, The Center for Astrophysics, the IUE Regional Data Analysis Facility at the Goddard Space Flight Center and at the Colorado Astrophysics and Space Astronomy Facility, the Pennsylvania State University, and the Space Telescope Science Institute. Before the end of the year several additional nodes will be added. These will include the National Space Science Data Center, the GRO Science Center, the Extreme Ultraviolet Explorer Science Center, as well as the NASA Extra Galactic database. In addition, the databases themselves are expected to expand their on-line holdings to include abstracts and documents.

The ADS has been distributed to a limited number of test users in the United States and Europe. It is scheduled for wide distribution to the U.S. astronomical community following the American Astronomical Society meeting at the end of May.

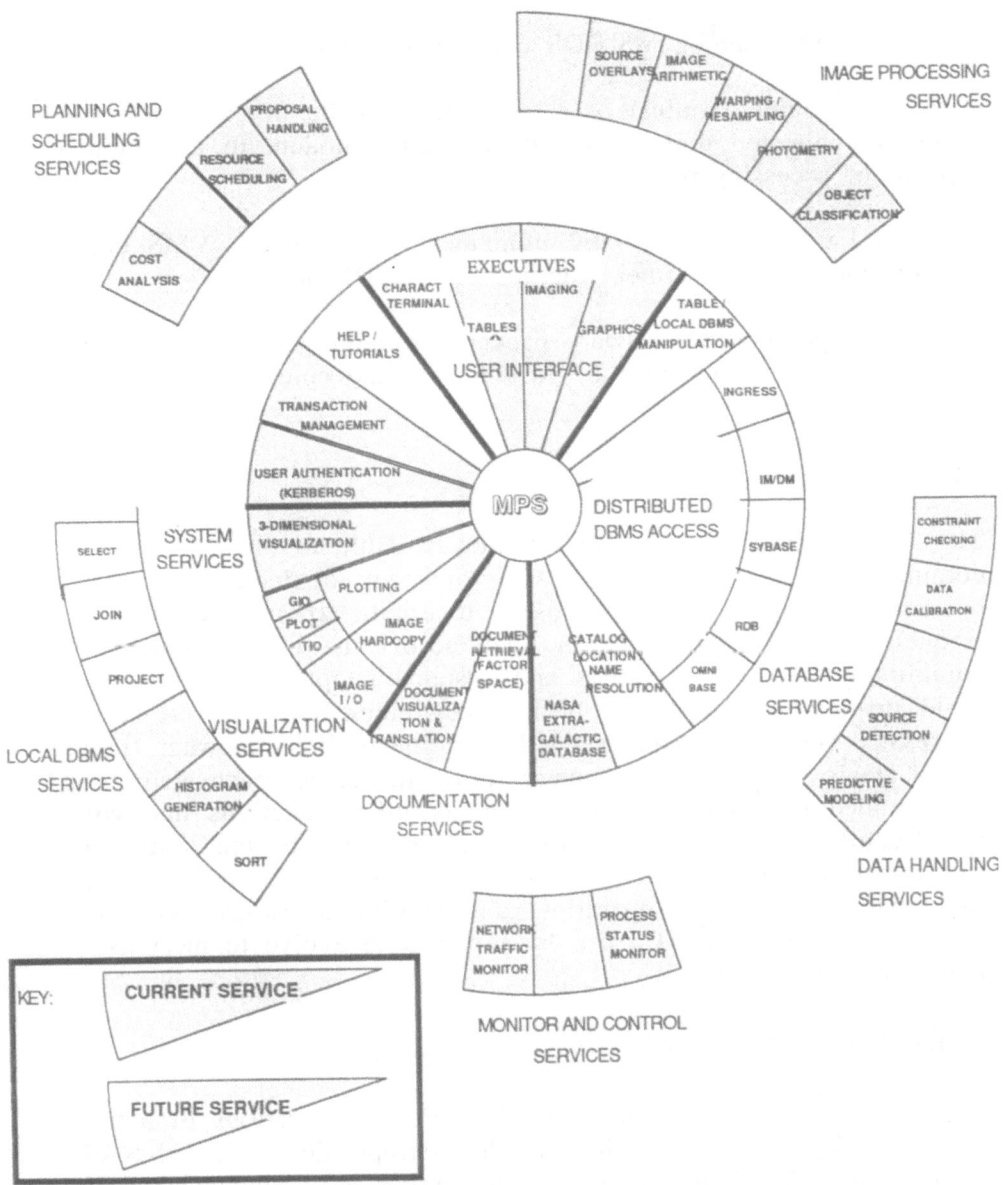

Figure 3. Astrophysics Data System Functional Architecture.

For a user to access the ADS he must have one of several commonly available computer systems:

1. A workstation running a UNIX operating systems with connection to Internet.

2. A PC running MS-DOS connected to Internet.

3. A dumb terminal or a computer capable of emulating a dumb terminal connected directly or via a dial-up modem to a computer capable of accessing the ADS.

4. Later in 1991, a DEC computer running under VMS operating system connected to either Internet or NSI/DECnet.

As well as a licenced software package containing the CUI and the KDS programs. This must be installed on the computer interfacing to the network.

CONCLUSIONS

The ADS is just beginning to become available to the astronomy community. It represents the work of many dedicated scientists and system developers, working closely together, although they were frequently widely distributed in body across the United States. The combination of both scientists and systems people, along with the rapid prototyping mode, has been a key element of the project's success. The mechanisms needed to achieve the somewhat limited initial goals of the project appear to be the same mechanisms needed to extended the system to include many other functions that will significantly improve the operation and usability of the system in the future. Thus, the evolution of the ADS system is far from complete, the basic system is in operation, and significant future enhancements are in the plan. The ADS can be expected to evolve to meet user needs and expectations.

ACKNOWLEDGEMENTS

The ADS Program is managed in the Science Operations Branch (Dr. G.uenter Riegler, Branch Chief), of the Astrophysics Division, NASA Headquarters, Washington, DC. The author has depended heavily on the advice and technical expertise of John Good (ADS Project manager), Rick Pomphrey, and Peter Shames of IPAC, and Steve Murray of the Center for Astrophysics (ADS Project Scientist). He also gratefully acknowledges the help of Sig Kutter of NASA Headquarters, Astrophysics Division, Science Operations Branch, and Elizabeth Giovane in the development of this paper.

MIDAS ON-LINE AT THE TELESCOPE

K. Banse, D. Baade, P. Biereichel, P. Grosbøl

European Southern Observatory
Karl-Schwarzschild-Straße 2
W-8046 Garching, Germany

Abstract

MIDAS will be the on-line image processing system used at ESO's Very large Telescope (VLT). As the first step on its way to the telescopes, MIDAS will soon replace the IHAP system at ESO's New Technology Telescope (NTT) when the NTT will be used also under remote control from ESO Headquarters in Garching. Observers will then have all utilities of a modern astronomical data analysis system at their finger tips, at the telescope.

This paper describes the modifications that have to be applied to MIDAS in order for it to run in a data acquisition environment. At the system level, they are rather few. An overview is given of the design goals for the functionality of the future, combined software environment at ESO telescopes: instrument control, data acquisition, 'quick-look' data analysis and full on-line data reduction. Possible implementations are discussed.

1 Introduction

At its site on La Silla in Chile, the European Southern Observatory (ESO) operates one of the largest assemblies of optical telescopes with an instrument park of considerable diversity. The biggest project which ESO is currently pursuing for the 1990's is the Very Large Telescope with an equivalent aperture of 16 m. The size of this investment and the potential data acquisition rate make it mandatory that adequate software support is given to

Data Analysis in Astronomy IV, Edited by V. Di Gesù et al.
Plenum Press, New York, 1992

the observer. Among other areas, this in particular includes efficient and versatile tools for 'quick-look' data analysis and full on-line data reduction.

Many years of experience within the ESO community have shown the invaluable advantage of using one and the same system for both on-line and off-line image processing. However, IHAP (Image Handling And Processing System [1]) which is presently installed at all major telescopes on La Silla is now reaching some limits. At the same time, the off-line processing of the majority of the observations obtained at La Silla is now done with MIDAS. Therefore, MIDAS was the obvious choice for the on-line data processing system at the VLT.

Starting from an analysis of the present situation and experiences, this paper attempts to project the resulting requirements into the VLT era and describes the efforts which are currently undertaken to enhance MIDAS for on-line usage on La Silla.

2 Current Set-up

At the moment, all telescopes on La Silla where instruments with solid-state detectors are installed, are equipped with computers of the HP 1000 family. The CPU is used for instrument control, data acquisition, and on-line data reduction. The standard hardware configuration consists of an ASCII terminal, a separate graphics terminal, and a color image display.

For the purposes of this paper it is sufficient to distinguish between only two major software components, the data acquisition program (DAP) and IHAP. The two share the graphics terminal, mainly for plots and instrument status display, and the character terminal. On the latter, commands to both DAP and IHAP are entered, and either system returns numerical results or status messages to it. The DAP can schedule macros (batch programs) in IHAP whereas IHAP has no control over the DAP. Both system share the same data base [11].

The design of this configuration was made in connection with the introduction of the Image Dissector Scanner (IDS) which was the first instrument at La Silla with a digital detector. Two requirements had to be simultaneously fulfilled: computer-aided instrument control (the DAP) and data visualization. Not surprisingly, the necessity and power of fully-fledged on-line data processing were realized very early on, and the data display tools expanded into what is now known as the IHAP system. Because of this historical development but also because 20 years ago computer equipment was still much more expensive than it is today and adequate network support did not exist, DAP and IHAP were implemented such that they shared most of the hardware.

IHAP achieves an excellent performance with its database, however it is a "single user" database only. For example, space for acquisition data can only be allocated through IHAP. When the DAP requests data space from IHAP (e.g., before starting a new exposure) it can happen that IHAP is just executing a command so that the DAP has to wait. This situation occurred with the introduction of new automatic observing modes (multiple exposures and exposure sequences).

Another disadvantage of the IHAP database is the maximum size of an acquisition file, namely 2000 x 2000 16-bit pixels, and its maximum total size (approximately 100 Mbytes). With some effort this size could be increased but even newer HP-1000 systems (e.g. A990 with 3-4 MIPS) could not handle such an amount of data in a reasonable time.

In short, IHAP is indeed very dependent on HP-1000 systems and a conversion to UNIX would lead to a system very similar to MIDAS, which exists already today.

Despite this list of reasons for the development of a new on-line system, the achievements of IHAP as an astronomical on-line data reduction system should not be forgotten. IHAP has pioneered this field, and many of its concepts will be copied by future systems. Far more than 10^5 observations of astronomical objects have been obtained with the present system, and it still supports several hundred observing programs which are carried out by visiting astronomers on La Silla every year. It could even be easily adapted to the needs of remote control observations from Garching [5]. Probably the most important asset of the present data acquisition system proved to be the homogeneous user interface that it offers at all telescopes. This cannot be overemphasized because visiting observers who typically have access to ESO telescopes and instruments for only few nights per year do not want to waste time by having to learn a new software system each time they use a new instrument. A single, modular system of course also facilitates maintenance.

3   Requirements

Analysis of the current situation, its strong points and its shortcomings leads to a number of requirements for an on-line Image Processing System (IPS). They are listed below, along with additional needs arising from possible new telescope operations concepts (cf. [6, 8]):

1. The interferences between DAP and IPS must be kept to a minimum.

   – It must be possible to upgrade the hardware in order to take advantage of technological progress (instruments, detectors, computers, etc.) without rendering the IPS useless.

- Under no circumstances may the acquisition of data by the DAP be blocked because of parallel work with the IPS (naturally, the IPS should also not depend on the DAP, except for its data output). Any possible locking situation must at least be resolved by a time-out.

- DAP and IPS have to share the data base. All data files created by the DAP should be available to the IPS. All acquired data must also be saved on a separate back-up medium, so that the IPS cannot inadvertantly corrupt the new data.

- DAP and IPS should not depend on common resources. For instance, if the DAP needs to display data, it has to do so on its 'own' display device and cannot rely on the IPS.

2. It must be possible for the DAP to also provide input for the IPS so that, *e.g.*, on-line calibration can automatically be initiated from the DAP right after data acquisition.

3. It must be easy to feed back to the DAP the results of an analysis carried out with the IPS, including computer readable error codes. An example is the astrometry of sources which subsequently are to be observed with a fiber-fed multi-object spectrograph.

4. There will be facilities in the IPS to support user input via a single key or a mouse click in order to minimize typing.

5. The IPS must be exportable so that observers can familiarize themselves with it off-line and even develop their own calibration software which they can also easily implement and run at the telescope during their observations.

6. The IPS must be portable to different computer systems. This will enable users to implement the IPS at their own site even if they do not have the same hardware as the one used at the observatory. Furthermore, it eases the migration to faster CPUs if on-line reduction needs increase.

7. The IPS must support a data base management system which does not have to be very sophisticated but must be very strong in its functionality. The needs for such a system are manifold:

- Calibration requires astronomical (*e.g.*, flux distributions of spectrophotometric standard stars), instrumental (*e.g.*, flat fields, approximate dispersion coefficients, etc.), and auxiliary (*e.g.*, wavelength tables) data.

– Observers may wish to bring and use target lists in machine readable form. Especially for service observations carried out in a flexible scheduling scheme, but also for efficient usage of a well pointing telescope, the target lists will, besides coordinates and identifiers, contain instrument settings such as exposure times, calibration exposures, etc. Such an enlarged target list can form the basis for an automatic, but supervised, batch observing mode.

8. The application software in the IPS must support the calibration of all data types (long-slit spectra, echelle spectra, images, etc.) generated by the instruments used with a given telescope and also different data formats (16-bit integer, 32-bit floating point, etc.).

9. Many of the VLT instruments, including their control and calibration software, will be contributed by the European astronomical community. Therefore the IPS must offer well documented, stable and efficient programmers' interfaces.

10. The interface between IPS and DAP for acquired data must be FITS format conforming to the new ESO archive data format [12]. It includes all information on the observation and permits on-line pipeline processing which is extremely important for efficient quality control.

11. High-quality hardcopies must be easily obtainable. The IPS must have a well defined interface to hardcopy device drivers because under the extreme climatic conditions of the observatory sites, special devices may have to be used.

12. The hardware and software architecture must be highly modular to make maintenance easy.

This list is not meant to be exhaustive nor a firm commitment that all the features will be implemented exactly as stated. But it sketches part of the path which MIDAS will probably follow over the next decade.

4 Possible Implementations

The telescopes will be equipped with workstations running Unix with NFS and the X11 Window system which form a local area network (LAN). This hardware greatly facilitates the implementation of the previously defined requirements.

The image display and graphics terminal are then realized as different windows on a bit mapped screen. The IPS will be a normal, interactive MIDAS session activated by the observer. The DAP will be completely independent of MIDAS but can use the ST/TB-interface routine [13] if direct access to MIDAS data structures is required. If the DAP needs image display capabilities these are obtained through a non-interactive MIDAS process. Then, there will be two independent MIDAS processes.

With that system architectur, the user is free to analyze his/her data in as much detail as he/she wishes without having to worry about the amount of time the required operations will take. When a new data frame has been obtained, the DAP stores it in the ESO Archive format on disk using the standard file management mechanisms provided by the operating system and saves it also on a separate disk or tape. The DAP (or the IPS) uses MIDAS to convert this disk file into a MIDAS file. The DAP may execute a separate, non-interactive MIDAS process to display the newly acquired frames in a window (but not in the display window of the IPS!) and to do any predefined preprocessing of the data or use any other utilities to indicate that new data has been acquired, *e.g.* just display a message. The IPS and DAP may use different windows on the same workstation screen or on a different monitor (*e.g.* another X-Terminal). Also, in a networked environment, the IPS and DAP may run on the same or on different CPUs still using windows on the same or separate screens. It is then up to the observer to use the interactive MIDAS session to load that new image into his display and/or do any other processing of the data. The DAP can additionally provide a choice of predefined procedures for specific instruments and detectors via a menu where the user just clicks a given field to activate the related command in the IPS. However, the actual user interface presented by the DAP is independent of MIDAS.

The DAP may also provide a live display of the data if the integration time is very short (like for infrared observations) and just indicate the directory where the result frames are to be found.

5   Implications for future development of MIDAS

Surprisingly few additional capabilities have to be considered with respect to the current version of MIDAS in order to offer the functionality outlined above.

Since there will be different display windows for the IPS and DAP, special measures to avoid interferences of the two domains are not needed anymore.

For the other requirements the following list specifies, point by point, what is still needed and what is available already now.

1. Contrary to IHAP, which out of necessity supports its own file management system, MIDAS uses the standard file system of Unix. Therefore, special interfaces to exchange information about physical locations of files become obsolete. That also opens up many choices for the implementation of the acquisition data base. The files with raw and preprocessed data coming from the DAP can but do not have to be on the same disk as the one the IPS is working on. Since the IPS has to fetch the data file from the DAP in any case, it can do so from anywhere in the network.

2. It is already possible to feed input into a MIDAS process from a different process (that may be another MIDAS process or any other process executing in the network). This functionality has been implemented on a test basis with encouraging results. But further work is necessary to enable these features also in an asynchronous way.

   We do not foresee, that it is necessary to interrupt an active MIDAS command, but in between two user commands input may also come from the DAP (though actually initiated by the user, e.g., by mouse clicking in a DAP provided menu).

3. Prototype interfaces have already been added which support communication not only between different MIDAS processes, but also between MIDAS and a program which is written to run outside the MIDAS environment, i.e. the DAP.

   For actual results in form of data files the standard file system is used, so there is no need for additional functionality in MIDAS.

4. Menues and other forms of advanced user interfaces have to be developed for MIDAS. First tests using just the standard MIDAS utilities to create menues for different purposes have already been done.

5. Extensive experience has been gained with exporting MIDAS in a rigorous 6-month cycle to more than 60 institutes in the ESO member countries. Procedures for preparing and shipping the system are in place and constantly refined.

6. MIDAS is a portable system in the sense, that Digital Equipment's familiar VMS operating system is supported as well as all major flavours of the UNIX operating system.

7. The MIDAS table system has been designed to provide all the basic functions of a "real" database system and in addition to offer features which are especially suited for the astronomical user ([3]).

8. Data files from different instruments must usually be preprocessed in-

dividually for each instrument to remove the instrumental signatures. After that, they can be treated alike by MIDAS applications.

MIDAS supports already several different formats for data representation. Since different data formats are foreseen in the basic building blocks of MIDAS, additional formats can be implemented quickly and easily.

9. There exists a well tested and stable set of interfaces for the MIDAS programming environment [13]. These interfaces cover most of the needs of a programmer who wants to integrate applications into MIDAS. However, additional interfaces are needed to support basic image display and graphics functions.

10. MIDAS supports the latest standard of the FITS format. Hierarchical keywords which are used in the ESO Archive format are also fully implemented.

11. The MIDAS image display applications are based on the IDI (Image Display Interfaces) routines which provide a portable interface to image displays. The definition of theses interfaces was a joint collaboration between the ESO Image Processing Group and major astronomical institutes ([7]). For plotting, MIDAS is based on AGL (ASTRONET Graphic Library [4]), the graphics package used by ASTRONET in Italy.

12. Workstations connected via a LAN offer a highly distributed computing environment. Malfunctions of a single workstation (node) in the network do not affect the whole system. MIDAS is a modular system where applications can be easily added or removed.

6 Conclusions and first actions

With the implementation of MIDAS as the on-line system in La Silla, ESO will again offer to its users a homogeneous environment for acquisition as well as for the reduction of data. The commands which the users already know from their off-line data reduction with MIDAS will also be applicable at the telescopes.

New instruments at La Silla will probably all be operated in the new data acquisition environment. The first such case is IRAC-2 on the 2.2m telescope for infrared astronomy where a Unix workstation will be installed.

The NTT will serve as a testbed for the operational concepts to be used with the VLT. The NTT is therefore the first telescope where the new MI-

DAS/DAP system will be employed exclusively. The possibility of remote observing, in principle also from other sites than Garching, forms an integral part of all NTT instruments. This means, MIDAS will be available on-line not only at the telescope but also at the remote control station.

References

[1] F. Middelburg: *IHAP Users Guide*, ESO Operating Manual (1983), P. Biereichel (1990 rev.).

[2] P. Crane, K. Banse, P. Grosbøl, C. Ounnas, D. Ponz: *MIDAS*, in Data Analysis in Astronomy, p. 183. Di Gesu *et al.* eds., Plenum Press (1985).

[3] P. Grosbøl, D. Ponz: *The MIDAS Table File System*, in Mem. Soc. astron. Italiana, **56**, p. 429 (1985).

[4] L. Fini: *ASTRONET Graphic Library User's Guide*, Astronet Documentation, AUM2 (1985).

[5] G. Raffi, M. Ziebell: *The ESO Messenger*, **44**, p. 26, Garching (1986).

[6] D. Baade: *From Low-Noise Spectra to High-Precision Data*, in Observational Astrophysics with High Precision Data, Proc. 27th Liège International Astrophysical Colloquium, p. 1 (1987).

[7] D.L. Terret et al.: *An Image Display Interface for Astronomical Image Processing*, Astronomy and Astrophysics Suppl., **76**, p. 263 (1988).

[8] European Southern Observatory, *Report by the VLT Operations Working Group*, Garching (1989).

[9] European Southern Observatory - *The ESO Users Manual*, Garching (1990).

[10] European Southern Observatory - Image Processing Group: *The MIDAS Users Guide*, ESO Operating Manual No. 1, Garching (1990).

[11] P. Biereichel: *IHAP Engineering Manual*, Garching (1990).

[12] F. Ochsenbein and P. Grosbøl: *ESO Archive - Acquisition Requirements*, Garching (1990).

[13] European Southern Observatory - Image Processing Group: *The MIDAS Environment*, Garching (1991).

ASTRONOMICAL DATABASES: A USER APPROACH

André Heck

Observatoire Astronomique
11 rue de l'Université
F–67000 Strasbourg
France

Data Archiving was for many years the most disregarded aspect of all data systems.

(Albrecht & Egret, 1991)

[According to NASA], scientists have looked at only 10 percent of the data sent back to Earth in the past 20 years, and have analyzed only 1 percent.

(Newsweek, 7 May 1990)

The Hubble Space Telescope is expected to send down yearly to Earth the equivalent of about 500 Gbyte of data.

(see e.g. Russo et al., 1986)

Abstract

A few thoughts are presented on astronomical databases and archives from a user point of view. Shortcomings and possible future trends are also mentioned, as well as some considerations on a future ideal environment for direct home applications on real and remotely located astronomical data.

1. Introduction

Astronomical theory and methodology would remain sheer rhetoric without the observational data with which they can be confronted and to which they can be applied. It is however not our aim to describe here all databases and archives available presently or in the near future to the astronomical community.

Various speakers at this meeting are describing better that I would do this part of the particular experiment they are involved with. The situation is also evolving quite rapidly in this field. The most recent and up–to–date picture of the situation can be found in a compilation by Albrecht & Egret (1991).

I would rather concentrate on a few aspects from a user point of view. A few shortcomings will be mentioned together with possible solutions and future trends that should ease their resulting constraints. A touch of dream will conclude this paper.

Databases are now linked to multiple aspects of our activities: from the observing proposal elaboration, submission and evaluation, to the reduction of new data with or without integration of data obtained previously or with different instrumentations, and to the finalization of papers implying at each step the collection of a maximum of isolated data of various types and from different sources, without forgetting the cross–checks with what has already been published on the objects studied or on similar ones.

All this has to be put in parallel with the dramatic revolution in scientific communications, expressed recently by the explosion of electronic mail and networks, as well as with the desktop publishing facilities, which were not at all in practice only a few years ago. This allows quick, efficient and high–quality dialogues, and consequently publishing, with remotely located collaborators.

2. Databases, archives and networks

The general understanding for an *archive* is in a context of a set of data obtained through a given experiment, be they on a classical support as photographic plates or digitalized on magnetic tapes or other supports. These data might not be anymore quite *raw*, in the sense that they might have undergone some amount of treatment or *processing*.

Typically archives would be for example:
- the set of ultraviolet spectra collected by the International Ultraviolet Explorer (IUE) (see Wamsteker, 1991);
- the European Southern Observatory (ESO) archiving project which has started systematically with the New Technology Telescope (NTT) operations (Ochsenbein, 1991);
- the National Radio Astronomical Observatory (NRAO) archives from its various facilities (Wells, 1991).

These are just three examples relative to different wavelength ranges (respectively ultraviolet, visible, and radio), but the list could be much longer and include the archiving aspects of other experiments such as COBE (White & Mather, 1991), EXOSAT (White & Giommi, 1991), GRO (den Helder, 1991), HIPPARCOS (Turon *et al.*, 1991), HST (Schreier *et al.*, 1991), IRAS (Walker, 1991), ROSAT (Zimmermann & Harris, 1991), and so on.

The term *database* could be generic and include archives as well. A stricter meaning however would link it to a context of scientific data extracted or derived from observational material after reduction and/or analysis, although a database can also be made of non–observational data. Contrary to a *databank*, a database would also generally contain some software to retrieve the data and possibly work on them.

A typical database is the pionneering SIMBAD hosted by the Astronomical Data Centre (CDS) at Strasbourg Observatory (Egret *et al.*, 1991).

Some experiments have got a database built around their archive and it is sometimes difficult to distinguish the two aspects. This is particularly the case for the EXOSAT Database (White & Giommi, 1991) and the NASA/IPAC Database (Helou *et al.*, 1991).

Pionneering networks such as ESA's ESIS (Albrecht, 1991) and NASA's ADS (Weiss & Good, 1991) aim at interconnecting and making easily available the most relevant archives and databases to the astronomical community. See Murtagh (1988 & 1991) for reviews on current issues and status.

The line 'speed' (or better 'data flux capacity') should still be increased in order to accomodate quick transfers of huge amount of data. The transfer of bi–dimensional images remain still prohibitive nowadays, but similar requirements in medical communications might have a stronger push on an advance in that respect than ours. Networks should also absorb more and more databases and archives, with more and more diversified types of data (see Sect. 8).

3. Provisions for databases and archives

Here are a few recommendations that could be formulated for ground and space experiments in relation with the data they would collect (already partially formulated in Heck, 1986):

- no project should be allowed to go ahead without proper provisions for an archive including data retrieval and dissemination at the end of a possible proprietary period;
- this possible proprietary period should be kept as reasonably short as possible;
- data availability should be as flexible as possible, as far as means of retrieval and supports are concerned; standard format (such as FITS – see Grosbøl, 1991) should be highly recommended;
- data processing for plain astronomers (i.e. other than principal investigators, project designers, staff members, and so on) should be possible at the archive centers, possibly by remote logon and through decentralized pieces of software (see Sect. 5);
- adequate investigations and simulations should be run in advance to identify, as far as feasible, the appropriate pieces of data (or telemetry in case of space experiments) to be retained;
- the importance of a clean, complete and homogeneous log of observations cannot be stressed enough, as well as its easy integration in a network of databases;
- standard and calibration data should be made routinely available;
- special attention should be paid to a good representation (in the statistical sense) of all target categories observable with a specific experiment;
- before the possible termination of an experiment, plans should have been made for a complete reprocessing of all data collected with the latest and supposedly best version of the corresponding image processing software (IPS);
- the end of an experiment should not mean the turnoff of all activities related to the project as the corresponding database should be maintained with an appropriate service to the astronomical community;
- budget provisions should be made accordingly.

Last, but not least, one should not forget the invaluable service that a minimum quantity of software could provide to database or archive users. If each user cannot presently host in his/her local system an extensive IPS, a minimum processing capability and some mathematical – essentially statistical – methodological facilities should be implemented.

A statistical package such as the one available in MIDAS, the image processing system developed at ESO, has been derived from Murtagh & Heck's (1987) set of routines. It could be considered for a minimum exploitation of the data at hand. Other environments are described in Murtagh & Heck (1988).

4. Maintenance of databases and archives

The recommendations above imply a number of measures that go beyond mere archiving and dissemination of data: one could never stress enough the necessity of a good maintenance of databases and archives.

As indicated in the previous section, this leads to a number of reprocessings (whenever bugs have been detected or refinements can improve the IPS) and to the availability of a dedicated staff, even after the possible termination of the data collection period for an experiment with a limited lifetime.

Quality controls have to be carried out at various levels. In the database such as SIMBAD, the catalogues integrated undergo first a screening by a specialist or a team of specialists in the field. However even after integration of data checked in such a way, it is necessary to have a number of people controlling the databases regularly for purposes such as homogeneity, errors detected by users, and so on.

Artificial intelligence, or more generally knowledge–based, procedures could be of a great help in this. Refer for instance to various papers in Heck & Murtagh (1989).

It is important that the database and archive managers stay in touch with their users. It might be appropriate to set up specific users committees and to run surveys regularly. Bulletin board services would be recommendable as well as mailbox systems to allow casual dialogues with users.

It might be the place to point out here the existence of a *Working Group on Modern Astronomical Methodology* and of a *Technical Committee* of the *International Association for Pattern Recognition* (TC 13 – Astronomy and Astrophysics) (refer i.a. to Heck *et al.*, 1985 and to Murtagh *et al.*, 1987). The rôle of those groups is to set up links between interested people, to keep them posted on the advances in the corresponding methodologies, to suggest, support and help organizing meetings, workshops, schools, and so on.

5. Decentralized (sub-)databases and (sub-)archives

We have to face a paradox here. Indeed one could say that, because of the increasing facility to get connected to specific databases and archives, the idea of decentralized databases and archives (or subsets of these) could be completely discarded.

I believe this is wrong for at least two main reasons. First, networks are not yet fully satisfactory as far as accessibility and line speed are concerned (not to speak of the so–called *developing countries*). Second, with the present impressive power of desktop computing facilities (in steady improvement), the difficulties for accomodating locally databases are becoming negligible. Refer for instance to what has been done with 'Einstein' data (Garcia *et al.*, 1990; Plummer *et al.*, 1991).

In parallel, the cost of producing massively CD-ROMs has gone down a lot and is decreasing rapidly. According to Garcia (1991), a thousand CD-ROMs would presently cost about US $ 5000. In view of this, one could only encourage all databases and archives managers to make available their babies on CD-ROMs together with an adequate and minimum piece of software. The progressive relative cheapness of such as system would also allow to keep the distributed (sub-)databases frequently updated.

According to Adorf (1991), there are now on the market high–performance image archival system consisting of rewritable CD–drive and software by *GigaSearch*: the disk stores up to 635 Mbytes with an access time of 58 msec. The corresponding catalogue can encompass 32,000 images. Additionally, each entry can be indexed with keywords allowing an efficient retrieval through logical expressions. This is definitely a way that need to be investigated.

Clear MOUs between data centres setting up collaborations have to be drawn with a minimum of rules guaranteeing a copyright protection and a good policy of acknowledgements in publications based on the various databases and archives, in such a way that everybody be happy – and essentially the funding institutions. This can be considered as an aspect of the general security issue of databases.

6. Hierarchical structure of data

The situation in this matter is not yet quite satisfactory and significant steps should be made to improve it.

Let us take a simple example. If you are interested in a star in Orion, you surely want to know it belongs to this region, which is not obvious if you know the object only by its HD number or even more by a more obscure identification. Thus for a point–like object, you are keen to obtain data, such as those relative to the interstellar absorption, that correspond to an extended region it may belong to. For a star cluster, you also wish to go upwards and downwards: from individual stellar data to the cluster ones, and from the cluster to the individual stars.

This is not easy to solve because another feature has to be taken into account: the resolution of data relative to extended objects. An infrared or an X-ray survey of the sky cannot always match the resolution obtained in other wavelengths. Therefore the correlation between databases cannot always be built up easily.

7. Integration of wavelength ranges

Thanks to the use of balloons, rockets and spacecrafts, the part of the electromagnetic spectrum that has been more signicantly studied has been broadened. It is common now to carry out studies involving visible, UV, IR and X data, and to conduct joint observing campaigns involving ground–based instruments and spacecrafts working simultaneously in various ranges.

Therefore a coordination between archives and databases is also desired. A user must be able to mix data from various instrumentations in the most transparent possible way. Some fields of activity in astronomy, historically linked to specific wavelengths ranges, still remain too disconnected from the rest of the corporation.

8. Complementary services

A network of interconnected databases and archives should not only provide scientific data, but also some of another nature, less scientific ones, but not least useful and practical, such as bibliographic and 'corporatists' ones.

One has already access to various kinds of bibliographic data on paper, like the invaluable *Astronomy & Astrophysics Abstracts (AAA)* (see e.g. the last four annual volumes by Burkhardt *et al.*, 1990a,b,c&d), as well as bibliographic on–line databases (Watson, 1991). These are accessible through authors and/or keywords. The bibliographic part of SIMBAD is object-oriented and thus of a complementary nature. All these facilities should be integrated.

Institutional data such as those provided by directories like ASpScROW (Heck, 1991a) or individual data such as those listed in the directories published by the *American Astronomical Society (AAS)* (1990) or the *Société Française des Spécialistes d'Astronomie (SFSA)* (Egret, 1991) ones should also be connected, as well as the electronic–mail data listed in Benn & Martin (1991) files.

Indeed one can imagine that someone searching the bibliography in SIMBAD can immediately get the communication 'plugs' of one author he/she wants to get in touch with. The same could apply with someone detaining the proprietary rights of some images from a given archive. Scientific profiles as listed in the ASpScROW and SFSA directories would offer other possibilities such as selective mailings for organizing meetings and so on.

Going one step further, one could consider the inclusion of dictionaries of acronyms and abbreviations (see e.g. Heck, 1991b), as well as other tools that might come up as most useful in the everyday life (thesauri, S/W packages of various kinds, standardized desktop publishing macros or procedures possibly allowing direct connections to databases, and so on).

9. Conclusions or where are we going?

A general user does not really care in fact how the system he has at the tips of his/her fingers works, as long as it works well and give him/her as much as possible, and hopefully more than expected. This is why I did not enter here into technical details on databases structures, formats, protocols, and so on.

It is fair however to say the user interfaces are also in permanent progress (see i.a. Pasian & Richmond, 1991) allowing an ever more standardized access to heterogeneous sets of archives and databases. Hypertext and hypermedia systems have also been investigated as means for storage, retrieval and dissemination of information, for instance in the HST context (Adorf, 1989).

Refer also e.g. to Adorf & Busch (1988) for uses of 'intelligent' text retrieval systems, to Kurtz (1991) for bibliographic retrieval through factor spaces and to Parsaye *et al.* for 'intelligent' databases.

Of course, one is never happy and this is one of the major reasons behind the progress of sciences and techniques. Remember how we were only a few years ago. What an enormous progress in a short time. What will be the situation in only a decade from now?

Generally the evolution goes with quite a few surprises and it would be stupid to play here the game of guessing what they will be. What is sure is that the future of databases and archives with be mainly linked to technological advances, but also to the fact that astronomers will more and more realize how intimately this field is involved with their activities: observing (remote or traditional), data reduction, general research, publication, management of instrument time, and so on.

Ideally, everyone would like to have from his/her desk (or home?) access quickly and in the most friendly way to the maximum of data. Upper shells should be accessible to interrogate databases and archives and retrieve in a practically usable form the necessary material. Local and cheap hard–, firm– and software should be available in order to then process this stuff locally.

Means of communications should eliminate geographical separation. There might be a time when the keyboard will not be necessary anymore and when the publishing as such will disappear, replaced by some electronic form of dialogue with the machine and the *knowledge* bases, a concept that will progressively replace those of databases and archives.

The present way we work is basically oriented towards the need of recognition (essentially through traditional publishing, even if eased thanks to the desktop tools) for getting positions (salaries), acceptance of proposals (data) and funding of projects (materialization of ideas). Most of the underlying rules were designed at a time when the pen (and then the typewriter) and noisy telephone lines over manned switchboards were the basic tools for communicating, with the index card as the access key to the archive (or library) when it was existing at all.

Human narcisism apart, is really recognition the best motivation behind the progress of science? How many redundancies could be avoided, as well as time and energy wasted in these and in administrative reporting of little fundamental use! The new era in archiving, in database interconnectability and in knowledge access, as well as in communication means, policies and habits might open the doors to new behaviors in our corporation.

Acknowledgements

My special gratitude goes to Miguel A. Albrecht and Daniel Egret for letting me have in advance of publication a copy of the book they edited on *Databases & On–line Data in Astronomy*. It is also a very pleasant duty to acknowledge here discussions with and/or material received from Hans–Martin Adorf, Jan Willem den Helder, Michael R. Garcia, Michael J. Kurtz and H. Ulrich Zimmermann.

References

Adorf, H.M. 1989, *Space Inf. Syst. Newsl.* **1**, 7–14.

Adorf, H.M. 1991, private communication.

Adorf, H.M. & Busch, E.K. 1988, in *Astronomy from Large Databases – Scientific Objectives and Methodological Approaches*, Eds. F. Murtagh & A. Heck, *ESO Conf. & Workshop Proc.* **28**, 143–147.

Albrecht, M.A. 1991, in *Databases & On–line Data in Astronomy*, eds. M.A. Albrecht & D. Egret, Kluwer Acad. Publ., Dordrecht, in press.

Albrecht, M.A. & Egret, D. (Eds.) 1991, *Databases & On-line Data in Astronomy*, Kluwer Acad. Publ., Dordrecht, in press.

American Astronomical Society 1990, *1991 Membership Directory*, Amer. Astron. Soc., Washington, 184 p.

Burkhardt, G., Esser, U., Hefele, H., Heinrich, I., Hofmann, W., Krahn, D., Matas, V.R., Schmadel, L.D., Wielen, R. & Zech, G. (Eds.) 1990a, Astronomy & Astrophysics Abstracts, Vol. **49A**, Literature 1989, Part 1, Springer–Verlag, Heildeberg, xii + pp. 1–640.

Burkhardt, G., Esser, U., Hefele, H., Heinrich, I., Hofmann, W., Krahn, D., Matas, V.R., Schmadel, L.D., Wielen, R. & Zech, G. (Eds.) 1990b, Astronomy & Astrophysics Abstracts, Vol. **49B**, Literature 1989, Part 1, Springer–Verlag, Heildeberg, viii + pp. 641–1424.

Burkhardt, G., Esser, U., Hefele, H., Heinrich, I., Hofmann, W., Krahn, D., Matas, V.R., Schmadel, L.D., Wielen, R. & Zech, G. (Eds.) 1990c, Astronomy & Astrophysics Abstracts, Vol. **50A**, Literature 1989, Part 2, Springer–Verlag, Heildeberg, x + pp. 1–656.

Burkhardt, G., Esser, U., Hefele, H., Heinrich, I., Hofmann, W., Krahn, D., Matas, V.R., Schmadel, L.D., Wielen, R. & Zech, G. (Eds.) 1990d, Astronomy & Astrophysics Abstracts, Vol. **50B**, Literature 1989, Part 2, Springer–Verlag, Heildeberg, viii + pp. 657–1430.

Benn, Chr. & Martin, R. 1991, *Electronic–Mail Directory 1991*, Roy. Greenwich Obs., Cambridge, e-mail files.

den Helder, J.W. 1991, private communication

Egret, D. 1991, *Annuaire de l'Astronomie Française 1991*, Soc. Française Spécialistes Astron., Paris, in press.

Egret, D., Wenger, M. & Dubois, P. 1991, in *Databases & On-line Data in Astronomy*, eds. M.A. Albrecht & D. Egret, Kluwer Acad. Publ., Dordrecht, in press.

Garcia, M.R. 1991, private communication.

Garcia, M.R., McSweeney, J.D., Karakashian, T., Thurman, J., Primini, F.A., Wilkes, B.J. & Elvis, M. 1990, *The Einstein Observatory Database of HRI X-Ray Images (FITS/CD-ROM Version)*, Smithsonian Institution Astrophys. Observ., Cambridge.

Grosbøl, P. 1991, in *Databases & On-line Data in Astronomy*, eds. M.A. Albrecht & D. Egret, Kluwer Acad. Publ., Dordrecht, in press.

Heck, A. 1986, *Bull. Inform. Strasbourg Data Center* **31**, 31–33.

Heck, A. 1991a, *Astronomy, Space Sciences and Related Organizations of the World 1991 – ASpScROW 1991*, CDS Spec. Publ. **16**, x + 1182 p. (two volumes).

Heck, A. 1991b, *Acronyms & Abbreviations in Astronomy, Space Sciences & Related Fields*, CDS Spec. Publ. **18**, xii + 530 p.

Heck, A. & Murtagh, F. (Eds.) 1989, *Knowledge–Based Systems in Astronomy, Lecture Notes in Physics* **329**, Springer–Verlag, Heidelberg, v + 280 p.

Heck, A., Murtagh, F. & Ponz, D.J. 1985, *Messenger* **41**, 22–25.

Helou, G., Madore, B.F., Schmitz, M., Bicay, M.D., Wu, X. & Bennett, J. 1991, in *Databases & On-line Data in Astronomy*, eds. M.A. Albrecht & D. Egret, Kluwer Acad. Publ., Dordrecht, in press.

Kurtz, M.J. 1991, in *On-line Astronomy Documentation and Literature*, Eds. F. Giovane & C. Pilachowski, NASA Conf. Proc., in press

Murtagh, F. 1988, *Bull. Inform. Strasbourg Data Center* **34**, 3–33.

Murtagh, F. 1991, in *Databases & On-line Data in Astronomy*, eds. M.A. Albrecht & D. Egret, Kluwer Acad. Publ., Dordrecht, in press.

Murtagh, F. & Heck, A. 1987, *Multivariate Data Analysis*, D. Reidel Publ. Co., Dordrecht, xvi + 210 p.

Murtagh, F. & Heck, A. (Eds.) 1988, *Astronomy from Large Databases – Scientific Objectives and Methodological Approaches*, ESO Conf. & Workshop Proc. **28**, xiv + 512 p.

Murtagh, F., Heck, A. & di Gesù, V. 1987, *Messenger* **47**, 23–24.

Ochsenbein, F. 1991, in *Databases & On–line Data in Astronomy*, eds. M.A. Albrecht & D. Egret, Kluwer Acad. Publ., Dordrecht, in press.

Parsaye, K., Chignell, M., Khoshafian, S. & Wong, H. 1989, *Intelligent Databases*, J. Wiley & Sons, New York, xvi + 480 p.

Pasian, F. & Richmond, A. 1991, in *Databases & On–line Data in Astronomy*, eds. M.A. Albrecht & D. Egret, Kluwer Acad. Publ., Dordrecht, in press.

Plummer, D., Schachter, J., Garcia, M. & Elvis, M. 1991, *The Einstein Observatory IPC Slew Survey (FITSA3D/CD-ROM Version)*, Smithsonian Institution Astrophys. Obs., Cambridge.

Rolfe, E.J. (Ed.) 1983, *Statistical Methods in Astronomy*, European Space Agency Spec. Publ. **201**, x + 262 p.

Russo, G., Richmond, A. & Albrecht, R. 1986, in *Data Analysis in Astronomy*, eds. V. Di Gesù, L. Scarsi, P. Crane, J.H. Friedman S. Levialdi, Plenum Press, New York, pp. 193–200

Schreier, E., Benvenuti, P. & Pasian, F. 1991, in *Databases & On–line Data in Astronomy*, eds. M.A. Albrecht & D. Egret, Kluwer Acad. Publ., Dordrecht, in press.

Turon, C., Arenou, F., Baylac, M.O., Boumghar, D., Crifo, F., Gómez, A., Marouard, M., Mekkas, M., Morin, D. & Sellier, A. 1991, in *Databases & On–line Data in Astronomy*, eds. M.A. Albrecht & D. Egret, Kluwer Acad. Publ., Dordrecht, in press.

Walker, H.J. 1991, in *Databases & On–line Data in Astronomy*, eds. M.A. Albrecht & D. Egret, Kluwer Acad. Publ., Dordrecht, in press.

Wamsteker, W. 1991, in *Databases & On–line Data in Astronomy*, eds. M.A. Albrecht & D. Egret, Kluwer Acad. Publ., Dordrecht, in press.

Watson, J.M. 1991, in *Databases & On–line Data in Astronomy*, eds. M.A. Albrecht & D. Egret, Kluwer Acad. Publ., Dordrecht, in press.

Weiss J.R. & Good, J.C. 1991, in *Databases & On–line Data in Astronomy*, eds. M.A. Albrecht & D. Egret, Kluwer Acad. Publ., Dordrecht, in press.

Wells, D.C. 1991, in *Databases & On–line Data in Astronomy*, eds. M.A. Albrecht & D. Egret, Kluwer Acad. Publ., Dordrecht, in press.

White, N.E. & Giommi, P. 1991, in *Databases & On–line Data in Astronomy*, eds. M.A. Albrecht & D. Egret, Kluwer Acad. Publ., Dordrecht, in press.

White, R.A. & Mather, J.C. 1991, in *Databases & On–line Data in Astronomy*, eds. M.A. Albrecht & D. Egret, Kluwer Acad. Publ., Dordrecht, in press.

Zimmermann, H.U. & Harris, A.W. 1991, in *Databases & On–line Data in Astronomy*, eds. M.A. Albrecht & D. Egret, Kluwer Acad. Publ., Dordrecht, in press.

ESIS: SCIENCE INFORMATION RETRIEVAL

Miguel A. Albrecht

ESA/ESRIN
Via Galileo Galilei
I-00044 Frascati
Italy[1]

INTRODUCTION

This chapter describes the work being currently undertaken to define and develop the European Space Information System (ESIS). We will concentrate here on the user functionality aspects of the project, and refer the reader, for further detail, to Albrecht et al., 1988.

ESIS aims at providing the Space Science community with a set of tools to access, exchange and handle information from a great variety of sources including space mission archives, databases of scientific results, bibliographical references, "white" - and "yellow"- page directories, and other information services. Access to all information contained in ESIS shall be possible through a homogeneous interrogation and data managing language. This language will support queries formulated in "discipline oriented" rather than "computer oriented" terms. A "personal database" environment will support the retrieval of data from different sources into a common homogeneous format: this constitutes the basis for cross correlations and other further processing.

A Pilot Project running in the timeframe 1988-1993 will demonstrate the validity of the concepts adopted and will establish a basic infrastructure for further developments.

USER NEEDS

The general requirements of space scientists on ESIS can be summarized in the points listed below.

(a) Open existing space data holdings to non-experts.

Due to the high level of specialisation required to access current archiving facilities, data is being utilized only by small "mission-specific" user communities. A global reference database could provide non-experts with an overview of available data. This requirement calls for an integrated data view model and a layered access structure.

---

[1] Current address:
European Southern Observatory, Karl-Schwarschild-Str. 2, D-8046 Garching bei München, Fed. Rep. of Germany

Data Analysis in Astronomy IV, Edited by V. Di Gesù *et al.*
Plenum Press, New York, 1992

(b) Integrate past and future mission archives.

Existing and future archiving facilities should be integrated into the system without losing their identity and/or independence. For past missions this means that current investments are preserved and that local user communities are not concerned by the integration. For future missions, independence from the system means the flexibility to set up their facilities taking advantage of state-of-the-art technology without making compromises. This requirement calls for an open architecture.

(c) Integrate data retrieval with information exchange

Finding out what data sets are of relevance for a particular research project, retrieving these data sets, processing them and possibly exchanging them with other collaborators should all be activities integrated into one single environment. Provision should also be made to interface in a transparent way other information systems and telecommunication networks. This requirement calls for a uniform access method to all services available through ESIS.

THE PILOT PROJECT

In order to test out feasible concepts both on the system engineering as well as user interface sides, a pilot project has been set up to run over the period 1988-1993. During this period the system functionality and its architecture is defined and tested in a "real-life" environment, i.e. it is actually used in day-to-day scientific work. It is assumed that a science information system can be designed independently of the science domains that is serves: being the discipline specific aspects confined into well defined, exchangeable layers. In fact, ESIS is meant to serve the whole range of space science domains, beginning in the pilot phase with Astronomy & Astrophysics and Space Physics but expected to include Earth Observation and Microgravity sciences later on. At the time this article is written, the pilot phase has gone half way through. The system architecture has been designed, a first set of user functions have been defined and the system is entering the implementation phase in the first half of 1991.

Pilot Project Configuration

The elements that form the complete system will be brifly described below. A number of decisions had to be made in the configuration of the baseline system in order to keep it sizeable and manageable. The number of databases to include, the number of user hardware platforms, the associated network protocols to support, etc., all define the starting system, but great care has been taken to isolate dependencies in order to ensure expansibility and further development.

Databases included in the pilot project

The archiving facilities and databases to be included in the pilot phase are:

- the IUE Vilspa DB, in Villafranca, Spain;
- the EXOSAT Database, at ESTEC, Noordwijk, theNetherlands;
- the SIMBAD Database, at the Centre de Données astronomiques de Strasbourg (CDS), France;
- the Space Telescope archive, at the ST-ECF, Garching, Fed. Rep. of Germany;
- the World Data Centre C1 and the Geophysical Data Facility managed by the Rutherford Appleton Laboratory, Chilton, Didcot, United Kingdom;
- relevant bibliographical databases offered on-line by the Information Retrieval Service (IRS) at ESRIN, Frascati, Italy.

Information services.

Value added information services will also be included. They comprise the "classical" set of telecommunication functions like electronic mail, file transfer and management, remote login, as well as group communication tools like electronic conferencing, bulletin boards, etc. A directory service for users, institutions, instrumentation and software packages will also be established that will allow both "white" - and "yellow" - page searches.

The Backbone network

Since ESIS has to connect distributed, heterogeneous data systems all over Europe, the underlaying telecommunication infrastructure plays a key role. The ESIS backbone network has been established making use of the telecommunication facilities provided by ESANET. In addition to the telecommunication lines, two kinds of elements form the backbone network: the ESIS access points and the service front-ends (also called "service shells"). A number of Vax computers have been procured to host these elements covering today six sites, and it is expected that within 1992 a series of further procurements will complement this infrastructure to cover all ESA member states.

SYSTEM ARCHITECTURE

ESIS is defined through a modular architecture that organizes the system functions into the following three groups: the Query Environment (QE), the Correlation Environment (CE) and the Information Services (IS); they are "glued" together with a User Management System (UMS) (see figure 1). Each module requests services from others via a set of interfaces. All functions, regardless of their location, both architectural and geographical, are accessed via a unique environment: the "User Shell".

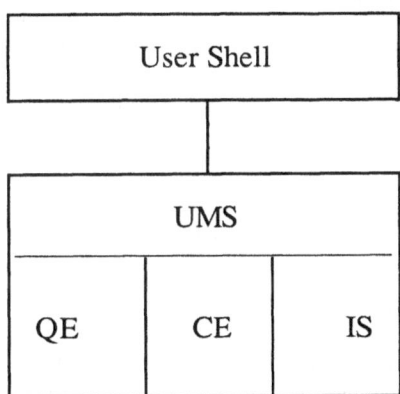

Fig.1. A user's perspective of the ESIS environments

The User Shell consists of the collection of "clients" that interact with the corresponding "servers" in the system. It is a software package to be distributed to end users for installation on their local computing infrastructure. The user interface is designed to support a large number of devices including terminals, workstations and personal computers, and offers a selection of dialogue modes such as menus, commands and forms. A user activating the User Shell will reach all ESIS functions through a uniform view of the system, independent of the local computing environment and of the networking method of accessing ESIS. On-line help and tutorial services will be provided locally, in addition to any similar services offered by the data facility, allowing for a quicker response and avoiding unecessary network load.

The Information Services include typical electronic "group-ware" services such as electronic mail, bulletin boards, file transfer, etc. The User Management System is concerned with providing controlled access to archiving facilities, and more generally provides a standard access paradigm between generic clients and servers for all information systems developed by ESRIN's Information Systems Division.

The ESIS Query and Correlation Environments will provide information retrieval, handling and storage functions, as well as the possibility to cross correlate data from different sources. The user functionality defined so far within these two environments will be described in better detail.

SCIENCE ORIENTED INFORMATION RETRIEVAL

A key issue in ESIS's capability to integrate heterogeneous information sources is the definition of a "unified data model". In this section we will describe how such a model has been built, what are its key components, and what user functions will eventually make ESIS a science information system.

The Heterogeneity Factor

In order to understand how a unified data model can be designed, it is first necessary to briefly summarize the levels of heterogeneity that are to be overcome. Heterogeneity is reflected at many levels:

PHYSICAL HETEROGENEITY

Data stored by different database management systems (DBMS) or file systems on a variety of different hardware platforms generally have different file formats, index mechanisms, record lengths and so on.

SYNTATIC HETEROGENEITY

The different catalogues and archives have their own denomination of objects, their different spelling rules, different abbreviations, different fault tolerance and, of course, their own query language. The result is that it is practically impossible for a novice user to remember exactly how to syntactically formulate his database queries.
Also different (mis) spelling causes problems. In the INSPEC database of bibliograph references, for instance, "The Monthly Notices of the Royal Astronomical Society of the UK" is spelled in (minimum) 10 different ways. In general, spelling mistakes resulting in uncontrolled keywords, may lead to loss of information.

SEMANTIC HETEROGENEITY

On the semantical level, heterogeneity appears when different meaning is attached to one single term (homonyms), or inversely when different terms express the same meaning (synonyms).

CONTEXTUAL HETEROGENEITY

Contextual heterogeneity primarily refers to the fact that equal terms can be interpreted differently in different science disciplines.

A Layered Data View Model

Generally speaking, connecting or combining information from different sources implies providing an overview of the available information, transparent to "where" and "how" it is stored. On the other hand one surely wants to have access to the data records themselves. These requirements call for a structure of a layers that provide different levels of abstaction. In fact, this is how ESIS overcomes all levels of data heterogeneity, the upper layer provides a unified view while the lower layers contain archive specific detail information.

SUMMARY DESCRIPTION LAYER

The summary description layer contains high-level key information about all accessible information in a uniform, homogenised way. The conceptual schema for this layer will be drawn from the union of all semantic definitions for the disciplines involved in ESIS. This key information should be suitable for making an initial determination of the potential usefulness of an information object (e.g. a data granule).
In this layer all heterogeneity levels are overcome except contextual heterogeneity, for there will be one of these layers for each discipline.

DETAIL DESCRIPTION LAYER

This is primarily what the ESIS pilot project databases offer today. It is still information about something, but a more detailed level than above.

There may be different kinds of detail descriptions:

- "Database catalogue entries" (e.g. an observation log), a detailed description of lower level data available at that data centre.

- "Quick-look" products, i.e. data sets of limited size and resolution created to provide an understanding of the type and quality of available full resolution data, and/or to enable the selection of intervals for the analysis of physical events.

- "Examples", simply limited-in-size examples of data instances, which could give a clue to the kind of data in question. Instead of, for instance retrieving a full data set, the user can request an example, in which case the system could provide 1-2 pages of the full data instance.

- "Abstracts", a textual summary of the contents of the actual information item. These could be descriptions of publications, quantitative results, catalogues, databases, data systems, etc.

DATA INSTANCES LAYER

At this layer we are confronted with the actual data themselves (and not descriptions thereof). In the case of observations this may be an image, a spectrum, etc., while in the case of publications, it could be a hardcopy of the actual article.
Naturally, information from the last two layers is specific to each database, and so of heterogeneous nature. However, when retrieved over ESIS, physical formats will be converted to an internal common standard allowing further manipulation at the physical level.

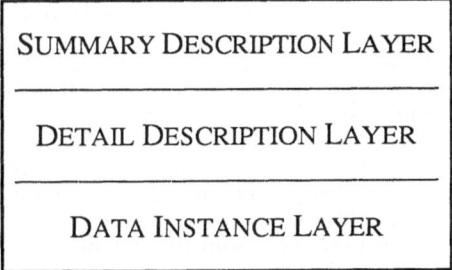

SUMMARY DESCRIPTION LAYER

DETAIL DESCRIPTION LAYER

DATA INSTANCE LAYER

Fig.2. ESIS Query Environment Information Access Layers

The ESIS Reference Directory

The summary description layer mentioned above reflects the unified data model for a particular discipline. In order to model the way scientists from different science domains see their "reality of interest", semantic networks have been designed that constitute the basis for the eventual retrieval system. Since each discipline has its own semantic denominations, one such network has been worked out for each of the disciplines that ESIS will serve: Astronomy & Astrophyisics, and Space Physics. The design of these schemata has been conducted by the Centre de Données astronomiques de Strabourg (CDS) for astronomy and by the Rutherford Appleton Laboratory (RAL) in collaboration with the Sheffield University for space physics, under contract for ESA, and has involved the participation of users through workshops and

Appleton Laboratory (RAL) in collaboration with the Sheffield University for space physics, under contract for ESA, and has involved the participation of users through workshops and working groups (Egret et al., 1990). Figure 3 shows the latest available version of the astronomical semantic network drawn using the Entity-Relationship diagramming technique. Boxes represent entities, diamonds represent relationship among entities; attributes for both are given when defined. A query is composed by selecting entities and/or relationships possibly qualifying them by giving values to some attributes.

In order to be able to answer user queries, the retrieval system must take into consideration both the user data models, as defined by the discipline semantic networks and the union of all local archive data descriptions. The database in ESIS that integrates these two aspects is called the "Reference Directory" and it includes both the global data dictionary as well as the summary description layer. The Reference Directory will play a fundamental rôle in the system's ability to provide quick responses to user requests.

The key difference between the Reference Directory and "classic" approaches (see e.g. Larson, 1989) solving the problem of heterogeneous, distributed databases, is that in ESIS heterogeneity is overcome not only at the level of the data description, but also at the level of their "semantic" contents (see section 5.1 above). Another distinctive feature of the Reference Directory is that the system can take full advantage of the possibility to resolve queries at a "summary information" level, i.e. needing only a small subset of the global data contents available in the system. After the user has gained a clear idea of the datasets of interest to his/her needs through inspecting summary descriptions thereof, a highly efficient fetching method can be applied to actually retrieve the datasets in question. In fact, not only datasets but also detailed descriptions, quick-looks or other information items can be retrieved in the same way. The method used here is that of "intelligent pointers" that are embedded in the Reference Directory for each datum described therein. These pointers can be thought of objects that are activated upon request and that contain enough knowledge to perform an information retrieval operation. The details of this operation are concealed within the object's interface: they could be small SQL programs or query language constructs, they could also be arithmetic operations, requests to a server on the network or simply file fetching commands.

Discipline Specific Query Languages

The superset of functions available to the user can be described in terms of what can be called a "Discipline Query Language". In fact, it has resulted from the definition of the system that each discipline has not only its own semantic definitions but also specific query primitives, e.g. {select-by-cone} in astronomy.

In general we envisage that through the Query and Correlation Environments, users will gain access to the global information contents of ESIS as well as access to personal databases. Personal databases could contain data imported from external sources or data retrieved from the global database. Also the exchange of databases with other users will be supported.

The user will interact with the environment through a suite of so-called "interaction paradigms". These paradigms will support command or query input in a variety of ways and will display results in a form suitable to the data. Database "views", i.e. the capability to tailor the amount of information the user sees from the database, will play an important rôle. Personal "views" of databases will allow a customized access to the information stored. Globally pre-defined "views" will allow groups of individuals to share the same way of accessing data. The formulation of queries will support the combination of any of the interaction paradigms with any user defined view. The overall environment will be enhanced by including standard support tools and discipline specific ones such as coordinate-system transformations and physical-unit conversions.

Among the functions that will be available through the User Shell the following should be highlighted:

QUERY AND COMMAND FORMULATION

Command or query formulation will be supported through a set of interaction paradigms that will be used in any combination at any given time. Depending on the kind of input required, the most suitable method will be selected either by the user himself or by the application as the default setting. The interaction paradigms mentioned above will be complemented by discipline specific query primitives, such as search-by-cone, -ellipse,-polygon, etc.

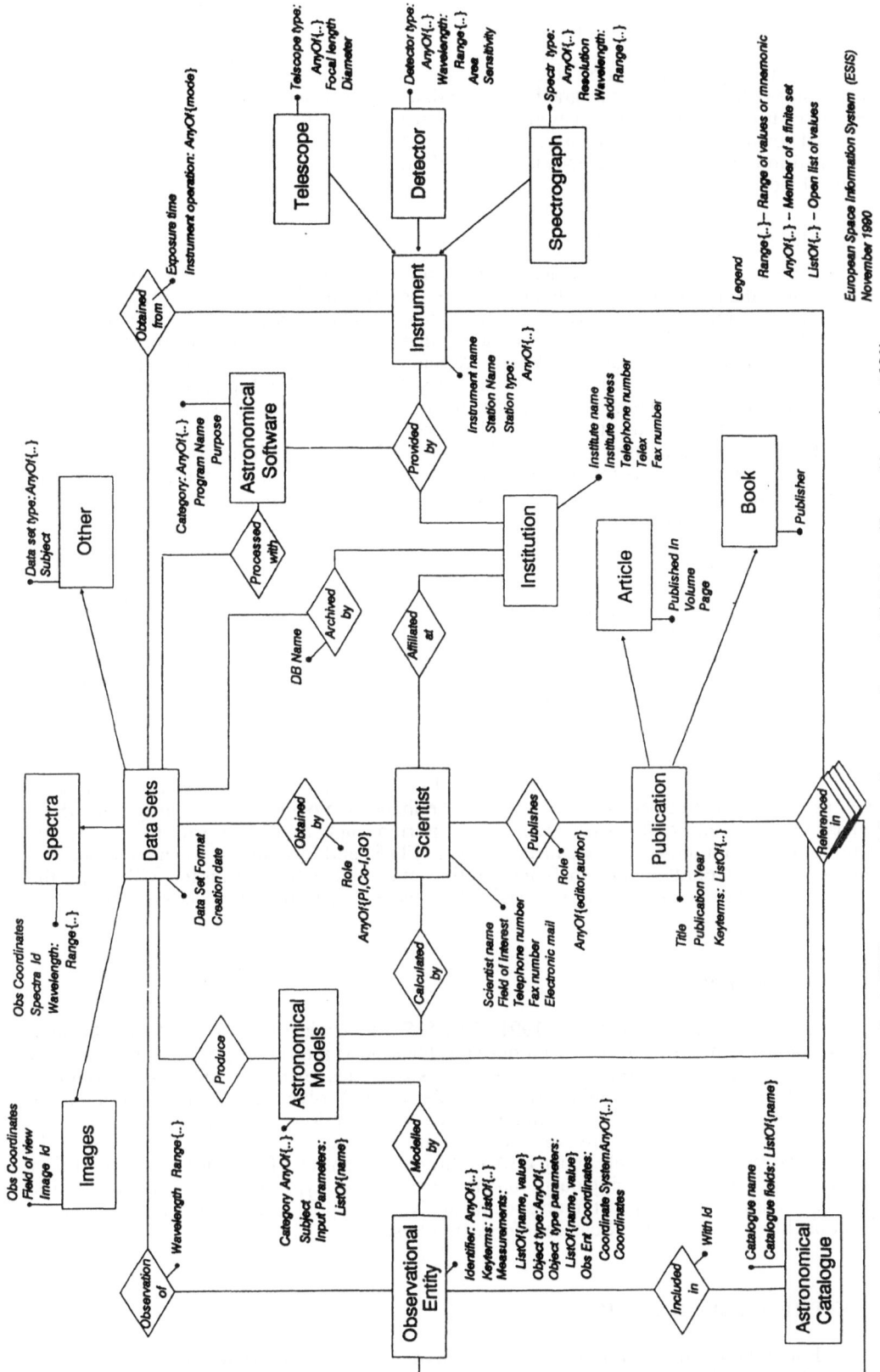

Fig. 3. ESIS Astronomical Query Language: Semantic Definitions (Status: November 1990).

OBJECT (DATA) VISUALIZATION ACCORDING TO STRUCTURE

The display of information items will be driven by their internal structure; in this way a vector plot will be visualized as graphs while a database record could be displayed in a form. Database structure will be shown as entity-relationship diagrams by default. Using an object-oriented terminology, we can say that sending the "message" {display-yourself} to any information item, will produce a visualization of the object's contents in the form best suited to the object's structure.

PERSONAL VIEWS AND DATABASES

A basic requirement in science information system is the ability to manipulate data according to the needs of a research project. The basis for such a capability is implemented in ESIS both through functions to keep and manage personal databases and functions to look at the database contents in particular ways (views) that are created circumstantially. A graphical concept manipulation language (see Catarci & Santucci, 1988) will allow users to create and manage databases and views in a user friendly way. Entity-Relationship diagrams will be used as input paradigm for all functions related to the manipulation of database structure, i.e. view manipulation and management of personal databases.

The personal database capability will form the basis for multi-mission, multi-spectral analysis. It will allow users to retrieve data from different sources, that, when delivered will share the same physical, syntactic and semantic format, paving the way towards cross correlations accross spectral domains.

SUPPORT TOOLS

It is envisaged to provide within the system general support tools for information manipulation. These functions are not yet fully defined; however, at the time of writing, they will include a "scientific spreadsheet" capability, an astronomical calculator and means to support user notes, glossaries and lexica. These tools will support coordinate system and physical unit transformations, the handling of user catalogues and their manipulation, etc.

INFORMATION EXCHANGE

The possibility to export data from the ESIS environment onto external applications as well as the complementary import function will enhance the overall functionality of ESIS. A set of standard external formats will be supported such as the formats in use by the astronomical community for spectra and images (FITS) and formats for graphical metafiles. Further, fully integrated support functions like mail and data set transfer will allow the transparent exchange of data sets, personal databases, and other processed information.

CONCLUSIONS

At the time of writing (March 1991) the ESIS pilot project has gone half way through. An overall system structure has been defined, leading to a modular, open architecture. A science oriented information retrieval system has been designed that takes into account all levels of heterogeneity among the different databases, and solves the integration problem by defining a homogeneous layer containing summary descriptions of the data. The conceptual schema of this layer is given by semantic definitions, that, for each discipline involved in ESIS, have been worked out, under ESA contract, by centres of excellence with end-user participation. Discipline specific query functions enhance the system interrogation capabilities. Personal databases will support multi-spectral analysis. Support tools will enhance the overall functionality to make ESIS a Science Information System.

FURTHER INFORMATION

For further information and to receive the ESIS Newsletter and workshop announcements please contact:

The ESIS Project
ESA/ESRIN
Via Galileo Galalei
I-00044 Frascati, Italy
Tel: +39 (6) 941801, Fax: +39 (6) 94180361
ESIS::ESIS on SPAN
ESIS@IFRESA51 on EARN

ACKNOWLEDGMENTS

The author wishes to acknowledge the enthusiasm and support experienced from all individuals that have actively participated in the various phases of the project, in particular, from the ESIS teams at ESRIN, at the CDS, at RAL and at Sheffield University. All members of the ESIS Steering Committee as well as of the Astronomical Databases and Space Physics Working Groups have injected substantial momentum into the project, their contribution has been highly appreciated.

REFERENCES

1. Albrecht M.A., Russo G., Richmond A., Hapgood M.A. (1988), "Towards a European Space Information System, Vol. 1, General Review of User Requirements", ESA internal report.

2. Albrecht M.A. & Bodini A. (1988), "Towards a European Space Information System, Vol.2, Implementation Plan", ESA Internal Report.

3. Catarci C. & Santucci G. (1988), "Query by Diagram: A graphical query system", Proc. 7th Conf. on ER Approach, North Holland, Amsterdam.

4. Larson J.A. (1989), "Four Reference Architectures for Distributed Database Management Systems", Computer Standards & Interfaces, 8, 209.

5. Egret D., Ansari S.G., Denizman L., Preite-Martinez A. (1990), "The ESIS Query Environment: Final Report", ESA Internal Report.

MARKOV RANDOM FIELD APPLICATIONS IN IMAGE ANALYSIS[1]

Anil K. Jain and Sateesha G. Nadabar

Department of Computer Science
Michigan State University
East Lansing, MI 48824
U.S.A.

Introduction

Image analysis, or computer vision, deals with machine processing of visual (or pictorial) information. Visual sensory data are usually obtained in the form of two-dimensional images of the physical world and consists of measurements which are dependent on factors such as the imaging geometry, illumination, and structures present in the world. The goal of any vision system is to recognize familiar structures in the system's environment, and obtain concise symbolic descriptions of unfamiliar structures, using visual sensory data. The task of a vision system, then, can be described as a data reduction task which extracts meaningful descriptions from huge amounts of visual data. The description obtained from the visual data should remain invariant for certain changes in the physical world [3]. In many computer vision applications, invariance against the following factors is required [3]: (i) sensor noise; (ii) optical distortion; (iii) viewpoint; (iv) perspective distortion; and (v) variations in photometric conditions.

Figure 1 shows a block diagram of a computer vision system. This chapter deals primarily with the early vision modules whose task is to recover the physical or intrinsic properties, at each image location (pixel), of the corresponding 3D world point being sensed. That is to say, early vision modules attempt to invert the *world-image* mapping. However, the world-image mapping is a many-to-one mapping and is not invertible because, typically, there are many world configurations which result in the same sensory data. In addition to dealing with this non-invertible mapping problem, early vision processes need to deal effectively with the ubiquitous sensor noise.

It is now generally agreed that computer vision systems must restrict the set of possible solutions using various constraints in order to obtain a unique stable solution to the problem. The addition of these constraints converts an ill-posed problem into a well posed one. Such methods have been called regularization methods [17]. The type of constraints used depends on the environment in which the vision system operates and the type of information that it seeks in the environment. Some generic constraints that have been incorporated in early vision modules are: (i) neighboring pixels are likely to have similar image properties; (ii) close parallel edges are unlikely; (iii) isolated edge elements are improbable; and (iv) surfaces are smooth.

We are interested in regularization methods that take the Bayesian approach to solve a variety of image analysis problems. In this approach, the *a priori* knowledge about image structure is modeled by a Markov random field (MRF). This prior information is combined with the observed data, and statistical estimation techniques are used to obtain the solution. A Markov random field associates a probability value to each point in the solution space according

---

[1] This work was supported by NSF grant IRI-8901513

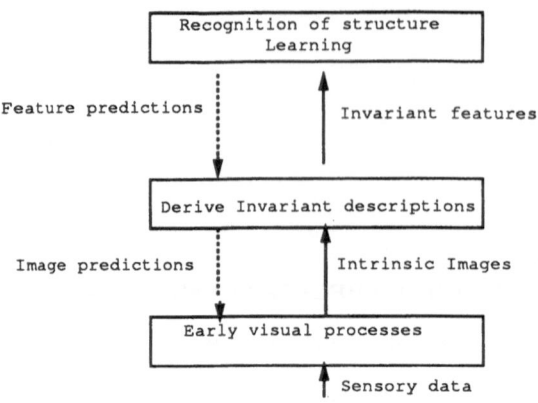

Figure 1. Block diagram of a typical vision system

to a Gibbs distribution. What is more important, a significant portion of the structure in the image is quantified in the form of statements such as "points that are close together in the image plane usually belong to the same surface" and this can be specified by a locally dependent MRF.

The computer vision literature contains several examples of problems for which MRF models have been used to obtain a solution, including texture analysis [14, 4, 9], image restoration [4, 18, 12], image segmentation [11, 9], and sensor fusion [19, 8]. These problems can, by and large, be grouped into two categories: (i) MRF is used to model the contextual information to enhance decision making algorithms, and (ii) MRF is used to describe the image using a few parameters so that the image can be synthesized. This chapter surveys some applications from both these categories.

Background

Let $\mathbf{X} = \{ X_1, X_2, ... X_M \}$ be the M-tuple random vector representing the colors or labels at the M sites of an image lattice S. Although the image is usually a 2-dimensional array, a linear ordering of the pixels (sites) of the image is assumed for notational convenience. The range of labels could be either discrete or continuous. Let the set of all possible labelings $\mathbf{x}$ be Ω. We now define a neighborhood system, Markov random field, and Gibbs random field.

A neighborhood system, $\mathcal{N}$, on a lattice, S, is a collection of subsets, $\mathcal{N}_t$ of S, where $\mathcal{N}_t$, the neighborhood of site t satisfies the following two criteria: (i) the site t is not a neighbor of itself ($t \notin \mathcal{N}_t$); and (ii) if s is a neighbor of t, then t is a neighbor of s (if $s \in \mathcal{N}_t$ then $t \in \mathcal{N}_s, \forall t$). First-order (four neighbors: East, West, North and South) and second-order (eight neighbors: four first-order neighbors plus four diagonals) neighborhoods are typically used. Consider the graph defined by $\{S, \mathcal{N}\}$. Here, the sites form the vertices of the graph and an edge is placed between two vertices if the corresponding sites are neighbors of each other. A *clique* is a subset $c \subset S$ in which every pair of sites are neighbors. That is, the subgraph formed by the sites in a clique along with the edges between them is completely connected. Let C denote the set of all cliques in $\{S, \mathcal{N}\}$. For the first- and second- order neighborhoods, the number of cliques in C are 5 and 21, respectively.

A *Markov Random Field* (MRF) is a joint probability distribution on a set of random variables which satisfies the following properties: (i) Positivity: $P(\mathbf{x}) > 0$ for all $\mathbf{x} \in \Omega$, and (ii) Markovianity: $P(x_t | \mathbf{x}_{S \setminus \{t\}}) = P(x_t | \mathbf{x}_{\mathcal{N}_t})$, where \ is the set difference operator.

While a MRF specifies the *local* properties of the image in terms of conditional distributions $P(x_t|\mathbf{x}_{\mathcal{N}_t})$, a Gibbs random field specifies the *global* properties of the image directly.

A *Gibbs Random Field* (GRF) on a lattice, S, for a given neighborhood system is a joint distribution which is specified by the following probability measure on the sample space Ω for the lattice.

$$P(\mathbf{x}) = e^{-U(\mathbf{x})}/Z,$$

where

$$U(\mathbf{x}) = \sum_{c \in C} V_c(\mathbf{x}),$$

$$Z = \sum_{\mathbf{x}} e^{-U(\mathbf{x})},$$

and $V_c(\mathbf{x})$ is the potential function (alternatively, clique function) associated with the clique c. The function Z is a normalizing constant and is known as the partition function. The summation in the definition of Z should be replaced by an integral when $\mathbf{x}$ is continuous. It should be noted that the computation of Z is not possible, except in trivial cases, due to extremely large cardinality of the set Ω.

One of the major difficulties with using MRFs is to verify whether a given conditional distribution is indeed valid or satisfies the definition of an MRF [1]. This makes it very difficult to specify an MRF in terms of conditional distributions. The Hammersley-Clifford theorem allows the specification of a MRF in terms of clique potentials [1]. Thus, for a given neighborhood system, a unique MRF exists for every GRF and vice versa. Typically, an image is modeled globally by a GRF using the associated clique potentials, and then the local dependence property of the MRF is used for computational advantage since only a small number of variables in the neighborhood of a site need to be used to compute the locally dependent conditional distribution at that site [12].

Texture Analysis

Texture is an important characteristic of many natural images. Following three problems are of interest in texture analysis studies: texture synthesis, texture classification, and texture segmentation. Gaussian Markov Random Field (GMRF) models are good candidates for modeling natural textures and have been used in all three aspects of texture analysis.

Texture synthesis

The goal of texture synthesis is to reconstruct the textured image from some texture models. We assume a non-causal finite lattice GMRF model with doubly periodic boundary condition. Under this assumption, the relationship between observations $y(s), s \in \Omega = \{(i,j)|i,j = 0, \cdots, M-1\}$ and their neighbors is represented as [4],

$$y(s) = \sum_{r \in \mathcal{N}} \theta_r y(s \oplus r) + e(s),$$

where the stationary Gaussian noise sequence $e(s)$ has the following properties:

$$E[e(s)e(r)] = \begin{cases} -\theta_{s-r}\nu, & (s-r) \in \mathcal{N} \\ \nu, & s = r \\ 0, & \text{otherwise.} \end{cases}$$

The notation $\oplus$ stands for sum modulo M along both coordinate axes, and the model parameters are θ_r and ν. The estimation of model parameters is a difficult problem [14]. Figure 2 shows two textured regions generated by two GMRFs with parameters listed below ($\nu = 1.1111$ for both regions): Left region: $\mathcal{N} = \{(1,0),(0,1),(-1,0),(0,-1)\}$, $\theta_{(1,0)} = \theta_{(-1,0)} = 0.2794$, $\theta_{(0,1)} = \theta_{(0,-1)} = 0.1825$; Right region: $\mathcal{N} = \{(1,0),(1,-1),(0,1),(1,1),(-1,0),(-1,1),(0,-1), (-1,-1)\}$, $\theta_{(1,0)} = \theta_{(-1,0)} = 0.3357$, $\theta_{(1,-1)} = \theta_{(-1,1)} = -0.25$, $\theta_{(0,1)} = \theta_{(0,-1)} = 0.3246$, $\theta_{(1,1)} = \theta_{(-1,-1)} = -0.2126$.

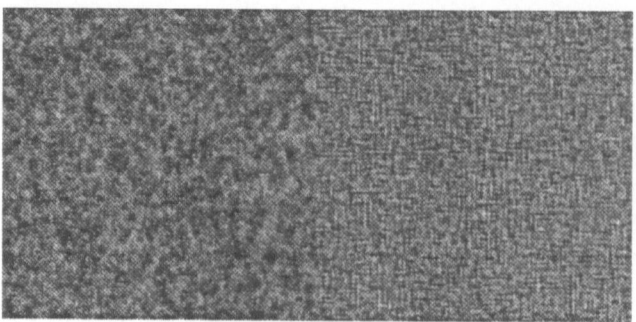

Figure 2. Two synthesized (128×128) textured regions.

Texture classification

The model parameters, θ_r, $r \in \mathcal{N}$, and ν capture the characteristics of the texture. The parameter θ_r specifies the dependence of a pixel on its neighboring pixels, and the standard deviation, ν, has a direct relationship with the visually perceived granularity or "business" of the texture. These model parameters, therefore, can be used as features for texture classification.

To improve the classification performance, we have introduced a multiresolution rotation-invariant simultaneous autoregressive (RISAR) model [14] which is basically a collection of RISAR models with the same neighborhood, but with different image resolutions. We observe that the classification accuracies using single-resolution images are not encouraging (60% – 89%), while the multiresolution RISAR model provides very high classification accuracies (94% – 100%).

Texture segmentation

The parameters of the GMRF and SAR models can also be used as features in texture segmentation algorithms. We introduced a multiresolution SAR model [14] which integrates the information captured by SAR models with different neighborhood sizes. The multiresolution SAR model can be written as,

$$y(s) = \mu_l + \sum_{r \in \mathcal{N}_l} \phi_l(r) y(s+r) + \varepsilon_l(s),$$

where $\mathcal{N}_l = \{(0,l),(0,-l),(l,0),(-l,0),(l,l),(l,-l),(-l,l),(-l,-l)\}, l = 1, \cdots, L$.

The value of l controls the neighborhood size. We used 25×25 overlapping windows moving every two pixels in both the horizontal and vertical directions to estimate the model parameters. Symmetric models are used, so each model has six parameters, five of which are used as features (we do not use μ_l due to its dependence on the mean gray value of the image). The parameters (features) from all the models, $\{\phi_l(r), \sigma_l \mid r \in \mathcal{N}_l, l = 1, ..., L\}$, are appended to form the feature vector at pixel s, $\{f_1(s), f_2(s), \cdots, f_d(s)\}$, where $d = 5L$. The k-means clustering algorithm is used for labeling each pixel. Segmentation results in Figure 3 indicate that the multiresolution SAR model provides better performance than the single-resolution SAR model.

MRFs as models of Context

Markov random field models have been successfully used to represent contextual information in many 'site' labeling problems [2, 5, 10, 11, 12, 15]. A site labeling problem involves classification of each site (pixel, edge element, region) into a certain number of classes based on an observed value (or vector) at each site. For example, in a typical image segmentation problem, pixels

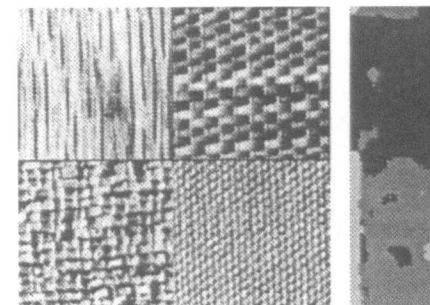

Figure 3. Segmentation results. Left column: input image (256 × 256); middle column: segmentation with $L = 1$; and right column: segmentation with $L = 3$.

O : Pixel site

— : Horizontal edge site

| : Vertical edge site

Figure 4. Pixel and line (edge) sites

are labeled based on observed value(s) at each pixel, where the number of labels correspond to the number of true segments or regions. We assume that the desired labeling is the true image. The goal of site labeling problem then is to recover the true image from the noisy observed image. Contextual information plays an important role here because the true label of a site is compatible in some sense with the labels of the neighboring sites. Context represents our *a priori* assumptions about the physical world such as continuity and smoothness. MRFs are appropriate models of context because they can be used to specify the spatial dependency.

Coupled Markov Random Fields

Geman and Geman [12] introduced the notion of line variables (sites) to augment usual pixel variables (sites) in order to model spatial discontinuities in the pixel values. The line variables are placed in a dual lattice (vertical and horizontal sites between the pixel sites) as shown in Figure 4. The stochastic process defined by the line variables is usually referred to as a *line process* or *edge process*.

A coupled MRF is defined on the coupled lattice of pixel and line variables as shown in Figure 4 [12, 15]. A commonly used neighborhood system for defining coupled MRF is shown in Figure 5. Coupled MRFs help in the specification of *a priori* constraints in two ways: (i) the values of line variables can be used to curtail undesired smoothing across discontinuities, and (ii) the cliques of the line process can be used to specify constraints on the structure of the discontinuities.

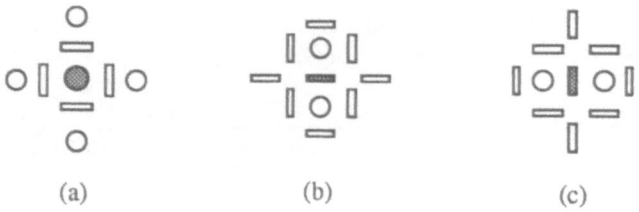

(a)   (b)   (c)

Figure 5. Neighborhood of a pixel and an edge site (a) pixel site (b) horizontal edge site (c) vertical edge site

MRFs as Models of Spatial Context: Bayesian Formulation

Let $(\mathbf{X}, \mathbf{L})$ be the process representing the 'true' labels at all the sites of a coupled lattice $(\mathcal{S}, \mathcal{L})$. The pixel label at site t, X_t, takes a value from the set of labels, $A = \{1,...,G\}$. The line label at site t, L_t, is typically binary indicating whether an edge is present at that location or not. Let $\mathbf{Y}$ represent the observations (e.g., gray values) at the pixel sites. Thus we have two types of information about the site labels. One is in the form of measurements $\mathbf{Y}$, and the other is in the form of spatial dependencies among the labels $(\mathbf{X}, \mathbf{L})$ represented by the probability distribution $P(\mathbf{x}, \mathbf{l})$. The *a posteriori* probability of $(\mathbf{X}, \mathbf{L})$ given $\mathbf{Y}$ is written as

$$P((\mathbf{x},\mathbf{l})|\mathbf{y}) = \frac{P(\mathbf{y}|(\mathbf{x},\mathbf{l}))P(\mathbf{x},\mathbf{l})}{P(\mathbf{y})} = \frac{P(\mathbf{y}|\mathbf{x})P(\mathbf{x},\mathbf{l})}{P(\mathbf{y})} \tag{1}$$

The *a priori* probability density function for $(\mathbf{X}, \mathbf{L})$ is assumed to be a Gibbs random field [1, 12] with respect to a neighborhood system of cliques

$$P(\mathbf{x},\mathbf{l}) = e^{-U(\mathbf{x},\mathbf{l})}/Z, \tag{2}$$

where Z is the partition function, and $U(.)$ is the energy function

$$U(\mathbf{x},\mathbf{l}) = \sum_{c \in C} V_c(\mathbf{x},\mathbf{l}), \tag{3}$$

C is the set of all cliques with respect to the neighborhood system, and $V_c(.)$ are the potential functions which map the local interactions of the sites in clique c to energy contributed by the clique towards the total energy. The functions $V_c(.)$ encode our *a priori* knowledge about the spatial dependence of labels at neighboring sites. For site labeling problems, natural constraints such as smoothness and continuity are enforced using $V_c(.)$.

For illustration purposes, we consider a simple degradation model, where the observed degraded image is written as

$$\mathbf{Y} = H\mathbf{X} + N$$

where H is the blurring matrix, and N is the Gaussian noise process which is independent of $\mathbf{X}$. Therefore, the distribution of $\mathbf{Y}$, given the true labeling, is

$$P(\mathbf{Y} = \mathbf{y}|\mathbf{X} = \mathbf{x}) = \left(\prod_{t \in \mathcal{S}}(2\pi\sigma^2)^{-1/2}\right) e^{-\sum_{t \in \mathcal{S}}\left[\frac{(y_t - y_t')^2}{2\sigma^2}\right]}$$

where, N is assumed to be i.i.d. Gaussian with mean zero and variance σ^2, and

$$H\mathbf{x} = \{y_t', t \in \mathcal{S}\}.$$

This conditional distribution is not valid when the image contains texture or correlated noise is present. For a formulation with more general degradation process, see [12].

In the above formulation, the *a posteriori* probability distribution for the site labels $\mathbf{X}$, given the observations $\mathbf{Y} = \mathbf{y}$ also has the form of a Gibbs random field. The neighborhood for the *a posteriori* GRF is determined by the size of the blurring matrix. If $\mathcal{H}_t$ denotes the set of pixels that affect the value y_t' at site t, then the neighborhood system for the posterior GRF is given by [12]

$$\mathcal{N}_t^P = \left\{ \begin{array}{ll} \mathcal{N}_t & t \in \mathcal{L} \\ \mathcal{N}_t \bigcup \mathcal{H}_t^2 \backslash \{t\} & t \in \mathcal{S} \end{array} \right.$$

where $\mathcal{H}_t^2 = \bigcup_{s \in \mathcal{H}_t} \mathcal{H}_s, t \in \mathcal{S}$. The *a posteriori* distribution of labels is given by,

$$P((\mathbf{x}, \mathbf{l})|\mathbf{y}) = e^{-U((\mathbf{x}, \mathbf{l})|\mathbf{y})}/Z_{\mathbf{y}}, \tag{4}$$

where $Z_{\mathbf{y}}$ is a normalizing constant, and the corresponding energy function is

$$U((\mathbf{x}, \mathbf{l})|\mathbf{y}) = \sum_{t \in \mathcal{S}} \left[\frac{(y_t - y_t')^2}{2\sigma^2} \right] + \sum_{c \in C} V_c(\mathbf{x}, \mathbf{l}). \tag{5}$$

Estimators for *a posteriori* labels

The site labeling problem can now be stated as follows: given the observation vector $\mathbf{y}$, estimate the label vector $\mathbf{x}$. Contextual information is represented by the rightmost term $(\sum_{c \in C} V_c(\mathbf{x}, \mathbf{l}))$ in Eq. (5). If this term was removed, energy minimization algorithms would assign the labels independently to each site. The MAP (Maximum *a posteriori*) estimate is the vector $\hat{\mathbf{x}}$ which maximizes $P(\mathbf{x}|\mathbf{y})$ with respect to $\mathbf{x}$. Maximizing this *a posteriori* probability mass function is a formidable task since the number of variables is very large even for images of moderate size. Geman and Geman [12] have used simulated annealing to obtain an approximation to the MAP estimate in the context of image restoration, but their procedure is also computationally expensive.

Several alternative estimates have been suggested for obtaining the *a posteriori* labeling. Marroquin *et al.* [15] have suggested that minimizing the expected value of an error functional is a better criterion for some problems. For this criterion, the optimal labeling is also the labeling that maximizes the *a posteriori* marginal probability (MPM) at each site. Besag [2] proposed a deterministic iterative scheme called *Iterated Conditional Modes* (ICM), in which, the labeling of a site is iteratively updated so as to maximize the local conditional probabilities. This algorithm finds a solution that corresponds to a local minimum of the energy function given in Eq. (5). For an experimental comparison of the simulated annealing, MPM, and ICM algorithms, in the context of intensity image segmentation, see Dubes *et al.* [11]. Recently Chou *et al.* [5] have proposed another estimation algorithm, known as the *Highest Confidence First* (HCF) algorithm. The HCF algorithm, like ICM, is deterministic and computationally attractive.

Image Segmentation

The problem of image segmentation is to partition an image into "homogeneous" regions. The homogeneity property could involve the pixel gray levels or some other property. The problem of partitioning an image based on homogeneity of texture is called the texture segmentation problem discussed earlier. Here, we will restrict the discussion to partitioning based on homogeneity of gray levels. This problem can be formulated as a site (pixel) labeling problem, where the desired labeling is the true region map. Contextual information is incorporated in the prior term in the Bayesian formulation and is modeled by a MRF. The MRF model-based segmentation algorithms perform well on regions of uniform intensity, but fail when the intensity in a region has a gradient as is often the case in real images [11]. Figure 6 illustrates this point.

We believe that this problem is better posed as an edge-based segmentation problem since discontinuities play the most important role in defining segments in such images. Most MRF

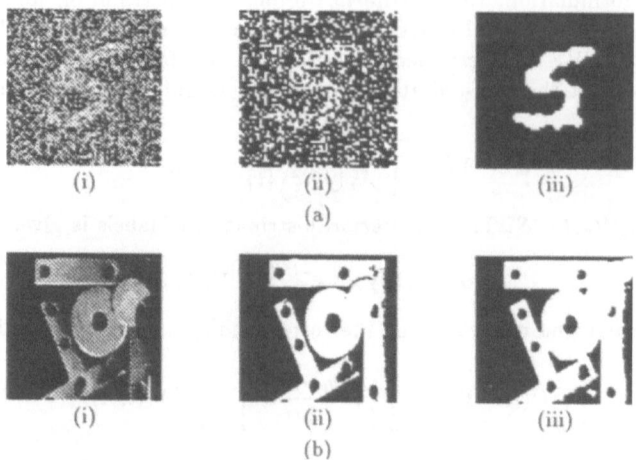

Figure 6. Segmentation results for (a) synthetic image (64 × 64) degraded with additive Gaussian noise and (b) real image (128 × 128) of industrial parts. (i) Noisy image, (ii) Maximum likelihood segmentation, and (iii) MRF segmentation using ICM algorithm

model-based segmentation algorithms require certain parameters to be specified [2, 9, 15, 12]. These include, the parameters of the prior model, the parameters of the degradation model, and the number of regions. Parameter estimation remains a difficult problem.

Image Restoration

Image restoration methods attempt to recover the "true" image from (noisy or degraded) observed image. A model of the degradation phenomenon is assumed to be available, and the contextual information is present in the form of spatial dependency of gray levels– nearby pixels tend to have similar gray levels. Image restoration is another instance of the site labeling problem. One significant difference between image segmentation and image restoration is that in restoration, the number of labels in the recovered image is of the same order as the number of gray levels in the corrupted image, whereas in segmentation, it is usually much less than the number of gray levels. A more fundamental difference is that segmentation labels usually are symbolic labels whereas labels in a restoration problem are numeric values corresponding to the gray levels in the image. Marroquin *et al.* [15] call the segmentation and restoration problems as reconstruction of piecewise constant and piecewise linear functions, respectively.

Discrete MRFs have been used for restoration by Geman and Geman [12], Besag [2], Marroquin *et al.* [15]. Geman and Geman used a hierarchical model consisting of the usual intensity process indexed by pixels, and an unobservable line process indexed by imaginary edge sites between pixels. The line process allows the formation of spatial discontinuities in the restored image. Further, line process allows for higher level geometric prior information such as encouraging continuous boundaries and discouraging bending and isolated edges. Jeng and Woods [13] have extended this work to continuous MRF models. Continuous MRFs are appealing for restoration, because usually, the undegraded pixel values are continuous in the real world. A popular class of continuous MRFs is the compound Gauss Markov random field model used by Jeng and Woods as well as by Simchony *et al.* [18] which has the advantage that sampling the conditional distribution is straightforward. Some impressive image restoration results have been obtained using MRF models [12, 18, 3, 15]. Figure 7 shows restoration results using a coupled MRF model for an image degraded by blur and additive Gaussian noise

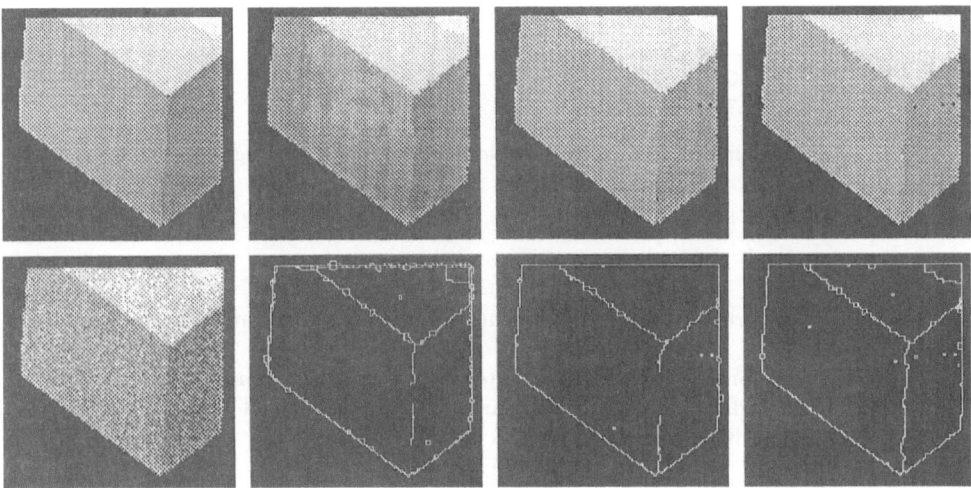

Figure 7. Restoration results using a coupled MRF model for an image degraded with blur and additive Gaussian noise: (a) true image (top) and blurred, noisy image (below); (b) restoration without using blurring model; (c) Restoration without using edge cliques of size 2 given in Figure 8; (d) Restoration using the complete model.

($\sigma = 15$). The blurring matrix H is

$$\begin{pmatrix} 1/8 & 1/8 & 1/8 \\ 1/8 & 1/2 & 1/8 \\ 1/8 & 1/8 & 1/8 \end{pmatrix}$$

The *a posteriori* energy function associated with this problem is

$$U((\mathbf{x}, \mathbf{l})|\mathbf{y}) = \sum_{t \in \mathcal{S}} \frac{(y_t - y_t')^2}{2\sigma^2} + \sum_{t \in \mathcal{S}} \sum_{\substack{s \in \mathcal{N}_t \bigcap \mathcal{S} \\ s > t}} \frac{(x_t - x_s)^2(1 - l_{ts})}{\nu} + \sum_{t \in \mathcal{L}} \sum_{\substack{s \in \mathcal{N}_t \bigcap \mathcal{L} \\ s > t}} V_{t,s}^L(l_t, l_s) + \sum_{t \in \mathcal{L}} \beta l_t$$

where ν and β are parameters, and $V_{t,s}^L(l_t, l_s)$ are as shown in Figure 8. Notice that the restored image is visually better when the complete model (edge constraints and blurring model) is used.

Multi-Sensor Image Analysis

Multi-sensor image analysis techniques deal with the fusion of information from different sources. Data fusion allows better decision making by relying more on sensory information and less on *a priori* assumptions which may or may not be true. By providing more sensory information, data fusion reduces the uncertainty in the decisions.

There are a variety of methods for achieving data fusion. These methods depend on the nature of the data sources, the relation between these sources and the desired output. For example, the type of processing required to combine two independent intensity images of the same scene to obtain a segmentation into uniform intensity regions is much different from that required for combining thermal and visual images of a scene to obtain segmentation in terms of uniform heat flux regions [16]. Clark and Yuille [6] give a good survey of various data fusion methods.

MRF models are appropriate for modeling the contextual information needed for image data fusion. In data fusion problems, a site is assumed to have a vector of observations.

Figure 8. Edge cliques used. Configurations not shown are assigned zero potential.

The labels at a site are univariate for segmentation problems and multivariate for restoration problems.

The data fusion problem can be formulated as a site labeling problem in much the same way as for univariate data. Assuming, there are k different sources of data, the energy function of Eq. (5) can be modified as follows.

$$U((\mathbf{x},\mathbf{l})|\mathbf{y}) = \sum_{t \in \mathcal{S}} (\mathbf{y}_t - \mathbf{y}_t')B^{-1}(\mathbf{y}_t - \mathbf{y}_t')^T + \sum_{c \in C} V_c(\mathbf{x},\mathbf{l}) \tag{6}$$

where $\mathbf{y}_t$ and $\mathbf{y}_t'$ are vectors of length k, and B is the $k \times k$ covariance matrix. Consider the problem of image restoration and edge detection in the presence of additive noise alone (no blurring assumed). Suppose k noisy images of the same scene are available. For simplicity, assume that the noisy images are statistically independent. The following energy function can be used to enforce the constraint that *edges in images from all sources occur at the same location* [19].

$$U((\mathbf{x},\mathbf{l})|\mathbf{y}) = \sum_{\alpha=1}^{k} \left[\sum_{t \in \mathcal{S}} \frac{(y_{t\alpha} - x_{t\alpha})^2}{2\sigma_\alpha^2} + \sum_{t \in \mathcal{S}} \sum_{\substack{s \in \mathcal{N}_t \cap \mathcal{S} \\ s > t}} \frac{(x_{t\alpha} - x_{s\alpha})^2(1 - l_{ts\alpha})}{\nu} \right.$$

$$\left. + \sum_{t \in \mathcal{L}} \sum_{\substack{s \in \mathcal{N}_t \cap \mathcal{L} \\ s > t}} V_{t,s}^L(l_{t\alpha}, l_{s\alpha}) + \sum_{t \in \mathcal{L}} \beta_\alpha l_{t\alpha} \right] + \sum_{t \in \mathcal{L}} V_t(\mathbf{l}_t)$$

where $y_{t\alpha}$ denotes the desired noisy value at site t in image no. α, $l_{ts\alpha}$ represents the line variable between pixel sites t and s in image no. α, $\mathbf{l}_t$ is the vector of line variables at edge site t, ν and β_α are parameters, $V_{t,s}^L(l_{t\alpha}, l_{s\alpha})$ are as shown in Figure 8, and $V_t(\mathbf{l}_t)$ is given by

$$V_t(\mathbf{l}_t) = \begin{cases} 0 & \text{if } \sum_{\alpha=1}^{k} l_{t\alpha} = 0 \text{ or } k \\ 1 & \text{otherwise} \end{cases}$$

The functions $V_t(\mathbf{l}_t)$ are used to incorporate the specified constraint since it penalizes configurations in which the values of all line variables at the same line site do not match. Figure 9 shows the results of using this approach to fuse two intensity images of the same scene with different amount of noise. It can be seen that edge map obtained by fusion is visually better than that obtained without fusion and illustrates the importance of fusion.

Summary

In this chapter we looked at two categories of image analysis problems in which MRF models have been applied to obtain a solution. In the first category, MRF models are used to

48

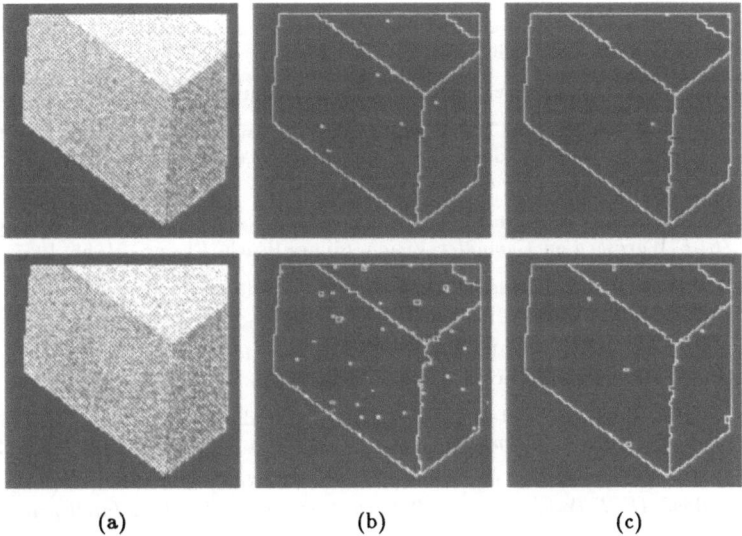

(a)   (b)   (c)

Figure 9. Edge detection results with and without fusion of intensity data from two sources: (a) noisy 100×100 images, $\sigma = 15$ (top) and $\sigma = 20$ (below); (b) Edges detected without fusion; (c) Edges detected with fusion.

mathematically describe image textures using only a few parameters. Significant success has been achieved in modeling image textures using MRF models. The issues in texture modeling using MRF are the specification of the model in terms of size of the neighborhood and clique parameters, robust estimation of parameters and test of goodness-of-fit between the data and the model.

MRF models have also been used to model the spatial *a priori* information or contextual information, in order to enhance the performance of decision making algorithms. This situation arises in a wide range of image analysis problems such as restoration, segmentation, edge detection, and multi-sensor fusion problems. Impressive results have been reported using methods that use MRF models in each of these problems. These results show that Markov random fields are useful as models of spatial dependency.

Despite these impressive results, MRF model-based algorithms are not widely used in practical applications because of some limitations in the specification of prior and degradation models, and the computational complexity associated with finding the optimal and even a suboptimal estimate. The specification of parameters for the prior contextual models is a difficult process and often the parameters are determined by trial and error. Also, an oversimplified degradation model is often assumed resulting in poor performance as illustrated by the segmentation example of Figure 6(b). The third and perhaps the most frequent criticism of MRF model-based algorithms is that they are computationally demanding. This is true for sequential implementations of (stochastic approximations of) MAP and MPM algorithms. These algorithms, however, are amenable to parallel implementation and could be efficiently implemented on massively parallel computers. Also, it is often sufficient to use the more efficient deterministic algorithms such as ICM, HCF, and Graduated Non-Convexity (GNC) [3] for restricted classes of energy functions.

References

[1] J. Besag, "Spatial Interaction and the Statistical Analysis of Lattice Systems," *Journal of the Royal Statistical Society, Ser. B*, vol. 36, pp. 192–236, 1974.

[2] J. Besag, "On the Statistical Analysis of Dirty Pictures," *Journal of the Royal Statistical Society, Ser. B*, vol. 48, pp. 259–302, 1986.

[3] A. Blake and A. Zisserman, *Visual Reconstruction*. Cambridge, MA: MIT Press, 1987.

[4] R. Chellappa. Two-Dimensional Discrete Gaussian Markov Random Field Models for Image Processing. In L.N. Kanal and A. Rosenfeld, editors, *Progress in Pattern Recognition 2*, pages 79–112. Elsevier Science Publishers B.V. (North-Holland), 1985.

[5] P. Chou, C. Brown, and R. Raman, "A Confidence-Based Approach to the Labeling Problem," in *Proc. of IEEE Workshop on Computer Vision*, Miami Beach, FL, pp. 51–56, 1987.

[6] J. Clark and A. Yuille, *Data Fusion for Sensory Information Processing Systems*. Boston, MA: Kluwer, 1990.

[7] G. Cross and A. Jain, "Markov Random Field Texture Models," *IEEE Transactions on Pattern Analysis and Machine Intelligence*, vol. 5, pp. 25–39, 1983.

[8] M. Daily, "Color Image Segmentation Using Markov Random Fields," in *Proc. 1989 IEEE Computer Society Conference on Computer Vision and Pattern Recognition*, pp. 304–312, 1989.

[9] H. Derin and H. Elliott, "Modeling and Segmentation of Noisy and Textured Images Using Gibbs Random Fields," *IEEE Transactions on Pattern Analysis and Machine Intelligence*, vol. 9, pp. 39–55, 1987.

[10] R. C. Dubes and A. K. Jain, "Random Field Models in Image Analysis," *Journal of Applied Statistics*, vol. 16, pp. 131–164, 1989.

[11] R. C. Dubes, A. K. Jain, S. G. Nadabar, and C. C. Chen, "MRF Model-Based Algorithms for Image Segmentation," in *Proc. Tenth International Conference on Pattern Recognition*, pp. 808–814, 1990.

[12] S. Geman and D. Geman, "Stochastic Relaxation, Gibbs Distributions, and the Bayesian Restoration of Images," *IEEE Transactions on Pattern Analysis and Machine Intelligence*, vol. 6, pp. 721–741, 1984.

[13] F. Jeng and J. Woods, "Simulated Annealing in Compound Gaussian Random Fields," *IEEE Transactions on Information Theory*, vol. 36, pp. 94–107, 1990.

[14] J.-C. Mao and A. K. Jain. Texture Classification and Segmentation Using Multiresolution Simultaneous Autoregressive Models. *Submitted to Pattern Recognition*, 1990.

[15] J. Marroquin, S. Mitter, and T. Poggio, "Probabilistic Solution of Ill-Posed Problems in Computational Vision," *Journal of the American Statistical Association*, vol. 82, pp. 76–89, 1987.

[16] N. Nandhakumar and J. Aggarwal, "Multisensor Integration — Experiments in Integrating Thermal and Visual Sensors," in *Proc. First IEEE International Conference on Computer Vision*, pp. 83–92, 1987.

[17] T. Poggio, V. Torre, and C. Koch, "Computational Vision and Regularization Theory," *Nature*, vol. 317, pp. 314–319, 1985.

[18] T. Simchony, R. Chellappa, and Z. Lichtenstein, "Pyramid Implementation of Optimal-Step Conjugate-Search Algorithms for Some Low-Level Vision Problems," *IEEE Trans. Systems, Man, and Cybernetics*, vol. 19, pp. 1408–278, 1989.

[19] W. Wright, "A Markov Random Field Approach to Data Fusion and Colour Segmentation," *Image and Vision Computing*, vol. 7, pp. 144–150, 1989.

SAST - A SOFTWARE PACKAGE FOR THE STATISTICAL ANALYSIS OF

SPARSE TIMES

R. Buccheri[1], M. Busetta[2], M.C. Maccarone[1]

[1]Istituto di Fisica Cosmica ed Applicazioni
  dell'Informatica, C.N.R., Palermo, Italy
[2]Space Science Department of ESA/ESTEC
  Noordwijk, The Netherlands

ABSTRACT

The paper describes the software package SAST developed for detection and description of periodicities and other irregular time structures in gamma-ray timing data. The package has been successfully used for the analysis of the photon arrival times measured by the COS-B satellite for gamma-ray astronomy and of the times derived from some observations of the Cerenkov telescope at the Potchefstroom University (South Africa).

INTRODUCTION

The analysis of gamma-ray photon arrival times for search of periodicities and other timing irregularities (i.e. short bursts of radiation) requires special attention due to the peculiarity of the experimental data whose main characteristics is the limited statistics connected with the low gamma-ray fluxes from celestial sources. In these conditions, the decision rules used by the experimenters for acceptation of a positive detection must be chosen with great care to avoid dubious results. The situation is generally much different in the X-ray range where the higher signals require more specialistic description techniques but allow little care in the decision phase. The standardization of the basic timing analysis tools for purposes of scientific comparison between different gamma-ray experiments adds therefore to the request by the gamma-ray scientific community of the implementation and use of a complete and self consistent software system of analysis because of the large amounts of data available per experiment.

We describe the analysis characteristics of the software package SAST, developed at IFCAI following many years of experience in the timing analysis of gamma-ray data. It presently consists of a set of FORTRAN77 programs which can be run in sequence in a VMS environment. A review of the

Data Analysis in Astronomy IV, Edited by V. Di Gesù *et al.*
Plenum Press, New York, 1992

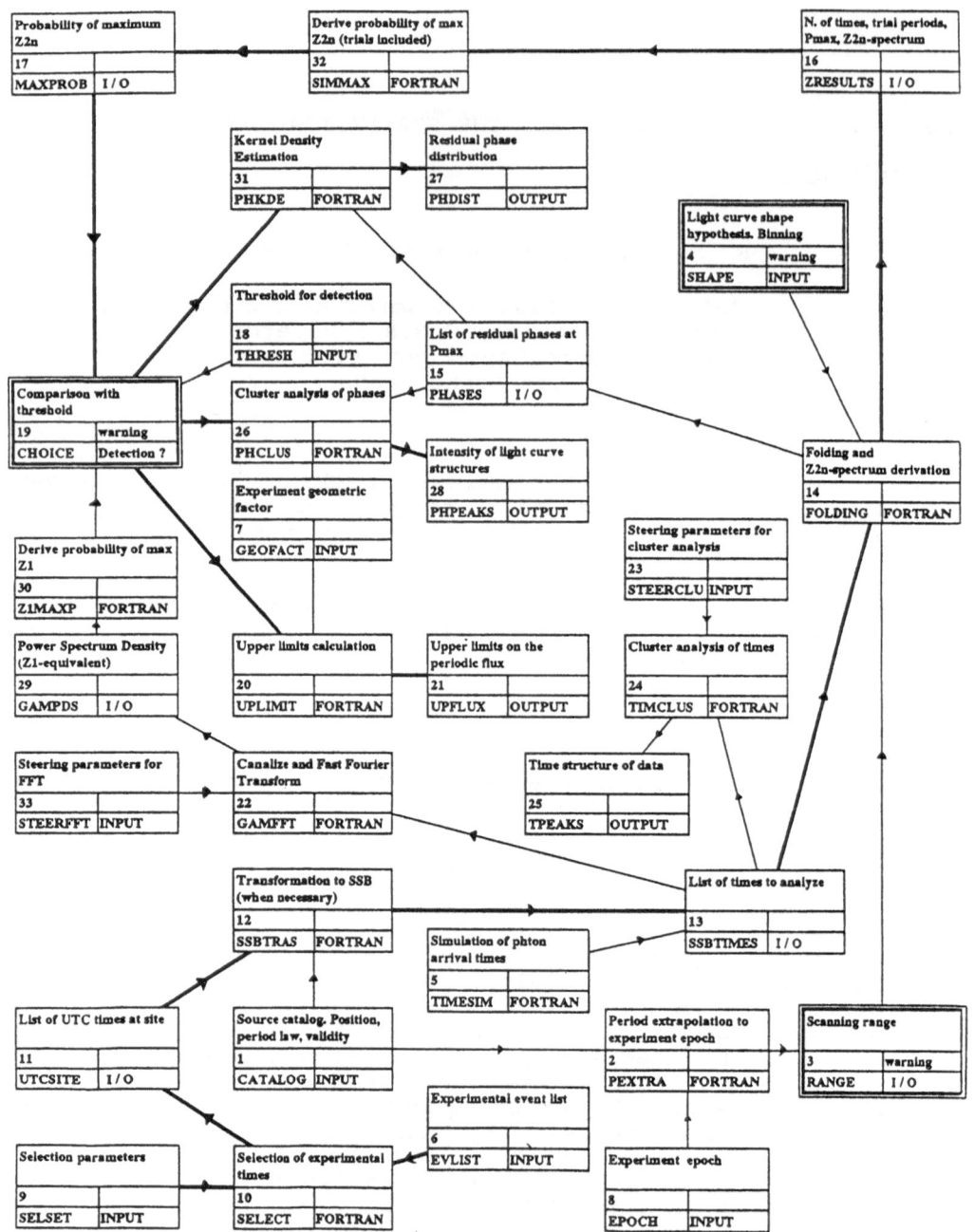

Probability of maximum Z2n			Derive probability of max Z2n (trials included)			N. of times, trial periods, Pmax, Z2n-spectrum		
17			32			16		
MAXPROB	I / O		SIMMAX	FORTRAN		ZRESULTS	I / O	

Kernel Density Estimation		Residual phase distribution	
31		27	
PHKDE	FORTRAN	PHDIST	OUTPUT

Light curve shape hypothesis. Binning	
4	warning
SHAPE	INPUT

Threshold for detection	
18	
THRESH	INPUT

List of residual phases at Pmax	
15	
PHASES	I / O

Comparison with threshold	
19	warning
CHOICE	Detection ?

Cluster analysis of phases	
26	
PHCLUS	FORTRAN

Intensity of light curve structures	
28	
PHPEAKS	OUTPUT

Folding and Z2n-spectrum derivation	
14	
FOLDING	FORTRAN

Experiment geometric factor	
7	
GEOFACT	INPUT

Steering parameters for cluster analysis	
23	
STEERCLU	INPUT

Derive probability of max Z1	
30	
Z1MAXP	FORTRAN

Power Spectrum Density (Z1-equivalent)	
29	
GAMPDS	I / O

Upper limits calculation	
20	
UPLIMIT	FORTRAN

Upper limits on the periodic flux	
21	
UPFLUX	OUTPUT

Cluster analysis of times	
24	
TIMCLUS	FORTRAN

Steering parameters for FFT	
33	
STEERFFT	INPUT

Canalize and Fast Fourier Transform	
22	
GAMFFT	FORTRAN

Time structure of data	
25	
TPEAKS	OUTPUT

Transformation to SSB (when necessary)	
12	
SSBTRAS	FORTRAN

Simulation of phton arrival times	
5	
TIMESIM	FORTRAN

List of times to analyze	
13	
SSBTIMES	I / O

List of UTC times at site	
11	
UTCSITE	I / O

Source catalog. Position, period law, validity	
1	
CATALOG	INPUT

Period extrapolation to experiment epoch	
2	
PEXTRA	FORTRAN

Scanning range	
3	warning
RANGE	I / O

Experimental event list	
6	
EVLIST	INPUT

Selection parameters	
9	
SELSET	INPUT

Selection of experimental times	
10	
SELECT	FORTRAN

Experiment epoch	
8	
EPOCH	INPUT

Figure 1

available literature and a discussion on the statistical aspects of the analysis for search of periodicities in gamma-ray data can be found in Buccheri (1991). The figure below shows the flow of operations followed by SAST; boxes show programs and data as described in the following.

INPUT DATA

Input data for SAST are information on the source to be investigated and on the experiment characteristics. Some parameters are also needed to steer the analysis and for decision on the acceptation of a positive signal.

Data on the sources investigated. These are contained in the source CATALOG describing the parameters of the sources to investigate (position, period law if known, epoch of measurement, flux, if known et.).

Data on the experiment. The experimental geometric factor (GEOFACT) is needed to evaluate the absolute intensity of the emission or its upper limit, the date of the experiment (EPOCH) for the extrapolation of the pulsation period and, of course, the experimental photon list (EVLIST) complete with all characterizing parameters (energy, position, time, etc.).

Data needed to steer the analysis. SELSET contains the parameters needed for the selection of the events to analyze (energy window, acceptance cone from the source, et.). In particular, to steer the FFT, the minimum period to investigate and, in connession, the bin width (both in STEERFFT) to canalize the times have to be given with the caution that a too short period could lead to an unrealisticly high computer time and to a too low detection efficiency. In STEERCLU it is provided the information on the size of the time structures looked for (ex. narrow structures like trojan matter emission). The steering set should be optimized at the beginning of the analysis process following the known characteristics of the experiment and the objective of the search.

Data needed for decision. The probability threshold (THRESH) for decision of positive acceptance of a pulsed signal has to be provided (see Maccarone & Buccheri, 1989 for a discussion on decision problems in the search for periodicities in gamma-ray astronomy). A word of caution must be spent in the choice of the threshold value; a reliable statistical confidence on the result of the analysis can only be obtained by including in its calculation the real period range investigated. In most cases, however, this is very difficult to achieve due to the intrinsic process of analysis in which some optimizations are done inconsciously (i.e. by considering, without any objective reason, for further analysis a "promising" result out of others not reaching the same consideration) and therefore, keeping too high the threshold for detection could lead to false alarms.

An hypothesis on the shape of the periodic light curve (SHAPE) is needed in order to optimize the choice of the statistical variable to use for testing the data. The only two known objects (the Crab and Vela pulsars) show a similar shape of two narrow peaks separated by 0.4 in phase with some emission between the two main pulses; on the other side, suspect pulsed emission from other sources has been indicated by TeV and PeV experiments, with a variety of shapes, from purely sinusoidal ones (see for ex. Chadwick et al., 1985) to triple peaks configurations (De Jager et al., 1988).

FORTRAN PROGRAMS AND RELATED ANALYSYS STEPS

We have roughly subdivided the analysis in three main logical steps. These are described in the following.

First level analysis: to get the list of times to analyze. The program SELECT derives the list of UTC times, selected in energy and direction, using the selection set described above. The selection procedures should be optimized following the instrument response (via simulations prior to the observations or by in-flight calibrations). The selected list of event times are transformed to Solar System Barycentre using SSBTRAS in order to correct for Doppler shift due to the motion of the earth around the sun and, as in the case of space experiments, due to the motion of the satellite around the earth.

In practically all the gamma-ray experiments, the basic data base searched for periodicities is constituted by a set of single event arrival times not uniformly distributed along the observation. In particular, the time distance between consecutive events does not follow, in general, an exponential distribution (as in the case of theoretical poissonian times) due to many reasons; the program TIMESIM is provided by SAST for producing simulated arrival times which can be analyzed by the same procedures for purposes of comparison with the analysis of experimental times.

Second level analysis: to analyze the list of times. The range of periods to search is first derived by the program PEXTRA; it uses the information (given by CATALOG) on the period of the source under investigation and the epoch of the experiment. When possible, oversampling (i.e. overlapping adjacent Independent Fourier Steps) is suggested for a better detectability of the underlying weak signal. In some cases the period trials can be too many and an operator decision may be required to reduce the operative scanning range in order to avoid a too low efficiency of detection and a too large analysis time. The program FOLDING then derives the list of residual phases per each value of the period investigated and computes the corresponding value of Z^2_n (Buccheri et al., 1983) and its maximum value over the scanning range. The number of harmonics n is input by SHAPE.

When no information is available on the source under investigation, one needs to analyze a very large range of periods. This can be done by using the program GAMFFT which employs a Fast Fourier Transform technique to reduce the computer time of the analysis (Protheroe, 1988). A word of caution: the algorithm gives the spectral power at each consecutive FFT channel which may result in a poorly sensitive analysis because i) the power of the harmonics of the fundamental (for non sinusoidal signals) are lost and ii) the maximum of the signal is not generally at the centre of the FFT channel.

The program TIMCLUS analyzes (via cluster techniques) the photon arrival times looking for detection of timing irregularities (ex. short bursts of emission, see Buccheri et al., 1989).

Third level analysis: to get the confidence level of the detection and describe the characteristics of the signal. The program SIMMAX derives, via computer simulation of the actual experiment, the probability to detect by chance the maximum measured value of Z^2_n . The real number of trials and, if needed, the oversampling are included in the simulation for a realistic estimate of the probability (see estimates of oversampling factor in De Jager, 1988 and Orford, 1991).

For analyses performed via FFT, the program Z1MAXP derives, using the known probability distributions of χ^2_2, the probability to detect by chance the measured maximum value of the Power Spectrum Density, including the the total number of FFT bins.

If detection is believed, according to the decision in CHOICE, the program PHKDE performs the analysis of the residual phases corresponding to the optimal value of the period for the derivation of the smoothed phase distribution representing the pulsar light curve. The technique used is the Kernel Density Estimator (see De Jager at al., 1988). In addition, a description of the single structures of the phase distribution representing the pulsar light curve is done by using the program PHCLUS which employs cluster techniques (Buccheri et al., 1988).

If the result of the search is negative (according to CHOICE), the program UPLIMIT is used to estimate the upper limit on the overall pulsed emission.

INTERMEDIATE DATA

As the result of the first level analysis, UTCSITE gives the list of selected UTC times to transform to SSB while the actual list of SSB times to analyze is contained in the file SSBTIMES.

The second level analysis produces:

the file RANGE giving information on the scanning range of periods where to look for periodicity;
GAMPDS with the Power Density Spectrum as resulting from the application of the FFT on the SSB times;
the file ZRESULTS containing the results of folding the lists of arrival times. In particular, the spectrum of Z^2_n values and its maximum in the scanning range, the period corresponding to $Z^2_n(max)$, the period trials and the total number of times used;
the list of residuals phases corresponding to $Z^2_n(max)$ or $\chi^2_2(max)$, contained in PHASES.

OUTPUT DATA

The final output of the analysis consists in:

MAXPROB containing the chance occurrence probability associated with the Z^2_n(max) or χ^2_2(max).

UPFLUX, giving the upper limit on the gamma-ray flux in case of negative detection;

PHDIST, with the residual phase distribution as resulting from the application of the Kernel Density analysis (for positive detections);

PHPEAKS, describing the light curve structures derived by appying cluster techniques to the residual phases;

TPEAKS, describing the aperiodic time structure of the data derived by appying cluster techniques to the list of SSB times.

REFERENCES

Buccheri R., 1991, Search for periodicities in high energy gamma-ray astrophysical data, Proc. of Conference on "Statistical Challenges in Modern Astronomy", Pennsylvania State University

Buccheri R., Bennett K., Bignami G.F., Bloemen J.B.G.M., Boriakoff V., Caraveo P.A., Hermsen W., Kanbach G., Manchester R.N., Masnou J.L., Mayer-Hasselwander H.A., Oezel M.E., Paul J.A., Sacco B., Scarsi L., Strong A.W., 1983, Astron. Astrophys., 128, 245-251

Buccheri R., Busetta M., Carollo C.M., 1989, Search for bursts of gamma-ray emission from the Crab pulsar in the COS-B data by adaptive cluster analysis, Proc. III Int. Workshop "Data Analysis in Astronomy", eds. V.Di Gesu', L.Scarsi, P.Crane, J.H. Friedman, S.Levialdi, M.C.Maccarone, Plenum Publ. Corp., p. 67-74

Buccheri R., Di Gesu' V., Maccarone M.C., Sacco B., 1988, High resolution cluster method for topological studies of the light curves of gamma-ray pulsars, Astron. Astrophys., 201, 194-198

Chadwick P.M., Dowthwaite J.C., Harrison A.B., Kirkman I.W., McComb T.J.L., Orford K.J., Turver K.E., 1985, Association of the 6-ms pulsar PSR1953 with the COS-B gamma-ray source 2CG065, Nature, 317, 236-238

De Jager O.C., 1987, The analysis and interpretation of VHE gamma-ray measurements, PhD thesis, University of Potchefstroom, South Africa

De Jager O.C., Raubenheimer B.C., North A.R., De Jager H.I., Van Urk G., 1988, A TeV triple peak from PSR1509-58?, Ap. J., 329,831-837

De Jager O.C., Swanepoel J.W.H., Raubenheimer B.C., 1986, Kernel Density Estimators applied to gamma-ray light curves; Astron. Astrophys., 170, 187-196

Maccarone M.C., Buccheri R., 1989, Decision problems in the search for periodicities in gamma-ray astronomy, Proc. III Int. Workshop "Data Analysis in Astronomy", eds. V.Di Gesu', L.Scarsi, P.Crane, J.H. Friedman, S.Levialdi, M.C.Maccarone, Plenum Publ. Corp., 237

Orford K.J., 1991, Analytical treatment of oversampling, Experimental Astronomy, vol. 1, n. 5, p. 305

Protheroe R.J., Hocking W.K., 1988, Period searches using the Fast Fourier Transform in X-ray and Gamma-ray Astronomy, Astron. Lett. and Communications, vol. 27, 237-240

XRONOS: A TIMING ANALYSIS SOFTWARE PACKAGE

L. Stella [1,2,*], L. Angelini [3,2,*]

[1] Osservatorio Astronomico di Brera, Via Brera 28, 20121 Milano, Italy
[2] EXOSAT Observatory, Astrophysics Division, Space Science Department
of ESA, ESTEC, Keplerlaan 1, 2200 AG Noordwijk, The Netherlands
[3] University Space Research Association, NASA GSFC, Greenbelt
MD 20771, USA - * Affiliated to ICRA

INTRODUCTION

XRONOS is a general purpose timing analysis package. Although it was designed primarily for X-ray astronomy, XRONOS is basically detector and wavelength-independent. It has been used to analyse data from the Einstein Observatory, EXOSAT and Ginga, as well as optical photometry and helioseismology data. In the currently available version (No. 2.0), XRONOS includes programs for: light curve(s), hardness ratio and colour-colour plotting, epoch folding, power spectrum, autocorrelation, crosscorrelation, time skewness and statistical analysis.

XRONOS consists of a collection of programs, each designated to one task, which can be run from the control environment provided by the xronos program. The program xronos is characterised by 3 different user interfaces: a Question/Answer user interface, a Partial Question/Answer user interface and a Command-Driven user interface. Command files are supported by the question/answer and command-driven user interfaces. XRONOS applications can be run in their command-driven fashion within the EXOSAT Database System[1,2]. XRONOS uses the QDP/PLT plotting package. Algorithms and features of XRONOS are described in detail in the XRONOS user's guide[3]. XRONOS is currently available for remote installation on VAX/VMS machines.

HISTORY

XRONOS is a descendant of a series of timing analysis programs written at the EXOSAT Observatory to analyse interactively the EXOSAT data on Hewlett Packard Series 1000 minicomputers. The development of XRONOS started in 1987 during the post-operational phase of the EXOSAT project. XRONOS was designed to provide a timing analysis package as independent of the detector specificity as possible so that it could be used for future missions with little or no changes. Some of the original design characteristics of XRONOS were greatly influenced by the experience with the analysis of large sets of high time resolution EXOSAT data. In order to limit the number of page folds, while retaining the array sizes needed e.g. for the fast Fourier transform (FFT) of large data

Data Analysis in Astronomy IV, Edited by V. Di Gesù et al.
Plenum Press, New York, 1992

59

sets, it was decided that XRONOS consist of a collection of programs, each designated to one task. Preference was initially given to a question/answer user interface.

A preliminary version of XRONOS was developed under UNIX on a Hewlett Packard 9000 computer. This was ported to VAX/VMS in 1988 to make XRONOS produce light curves within the EXOSAT database system[1,2]. A test version of XRONOS, including only a small number of applications, was presented at the the first EXOSAT Database Workshop in December 1988. In response to suggestions by participants it was decided to make XRONOS more user-friendly and to add a command/driven user interface which could speed up the work of experienced users. The number of XRONOS applications grew rapidly in 1989. Since early 1990 all XRONOS applications can be run from the control environment provided by the program xronos, which is characterised by 3 different user interfaces: a question/answer user interface, a simplified, "partial" question/answer user interface and a command-driven user interface. Moreover, all XRONOS applications are available in a command-driven fashion within the EXOSAT Database System. A version of XRONOS, similar to the current one, was presented at the Second EXOSAT Database Workshop in February 1990. XRONOS currently runs in six different institutes.

OVERVIEW

XRONOS consists of a set of programs which are accessed by the control program xronos. This is invoked from the operating system by typing xronos. There are 3 different user interfaces which can be used within XRONOS:

* QUESTION/ANSWER user interface, in which the user specifies input files and analysis parameters by answering individual questions;
* PARTIAL QUESTION/ANSWER user interface, in which input filenames are specified only once and analysis parameters are entered by replying to questions;
* COMMAND-DRIVEN user interface, in which applications are run by entering a command string followed by qualifiers and arguments.

Unexperienced users are advised to start with the question/answer user interfaces.

XRONOS issues a Xronos> prompt from which commands are entered. In the question/answer user interface (the default), after entering a command corresponding to a XRONOS application, each question is written on the screen followed by a => prompt. Defaults, where relevant, are indicated in square brackets. In the command-driven user interface the execution of a XRONOS application is controlled by specifying qualifiers in the command string. The XRONOS help file is accessed by entering help from any XRONOS prompt. Operating system commands can be given from within XRONOS by prefixing the commad with $. A set of XRONOS commands in a file can be input using @filename from the Xronos> prompt. Note that all commands can be abbreviated to unique strings.

READING DATA

XRONOS is designed to read in time series from many (up to 50) consecutive input files (this is to avoid wasting disk space by chaining many input files in a single file). Time series from up to 3 different energy ranges can be handled simultaneously within XRONOS. Additional flexibility is provided by "Input File Options", which are used, e.g. to perform algebraic operations on individual input files, shift times, apply barycentric and dead-time corrections etc.

In the question/answer user interface input data is specified by entering the input filenames, one by one, in response to the relevant questions. In the partial question/answer and command-driven user interfaces the input files are specified using the data command. Currently, two different kinds of input files are supported: "Rate Buffers" (with extension .rbf), which are binary files (4 different formats are available), and "QDP files", which are ASCII files written according to the QDP/PLT standard[4]. XRONOS input files can either contain binned data or unbinned data (i.e. input files containing the arrival times of individual photons).

APPLICATIONS

XRONOS applications are the subset of XRONOS commands which are used to carry out different types of analyses. The following XRONOS applications are available:

* lda - List DAta in input (header and points)
* asc - convert binary file to ASCII
* lc1 - plot Light Curve for 1 time series
* lc2 - plot Light Curve and ratio for 2 time series
* lc3 - plot Light Curve and ratios for 3 series (and colour-colour)
* sta - STatistical Analysis
* win - create/read/change a WINdow file
* ef1 - Epoch Folding for 1 time series
* ef2 - Epoch Folding and ratio for 2 time series
* ef3 - Epoch Folding and ratios for 3 series (and colour-colour)
* efs - Epoch Folding Search for periodicities
* qdp - plot a QDP file
* psd - Power Spectrum Density analysis with FFT
* pss - Power Spectrum density analysis with Slow direct FT
* acf - Auto Correlation Function with fast method
* acs - Auto Correlation function with Slow method
* ccf - Cross Correlation Function with fast method
* ccs - Cross Correlation function with Slow method
* tss - Time Skewness function with Slow method

Note that most applications include utilities to rebin the results of the analysis either by using a constant or a logarithmic step and to remove (or analyse) various types of trends from the input series.

TERMINOLOGY

Within XRONOS the user has full control of the binning to be used in the analysis, the way of subdividing a time series and/or grouping the results of the analysis. To this aim, four basic entities are used in most XRONOS applications:

* BINS : these are the time bins of the time series which is/are being analysed. Note that there 'can be more than one bin duration, e.g. two consecutive time series, one with 0.5 s bins and the other with 2 s bins, as in the example below.
* NEWBINS : these correspond to the time resolution with which the analysis is carried out. The example below demonstrates how to calculate the power spectrum using a 4 s resolution, i.e. newbins of 4 s. Note that: (i) newbins cannot be shorter than the longest bin duration of the time series which is/are being analysed; (ii) in many XRONOS applications (e.g. psd, pss, acf, ccf) the newbin duration is forced to be an integer multiple of the longest bin duration.

* INTERVALS : an interval is defined by the number of newbins over which the analysis is carried out. The example below shows how to calculate power spectra for intervals of 256 newbins, corresponding to 128 independent Fourier frequencies. Note that in applications using FFT algorithms (e.g. psd, acf, ccf) the number of newbins in an interval must be a power of 2.
* FRAMES : a frame consists of the average of the results of the analysis of one or more contiguous intervals. Each XRONOS QDP file and therefore each plot produced by XRONOS correspond to a frame. The example below shows how to average the power spectra from 5 intervals (each of 256 newbins of 4 s) in a frame. Note that in certain applications (e.g. 1c1, 1c2 and 1c3), a frame consists always of one interval.

The XRONOS terminology is illustrated in the following example (the command-driven user interface is used). Suppose that two input time series with 0.5 s bins and 10000 s duration and 2 s bins and 20000 s durations, respectively, are specified. In this case application psd will set the newbin duration by default such as to produce a power spectrum from a single interval of 4096 newbins at most. Therefore:

Xronos> psd

sets the newbin duration to 8 s, corresponding to 3750 newbins, calculates the power spectrum from a single interval of 4096 newbins (empty newbins are replaced with the average count rate), writes the file xronos.qpsd and plots the power spectrum with 2048 independent Fourier frequencies.

Xronos> psd/tnewbin=4.

or equivalently, (note that the longest bin in the time series is 2 s).

Xronos> psd/nint=2

sets the newbin duration to 4 s, corresponding to 7500 newbins. In this case psd calculates the power spectra from two intervals of 4096 newbins (the default), averages the results in a frame, writes the file xronos.qpsd and plots the average power spectrum with 2048 Fourier frequencies.

Xronos> psd/tnewbin=4./nbins=256

specifies also that each interval consists of 256 newbins; in this case the power spectra from 29 intervals are calculated (128 independent Fourier frequencies) and averaged by default in a single frame to produce a single output file xronos.qpsd and plot it.

Xronos> psd/tnewbin=4./nbins=256/navpsd=5

carries out the analysis as above, except that the power spectra from (up to) 5 consecutive intervals are averaged in a frame, written to the output file xronos.qpsd and plotted, therefore, producing a sequence of 6 average power spectra.

WINDOWS

There are 4 different types of windows which can be specified within XRONOS: time, phase, intensity and exposure windows. In applications requiring more than one input series (e.g. a cross-correlation analysis) intensity and exposure windows are specified separately for each series. Intensity and exposure windows can be specified separately for bins, newbins and intervals. Command win is to create, read or change XRONOS window files (default extension .wi). Many XRONOS applications use some default exposure windows, which are designed to avoid analysing data sets which are too inhomogeneous with respect to their statistical properties.

PLOTTING

Most XRONOS applications create a "XRONOS QDP File" containing the results of the analysis carried out with a XRONOS application. These files have a .qXXX extension, where XXX is the relevant XRONOS application (e.g. files created with application psd have extension .qpsd), and contain the results from ONE Frame only. Plotting within XRONOS is performed through QDP/PLT[4] by using the interface provided by these files (QDP/PLT uses the graphics package PGPLOT). By default XRONOS applications plot results from one frame automatically and return either to the execution of the rest of the analysis (if any) or to the XRONOS prompt. Alternatively, by replying **interactive** to the relevant question in the question/answer user interfaces or by using the /plt qualifier in the command-driven user interface, the user is left in the interactive plotting mode (PLT> prompt) to allow customization of the plot. A number of other plot options can be specified by replying to the relevant questions in the question/answer user interfaces, or by using designated qualifiers in the command-driven user interface. In particular, a stack of QDP/PLT commands written in a file can be applied to each plot. The plot device to be used is set at the beginning of a XRONOS session or by using the command cpd (change plot device) from the XRONOS prompt.

GLOBAL PARAMETERS

Additional flexibility is achieved by using a number of global parameters controlling the XRONOS session. These are specified by running command cpf ("change parameter file"). XRONOS global parameters are designed to:

* control the chattiness of XRONOS applications;
* write to a history file all the interaction with XRONOS applications and specify the history file chattiness;
* change the default output filename root and i/o status;
* replace data gaps with a moving average of the light curve(s);
* force a given start time in the analysis;
* calculate error bars for the results of various types of analysis by propagating theoretical error bars or by direct evaluation of the scatter around the mean values in a frame;
* analyse window functions;
* normalise the results in various ways;
* force strict simultaneousness in time series from two or three energy bands, e.g. for cross-correlation analysis;
* specify special windows for burst analysis;
* specify a rescaling factor and an additive constant for the results.

MISCELLANEOUS

XRONOS has an interactive help facility which is accessed by the **help** command from the XRONOS prompt, or by replying **help** to any question asked by a XRONOS application.

A log of the commands entered from the XRONOS prompt can be written to an ASCII file by using the **log** command. In the question/answer user interfaces, each application can be terminated by entering cntl z in response to any question, or by entering cntl y during execution. XRONOS can be terminated by the **exit** or **quit** commands.

FUTURE

A portable version of XRONOS (VAX/VMS and SUN and HP UNIX) is an advanced state of development and will be available during 1991. XRONOS will also be interfaced to FITS tables. Future applications will include:

* Searches for coherent periodicities both with power spectrum and epoch folding techniques;
* Display and analysis of "time images" (e.g. a large number of colour-coded power spectra is displayed versus frequency and time);
* Pulse phase determination for light curves with a coherent signal;
* Alternative Power spectrum and correlation techniques in the presence of sparse and/or non-equispaced data;
* Cross spectrum analysis.

CONTACT PERSON

Further information about XRONOS can be obtained from Luigi Stella at

* Brera Astronomical Observatory, Via Brera 28, 20121 Milano, Italy
* Telephone: +39-2-874444
* Fax: +39-2-72001600
* SPAN: ASTMIB::STELLA (or, equivalently, 39216::STELLA)

or from Lorella Angelini at

* University Space Research Association, NASA GSFC, Greenbelt, MD 20771, USA
* Telephone: +1-301-286-3607
* Fax: +1-301-286-9240
* SPAN: 6197::ANGELINI

ACKNOWLEDGEMENTS

P. Giommi, A. Parmar, A. Pollock, G. Tagliaferri, M. van der Klis and N. White provided several fortran routines and a number of suggestions. Useful suggestions and bug reports were also provided by a number of other people.

REFERENCES

1. EXOSAT Observatory Team, "The EXOSAT Database System: BROWSE Version 3.0 User's Guide", ESA TM-11 (1991).
2. EXOSAT Observatory Team, "The EXOSAT Database System: On-line User's Guide, ESA TM-12 (1991).
3. L. Stella and L. Angelini, "XRONOS: a Timing Analysis Software Package. User's Guide. Version 2.0", in press (1991).
4. A. Tennant, "The QDP/PLT User's Guide", NASA-MSFC preprint (1990)

AN AUTOMATIC SPECTROPHOTOMETRIC DATA REDUCTION SYSTEM

FOR QUANTITATIVE SPECTRAL CLASSIFICATION OF STARS

Valeri Malyuto, Jaan Pelt

Tartu Astrophysical Observatory
202444, Toravere, Estonia, USSR

Teimuraz Shvelidze

Abastumani Astrophysical Observatory
383762, Abastumani, Georgia, USSR

INTRODUCTION

Some years ago a complex programme of studying the main meridianal section of the Galaxy (MEGA = MEridian of GAlaxy) was started to improve our knowledge about spacial and kinematical characteristics of stellar populations. The tasks and the present state of the MEGA program has been described recently by Malyuto, Einasto, Kharchenko and Schilbach (1990). The use of absolute proper motions of stars (with respect to galaxies), photoelectric UBVR and Vilnius photometry together with automated quantitative spectral classification for large stellar-statistical samples are among the most essential features of the programme. The data are gathered in 47 areas within 30° from the main meridional section of the Galaxy.

To classify stars, objective prism stellar spectra of intermediate dispersion (D = 166 A/mm at H_γ), obtained with the 70-cm meniscus telescope at the Abastumani Astrophysical Observatory, are used. The field diameter$_m$ is about 5°, the limiting photographical stellar magnitude is about 12 .

Our automated quantitative spectral classification of G-K stars applies criteria evaluation and consists of two parts: a spectrophotometric data reduction system and a classification technique. Underlying the reduction system is the transformation of spectral densities into classification criteria values. Underlying the classification technique (Malyuto and Shvelidze, 1989) is the deduction of the main physical parameters of stars from the classification criteria values by means of a linear regression model (the same model and its coefficients were determined beforehand on the basis of standard stars).

DIGITIZATION OF OBSERVATIONAL DATA

The spectra widened to 0.15-0.25mm are obtained on Kodak IIaO or IIIaJ photographic emulsion. For their digitization on magnetic type we apply the PDS machine installed at the Tartu Astrophysical Observatory. The slit area corresponds to 5x100µ on the plate, one displacement step equals 5µ, the recorded linear distance is about 10mm (1801 pixels). The selection and positioning of the spectrum in the PDS machine are performed manually. To fix the approximate starting point for spectral recording, we use the line H CaII + H_ε (λ 3969 A), easily identifable visually in all well-exposed spectra. The calibration plates obtained with a laboratory spectrograph equipped with the step wedge, have been recorded with the same scanning parameters as for stellar spectra at four different wavelengths ($\lambda\lambda$ 4650, 4340, 4100 and 3970 AA).

PROCESSING OF CALIBRATION SPECTRA

We apply the method of constructing the calibration curves in an interactive mode elaborated by K. Annuk (1986) at the Tartu observatory. Each calibration curve approximates log I values in terms of the Baker density and wavelength by a polynomial up to the third degree.

REDUCTION OF THE STELLAR SPECTRA

The software package for stellar spectral data reduction together with the classification tecnique has been written in FORTRAN 77 language. The package has been ajusted to the IBM PC type computer.

Filtering the spectra

Filtering the spectra from the emulsion grain noise is performed by the proper subroutine by means of a Fourier transform. We use the method described by Lindgren (1975) but with small modifications. Differently from the Lindgren's technique, we apply the cut-off filter for power spectra using the exponential function.

The filtered spectral densities are transformed into intensities by means of the polynomial mentioned in Section 3.

Fitting of the continuum with the cubic spline technique

For automated spectrophotometric data reduction system created with the aim of classifying large samples of stars, a fast and impersonal fitting of a continuum is crucial. Our algorithm of continuum determination is based on a simple analytical model for the recorded spectral intensity distribution

$$X(i) = C(i) + I(i), \quad i=1,\ldots,n,$$

where $C(i)$ is the smooth continuum and $I(i)$ is the erratic line absorbtion spectrum, i is the pixel number and n is the total number of pixels. To determine the continuum we apply an iterative approach and a standard cubic splines technique to define the continuum at each iteration. We divide every spectrum into some equal intervals and define a current estimate of the continuum by the third degree polynomial at each interval. It yields the smoothed function $C^1(i)$ which is a first approximation to $C(i)$. At the next stage we take

$$X^1(i) = \max[X(i), C^1(i)]$$

as an initial spectral distribution instead of $X(i)$ and repeat the process. The number of iterations (put to equal 100) is included into free parameters of the procedure. The results of continuum line fitting are illustrated (together with subsequent stages of data reduction) for a representative star in Fig. 1. Using the continuum defined, the main programme transforms spectral intensities into residual spectral intensities.

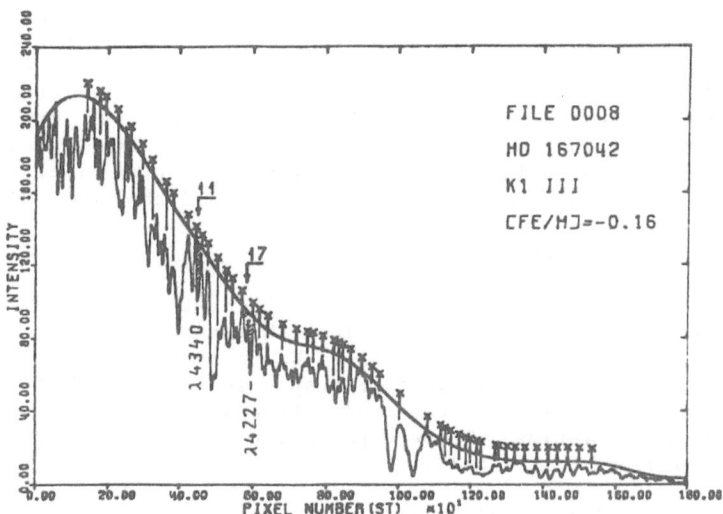

Fig. 1. The spectral intensities versus the standardized pixel numbers with the continuum line for the presentative K1 III star. The border pixel numbers are marked by vertical lines with crosses. The band areas 11 and 17 are shaded and marked with the corresponding wavelenghs for the lines characteristic to those areas.

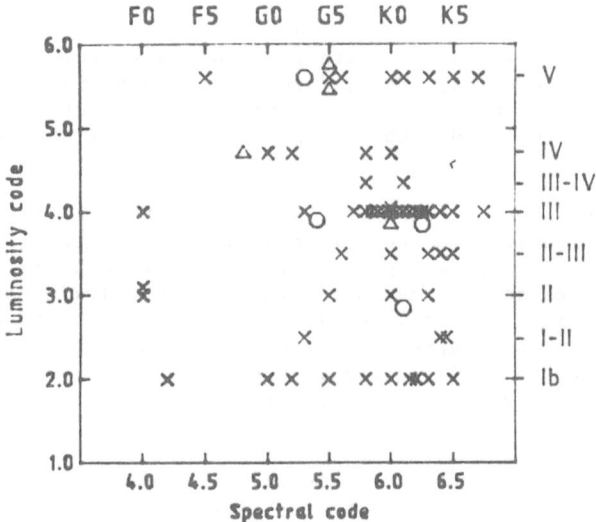

Fig. 2. The HR diagram for the standard stars. The designations used: crosses - normal metallicity stars (+0.4 > [Fe/H] > -0.4); triangles - stars with intermediate metal deficiency. (-0.4 > [Fe/H] > -1.0); circles - stars with low metal deficiency (-1.4 > [Fe/H] > -3.0).

67

Standardization of pixel numbers

Under the standardization of pixel numbers we mean matching the zero-point and the scale of all recordings of spectra. Standardization is naturally performed with the use of spectral lines. The strong lines are searched for by the computer in the spectrum under study and the positions of these lines are compared with the positions of the lines in standard lists with the aim of establishing common lines. As a rule, a computer find from 5 to 13 common lines within 1-2 pixel numbers from the expected positions allowing for the zero-point and the scale differences.

Measurements in the spectra

As classification criteria we measure the ratios of equivalent areas of bands confined by the continuum line, the residual spectral intensity distribution and the border wavelengths. We foresaw that the band area ratios would be less sensitive to atmospheric seeing and the graininess of photographical emulsion than conventional classification criteria (central intensity ratios). These expectations have been supported by the work of Malyuto and Pelt (1981).

We have compiled the list of 53 classification criteria chosen by West (1970, 1972), Shiukashvili (1969) and Malyuto (1977), central intensity ratios being replaced by corresponding band area ratios. Choosing the band borders we try to ensure that all those spectral lines whose central intensities are used in these classification criteria would be isolated in the separated border intervals. In the spectrum of a representative star with the marked band borders is given in Fig. 1. Two band areas in figure are shaded to underline the fact that these and other similar ratios of the band areas serve as classification criteria.

CRITERIA RELIABILITY

The classification techniques have been developed by Malyuto and Shvelidze (1989) to deduce the main physical parameters of stars (spectral and luminosity classes, [Fe/H] values) from spectral classification criteria by means of a linear regression model. The model for the standard stars has been defined with a stepping procedure based mainly on polynomial relations.

About 115 F0-K5 standard stars photographed by West (1970, 1972) and by us have been involved in the solution. The HR diagram for these standards is presented in Fig.2. Together with conventional discrete spectral and luminosity classes, corresponding spectral and luminosity codes were introduced in the form of continiously changing values. These codes have used in defining the regression model, a code unit approximately corresponds to each unit of spectral and luminosity classes, respectively. The lack of F0-G2 stars attracts attention in this Figure. The stars of normal metallicity have been picked out to demonstrate spectral and luminosity effects (examples see in Figs. 3-5) and the luminosity III stars have been picked out to demonstrate spectral and metallicity effects (example see in Fig. 6). We found that the effects are pronounced. However the scatter is significant for some figures. For reliable classification of G5-K5 stars we resort least to ten criteria per star.

Adopting the regression model we found that multiple correlation coefficients between the classification criteria and the the main physical parameters of stars were confined between 0.64 and 0.98.

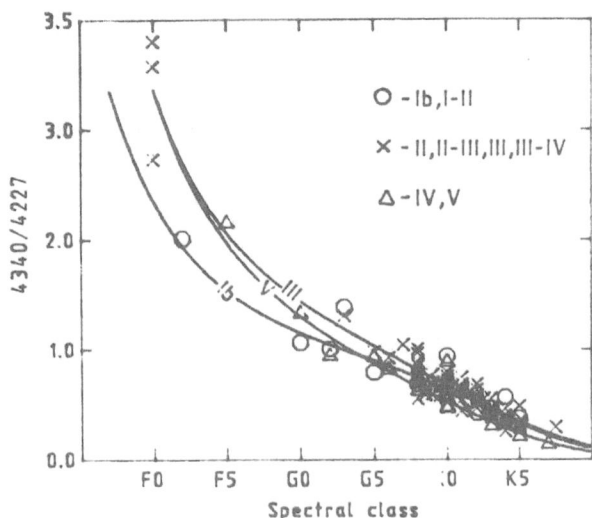

Fig. 3. The criterion 11/17 ($\lambda4340/\lambda4227$) versus the spectral classes for the standard stars with normal metallicity ($-0.4 >$ [Fe/H] $> +0.4$). For different luminosity classes and [Fe/H] $= 0.0$ the analytical regression model lines are imposed.

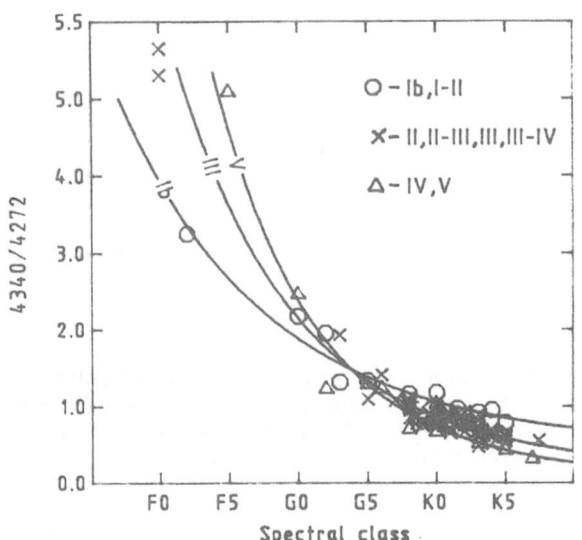

Fig. 4. The same as for Fig. 3 but for the criterion $\lambda4340/\lambda4272$.

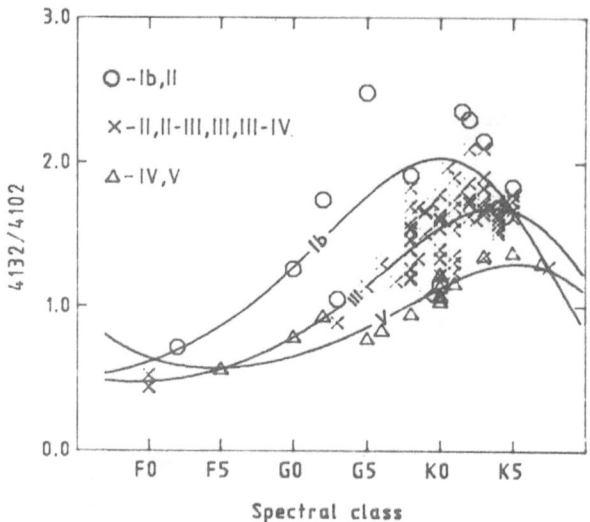

Fig. 5. The same as in Fig. 3 but for the criterion λ4132/λ4102.

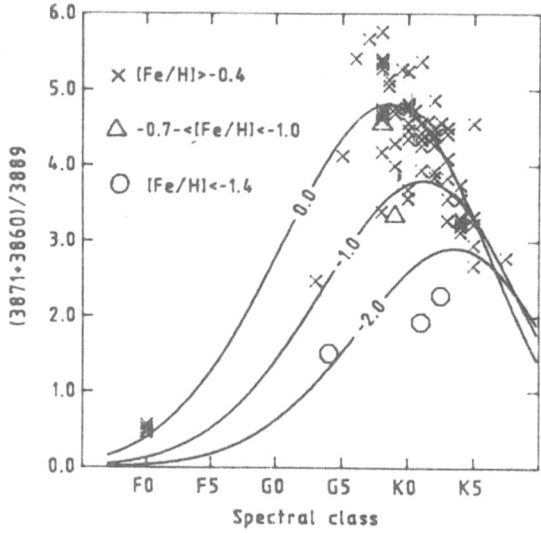

Fig. 6. The classification criterion ((λ3871+λ3860)/λ3889) versus the
spectral classes for the standard stars of luminosity class III.
For different [Fe/H] values and luminosity class III the analytical regression model lines are imposed.

ACCURACY OF CLASSIFICATION ESTIMATED FROM DATA COMPARISONS

The effeciency of the use of automatically defined classification
criteria in spectral classification is demonstrated by the results of the
automated classification. We found the r.m.s. differences between standard and calculated values of main physical parameters of standard stars.
After extracting the r.m.s. errors of the standard values quoted in

literature we found that the external r.m.s. errors of spectral classification of standard stars were about ± 0.06 for spectral classes, higher
than +-0.4 for luminosity classes and +-0.16 for [Fe/H]. These estimates
are comparable with accuracy with the appropriate results for the most
reliable data published.

Two areas in the well-studied North Galactic Pole region have been
treated which allowed us to deduce the accuracy of spectral classification of programme stars from a comparision with the DDO photoelectric
photometric classification (Hartkoph and Yoss, 1982) for about 60 G-K
common stars. From these comparisions we obtained the following external
r.m.s. errors of spectral classification for programme stars: about
+-0.1 for spectral classes, higher than +-0.5 for luminosity classes,
+-0.22 for [Fe/H], which are comparable with the accuracy of the DDO photoelectric photometric classification.

About 10 areas of the MEGA programme have been photographed and are
under treatment now. We hope to obtain classification results numerous
enough for statistical application by the date of the IAU Symposium 149
"The Stellar Populations of Galaxies".

FUTURE PROBLEMS

We suppose that the use of two-dimensional scanning of spectra will
enhance the classification accuracy thanks to detecting the plate faults
eluding visual control. The manual selection and positioning of the spectrum in the PDS machine have remained so far the most laborious steps in
classification which are to be automated in the future.

References

Annuk, K.: 1986, *Tartu Astrofüüs. Obs. Prepr.* **A-5,** 3
Harkopf, W. and Yoss, K.: 1982, *Astron. J.* **87,** 1679
Lindgren, H. : 1975, *Reports from the Observatory of Lund* **6,** 15
Malyuto, V.: 1977, *Tartu Astrofüüs. Obs. Publ.* **45,** 170
Malyuto, V., Einasto, J., Kharchenko, N., Schilbach, V: 1990, *Tartu
    Astrofüüs. Obs. Prepr.* **102,** 12
Malyuto, V., Pelt, J.: 1981, *Tartu Astrofüüs. Obs. Prepr.* **A-1,** 3
Malyuto, V. and Shvelidze, T.: 1989, *Astrophys. Space Sci.* **155,** 71
Shiukashvili, M.: 1969, *Abastumani Astrophys. Obs. Bull.* **37,** 43
West, R.: 1970, *Abastumani Astrophys. Obs. Bull.* **39,** 29
West, R.: 1972, *Abastumani Astrophys. Obs. Bull.* **43,** 109

THE HST MISSION

THE HUBBLE SPACE TELESCOPE: YEAR ONE

Robert J. Hanisch

Space Telescope Science Institute[1]
3700 San Martin Drive
Baltimore, Maryland 21218 USA

Abstract

The first year of in-orbit operations of the Hubble Space Telescope has produced a variety of exciting scientific discoveries, despite the fact that the primary mirror suffers from spherical aberration. The resulting image quality is considerably below the design specifications, and causes a loss of sensitivity of approximately 2.5 magnitudes. Fortunately, the HST point spread function retains a sharp core having almost the nominal diffraction-limited resolution. Through the judicious use of image restoration algorithms and model fitting, a number of the scientific objectives of the HST mission can still be met. Following the installation of corrective optics on the first shuttle servicing mission, with the subsequent installation of second-generation scientific instruments with built-in corrective optics, the full scientific potential of HST should be realized.

INTRODUCTION

The Hubble Space Telescope was carried into low earth orbit on the space shuttle Discovery (STS-31) on 24 April 1990. The launch of HST had been delayed for some seven years owing to a combination of technical problems with the spacecraft and the tragic accident of the space shuttle Challenger. The expectations for HST were tremendous, and there was almost a sense of disbelief amongst the staff at the Space Telescope Science Institute that after all these years, HST was actually in orbit.

Deployment of the spacecraft went relatively smoothly, with a minor glitch in the unrolling of the solar panels and a stuck telemetry antenna being the primary problems. After a period of orbital verification observations in which the spacecraft's electrical systems, attitude control systems, and data links were checked out, the first-light images from the Wide Field/Planetary Camera and ESA's Faint Object Camera were relayed to the ground.

---

[1]The Space Telescope Science Institute is operated by the Association of Universities for Research in Astronomy, Inc., for the National Aeronautics and Space Administration.

The first-light images were not the crystal-clear images everyone had hoped to see, and the poor quality was attributed to inadequate focus testing in orbit. Within two months, however, after numerous attempts to improve the focus, it was realized that stellar images showed the unambiguous signature of spherical aberration. The work of Chris Burrows at ST ScI, through the use of his HST optical modelling software, was fundamental in making this discovery.

Over the last year ST ScI has been involved in a number of efforts to deal with the aberration problem. An HST Strategy Panel reviewed a number of possible approaches, and ultimately proposed a system of corrective optics known as COSTAR (discussed further below). ST ScI established an Image Restoration Working Group, including members both within and outside the Institute, in order to pool the available expertise on image restoration algorithms and understanding the point spread function. Through a combination of hardware and software solutions, we are optimistic that the full scientific potential of the HST will be recovered.

THE HST AND SPACECRAFT OPERATIONS

Optical Telescope Assembly and Scientific Instruments

The design and construction of the HST has been thoroughly described elsewhere (see, for example, Hall 1982). Briefly, the telescope is a Ritchey-Chretien design with a 2.4-m primary. The nominal angular resolution is 0.1 arcsec, and limiting magnitude (without spherical aberration) is $29 M_v$. The telescope's Fine Guidance Sensors allow the spacecraft to be pointed to better than 0.01 arcsec accuracy. The entire telescope assembly has a mass of 11,500 kg.

There are five primary scientific instruments on the HST. The Wide Field/ Planetary Camera uses CCD detectors and has a field of view of 2.7 arcmin (WF mode) or 1.2 arcmin (PC mode). The Faint Object Camera uses photon-counting image intensifiers, and has a field of view of up to 22 arcsec in its f/48 mode. The Goddard High Resolution Spectrograph provides spectral resolution $\lambda/d\lambda$ of up to 100,000, and operates solely in the UV part of the spectrum, from approximately 110 to 320 nm. The Faint Object Spectrograph provides a spectral resolution of 1200 but operates from 110 to 930 nm. The High Speed Photometer has a time resolution of 10 μs for relatively bright objects, and is sensitive over a wavelength range from 120 to 700 nm. In addition, the Fine Guidance Sensors can be used as astrometric instruments, allowing precise measurements of star locations and the separation of close binaries through the analysis of the interferometric transfer function.

Spacecraft Operations

The operation of a low-earth orbit optical telescope is very complicated. Because there are major restrictions on spacecraft attitude (sun, earth, and moon avoidance angles, telemetry antenna pointing), power consumption, and telemetry rates and the availability of the Tracking and Data Relay Satellites (TDRS), it is a major effort to schedule scientific observations in an efficient manner. The majority of all HST observations are planned well in advance, and real-time observations are limited primarily to target acquisitions.

Figure 1 show the overall flow of data and other information for the Space Telescope. Guest Observers (GOs) and Archival Researchers (ARs) submit observing proposals via a Remote Proposal Submission System (RPSS). Proposals entered into RPSS are then sent to the Proposal Entry Processer (PEP), which is used to validate the observing program. A Time Allocation Committee (TAC) reviews and ranks proposals, using PEP as an aid in this review as necessary. Approved proposals are then scheduled using two scheduling tools: Spike, an AI-based tool which is used to define an optimum long-range schedule, and the Science Planning and Scheduling System (SPSS), which generates the detailed observing plan and spacecraft commanding in the form of a Science Mission Schedule (SMS). SPSS utilizes the Guide Star Selection System (GSSS) and the associated Guide Star Catalog (GSC) to choose suitable guide star candidates for target acquisition and pointing control.

The SMS (which typically covers several days of observing) is sent to the Payload Operations Control Center (POCC) for uplinking to the spacecraft. The mission schedule is executed by HST, and data is telemetered back to the ground via the TDRS satellites. During periods of non-availability of TDRS, data is stored on board the spacecraft on tape recorders for later down-linking. The telemetry data stream is received back at the POCC and the Data Capture Facility (DCF), which manages both the engineering and science data streams. Portions of the data are subsequently sent to ST ScI for on-line review with the Observation Support System (OSS). OSS is used to review spacecraft performance in real-time, and to control special target acquisitions and review data quality. All data are received by the Post-Observation Data Processing System (PODPS) for final reformatting and the application of the standard calibration processing. Data emerging from PODPS is fully calibrated using the best known calibration parameters at the time of the observation. Calibrated data are typically available within 24-48 hours of the time the data were taken.

All data are archived on optical disk, including the engineering data, raw science data, and calibrated science data. The current archive system is called the Date Management Facility (DMF), which is a prototype due to be replaced by the Data Archive and Distribution Facility (DADS). DADS is expected to store at least one-year's worth of HST data (1 TByte) on-line, with the remainder being stored 'near-line' for access within a few minutes. Over the expected mission lifetime of 15 years, some 15 TBytes of data should be accumulated in the archive. All data in the DADS archive will be stored in standard FITS formats.

Copies of the archive are created for the Space Telescope – European Coordinating Facility (ST–ECF) in Garching, Germany, and for the Canadian Astronomy Data Center (CADC) in Victoria, British Columbia. A copy of the complete archive is sent to the ST–ECF, and they impose the same access restrictions on proprietary data as ST ScI. The CADC receives only non-proprietary data. Data is retrieved from the archive either by operators working on behalf of GOs/ARs or by the GOs/ARs themselves through the DMF/DADS user interface. The archive limits access to data to each project's principal investigator and other persons authorized by the PI throughout the proprietary period of one year. Data is made available to GOs/ARs either in the FITS format, or in one of several host-dependent formats recognized by the STSDAS/IRAF data analysis software system.

Users may recalibrate their data and do their final data analysis using the facilities provided within the Space Telescope Science Data Analysis System (STSDAS). STS-DAS is an set of both general and HST-specific software which is fully layered on the Image Reduction and Analysis Facility (IRAF) from the National Optical Astronomy

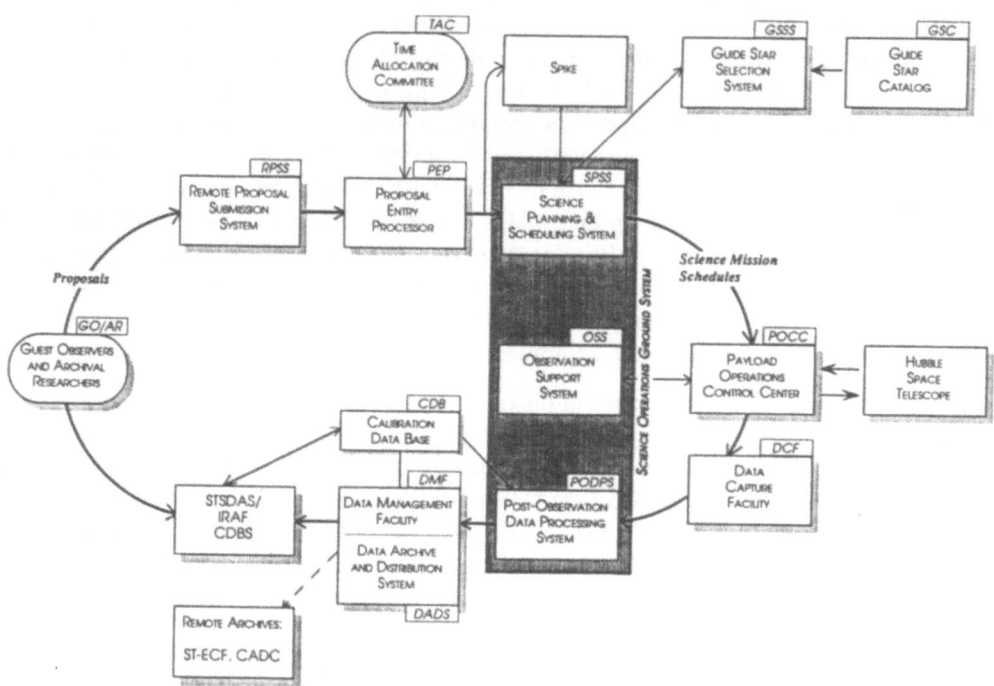

Figure 1. Overall Data Flow for HST

Observatories. These systems are described further in other papers (Hanisch, Tody) in this volume.

FIRST YEAR HIGHLIGHTS

During the summer and fall of 1990 a number of scientific observations were carried out with the HST in order to assess the performance of the telescope and the viability of the planned scientific program. This Science Assessment/Early Release Observations program produced a number of truly outstanding scientific results (see the 10 March 1991 issue of *The Astrophysical Journal Letters* for an overview). In addition, these data have helped us to understand the details of the spherical aberration (Burrows et al., 1991) and to begin to devise methods for correcting for the aberration using image processing software (see, for example, the paper by Adorf in this volume and references therein).

Solar System Studies

HST has turned out to be an excellent telescope for observations of the planets and comets. Saturn was first observed last summer, and when a storm was later observed from ground-based telescopes, HST was able to follow the progress of the Saturnian weather patterns. The WF/PC science team assembled multiple frames of Saturn into a movie, showing the evolution of the storm system with time (Westphal et al., 1991.) The spatial resolution of HST is comparable to Voyager images taken several days from closest approach. After image restoration the Encke division in Saturn's rings is clearly visible, a feature which has never been successfully recorded in ground-based photographs.

FOC observations of the planet Pluto show the well-resolved images of the planet and its moon Charon. At an angular separation of 0.9 arcsec, the planet and its moon are easily separated with HST imaging. Image restoration should permit observers to discern at least some surface features.

Recent WFPC observations of Mars taken at during its recent opposition again show the ability of HST to do detailed studies of planetary atmospheres and weather patterns. Major surface features are also visible in these images, which were part of the first major Guest Investigator program carried out with HST.

Comet Levy was observed as a target of opportunity, demonstrating the capability to reprogram HSTs observing schedule in a fairly short time. The comet was observed on 27 September 1990 at a distance of 160 million km from the Earth. At this distance the image pixel size corresponds to 78 kilometers.

Galactic Astronomy

The FOC was used to obtain images of the symbiotic star R Aqr, showing in remarkable detail the jet-like structures tracing the outflow of material (Paresce et al., 1991).

The WF/PC recorded images of a portion of the Orion nebula in several different filters (Hester et al., 1991). Emission in the light of S II is more filamentary and clumpy than that of H II and O III, indicating that the S II essentially traces the boundaries between hot and cool clouds. Also visible are bipolar flows from Herbig-Haro objects.

Extragalactic Astronomy

Perhaps the first image of real scientific interest taken with the HST was of the star cluster R136 in the 30 Doradus complex in the Large Magellanic Cloud. This image was actually only taken as a finder frame in order to help locate the apertures of the Goddard High Resolution Spectrograph. However, the original WF observation, followed up with PC and FOC images, indicates that R136 is certainly not the single supermassive star that was once postulated, but rather is a dense, young cluster of stars, including some members with masses of order $100 M_\odot$.

FOC observations of SN1987a show with unprecedented clarity a circumstellar ring of ionized gas (Jakobsen et al., 1991). This ring was produced prior to the supernova explosion during the star's red giant phase, but was ionized by the photons from the explosion. Image restoration techniques allow one to show from the lack of light interior to the ring that the feature is a true ring rather than a limb-brightened sphere. Comparisons of the SN image with a stellar image indicate that the SN is slightly resolved, making this the first image of the expanding shell of debris from the explosion. Over the course of the next ten years HST should be able to record the evolution of the SN remnant in great detail.

The normal elliptical galaxy NGC 7457 was observed with the WF/PC, and even without image restoration processing one can see an unresolved nucleus indicative of an extremely high concentration of mass at the center of the galaxy (Lauer et al., 1991). This, combined with additional planned observations of normal galaxy nuclei, should help to settle the question of whether black holes are commonly found at the centers of galaxies.

The radio galaxies 3C66B and PKS0521-36 were observed with the FOC, and direct imaging of the optical counterparts to the radio jets was obtained (Macchetto et al., 1991). The close correspondence in morphologies indicates that the radio and optical emission are both due to synchrotron radiation from relativistic electrons.

PLANS AND PROSPECTS

While it is now felt that image restoration software will make it possible to restore the spatial resolution in most HST images, nothing short of the installation of corrective optics will allow us to recover the loss in sensitivity. NASA is now making a final decision on whether to build an instrument called COSTAR – Corrective Optics Space Telescope Axial Replacement – and to include its installation on the first HST shuttle servicing mission in late 1993. A number of panels have recommended that COSTAR be built.

COSTAR will consist of a deployable optical bench; small mirrors that compensate for the spherical aberration in the primary will be positioned along the light path for the High Resolution Spectrograph, Faint Object Spectrograph, and Faint Object Camera. The High Speed Photometer will be removed in order to make room for COSTAR. The corrective mirrors are expected to return HST to nearly its original design specifications for sensitivity and spatial resolution.

Also on board the first servicing shuttle flight will be WF/PC II (an improved version of the Wide Field / Planetary Camera which has its own corrective optics), a new set of solar panels which will correct the problem with vibrations that are now induced at crossings from sunlight to shadow, a replacement gyro (one of the six failed a few months ago), and possibly a replacement Fine Guidance Sensor. Installation of these components will require an unprecedented number of spacewalks and a long shuttle mission.

Currently scheduled for the second servicing mission in 1997 is the installation of the Near Infrared Camera (NIC). NIC is a somewhat scaled back version of an earlier instrument called NICMOS (MOS for multiple object spectrograph). NIC will provide the first true IR imaging capabilities for HST, with three cameras of varying spatial resolution operating in the spectral range of 1 – 2.5 micron.

The Space Telescope Imaging Spectrograph (STIS) will probably be installed on the third servicing mission in 1999. STIS is an ambitious instrument, using both CCD and MAMA (multi-anode microchannel array) detectors to cover a spectral range from 105 – 1100 nm. The capabilities of STIS should exceed those of HRS, FOS, and FOC combined.

SUMMARY

HST has just begun what should be a long and productive mission, and as our understanding of the instruments and image processing techniques to remove the effects of the spherical aberration improve, the scientific return from the project will be even greater. The installation of COSTAR and the second-generation scientific instruments, with their own internal corrective optics, should return HST to its originally planned capabilities.

REFERENCES

Burrows, C. J., 1991. *Astrophys. J.* **369**, L21.

Hall, D. N. B., ed., 1982. *The Space Telescope Observatory*, Proceedings of IAU Comm. 44.

Hester, J. J., et al., 1991. *Astrophys. J.* **369**, L75.

Jakobsen, P. et al., 1991. *Astrophys. J.* **369**, L63.

Lauer, T. R. et al., 1991. *Astrophys. J.* **369**, L45.

Macchetto, F. et al., 1991. *Astrophys. J.* **369**, L55.

Paresce, F. et al., 1991. *Astrophys. J.* **369**, L67.

Westphal, J. A. et al., 1991. *Astrophys. J.* **369**, L51.

HST IMAGE RESTORATION - STATUS AND PROSPECTS

Hans-Martin Adorf

Space Telescope - European Coordinating Facility, European Southern Observatory
Karl-Schwarzschild-Straße 2, D-8046 Garching b. München, F.R. Germany
Tel.: +49 - 89 - 320 06 - 261
EARN/BITNET: adorf@dgaeso51.bitnet — SPAN: eso::adorf

Abstract: The spherical aberration of the otherwise diffraction limited Hubble Space Telescope optics has spurred a surge of interest in image restoration methods, hardly seen before in optical astronomy. The characteristics of the aberrated point-spread function together with the stochastic properties of quantum-limited data have lead to the recognition of statistical algorithms as the methods of choice for HST image restoration.

It is
very hazardous
to ignore
noise effects
when attempting
restoration.
— B.R. Frieden, D.C. Wells 1978

I have had no experience
in dealing with astronomy imagery,
but I have worked with medical imagery,
and the impression I get is
that all the push really comes from people
on our side of the fence.
— M.J.B. Duff 1986

1. INTRODUCTION

The spherical aberration of the Hubble Space Telescope optics, announced on 21 June 1990 is adversely affecting all science instruments aboard HST. The occurrence of such a fundamental error (see STScI Newsletter, Aug. 1990; Wilson 1990; Allen *et al.* 1990) has caused consternation and disappointment in the HST community, particularly since all science instruments are behaving so flawlessly, with no component failures and few operational problems. Although aberration is affecting the pointing control system — the fine guidance sensors are perturbed by jitter (Blades 1991), loss of fine lock during terminator transitions and a loss of fainter guide stars (Doxsey & Reppert 1990) — the initial Science Assessment and Early Release Observation (SAO/ERO) programmes, defined shortly after the discovery of the aberration, have been remarkably successful (see the special issue of the *Astrophys. J. Lett.* **369** (1991)).

The series of first observations showed that HST's resolution is less affected by the telescope's aberration than originally feared (Hanisch 1990); the spherically aberrated point-spread function (PSF, see e.g. Fig. 2 in White & Burrows 1991) does not act as a Gaussian blur suppressing all fine detail. Instead, it redistributes ("encodes") the image flux in a peculiar fashion retaining much of the fine details at high spatial frequencies. Thus, the current problem mainly affects the telescope's (i) *faint limit*, (ii) *dynamic range*, i.e. HST's highly esteemed capability to image high contrast scenes such as "quasar fuzz" or the proverbial "faint planet close to a bright star", and (iii) obtainable *photometric and astrometric precision*. However, even in its present state, HST remains a remarkable discovery instrument opening completely new spectral windows for high-resolution imaging.

Data Analysis in Astronomy IV, Edited by V. Di Gesù *et al.*
Plenum Press, New York, 1992

Soon after the discovery of the impaired HST-optics the usage of image restoration was suggested in the hope that it could alleviate some of the difficulties encountered and improve the scientific gain of HST, until a service-mission — currently planned for 1993 (see *STScI Newsletter*, Dec 1990; Brown & Ford 1991) — would restore much of the anticipated capabilities of HST. Image restoration, which in a narrower sense comprises only the removal of spreading effects caused by a non-ideal PSF, is considered here as a comprehensive form of *stochastic inversion*, a process that in principle can include (or replace) all standard data calibration steps. Restoration in this sense aims at estimating from the distorted, noisy and discretely sampled observation, and some auxiliary data, an ideal intensity distribution by means of some inferencing process capable of removing as much as possible the effects of imperfect observing conditions and the instruments.

HST image restoration is a case of an ill-posed stochastic inverse problem. As is well known, in the presence of noise, there is an infinite number of "solutions" that, within the limits set by noise, are consistent with the data. Regularization methods can be used to resolve the ambiguities inherent to the problem.

The choice of an appropriate restoration method is partially driven by the availability of prior knowledge about the imaged objects, the characteristics of the data and the data analysis goal. These three areas will be addressed below in reverse order, followed by brief discussions on how restoration works in practice, the kind of results obtained so far and an outlook on future work.

2. DATA ANALYSIS GOALS

HST image data are analysed in either of two modes which may be called "discovery" and "quantitative refinement".

In discovery mode one wishes to use as little general astronomical prior knowledge as possible. Here, image restoration serves the purpose of drawing the astronomer's attention to unusual details in the data deserving closer inspection. A restoration method suitable for discovery should be fast and easy to use, require little auxiliary data, be robust and as unconstrained as possible. Photometric and astrometric fidelity are less important. Discovery mode is the realm of (model-free) image restoration, particularly methods based on Fisher's (1922) maximum likelihood principle.

In refinement mode, on the other hand, one is interested in the quantitative estimations of effects which in principle are known prior to data intake. Prime examples are the improvement of the cosmic distance scale by determining the Hubble constant — one of the original goals for building HST — or the determination of chemical composition and age of globular clusters in a distant galaxy. In such cases HST is used as a measuring device supposed to deliver accurate and precise estimates of physical quantities. Every piece of available prior information can and should be used in order to guide and constrain the estimation process and thereby e.g. suppress potential instabilities. Refinement mode is the realm of the different variants of the Bayesian method.

When quantitative information is sought, *error assessment* becomes important. Unfortunately, the analysis of restoration errors is still in its infancy and for the time being one will often have to resort to a qualitative error assessment by looking for invariants in data frames restored using different methods.

3. DATA CHARACTERISTICS

3.1. The HST-OTA point-spread function

The spherically aberrated PSF (Burrows *et al.* 1991; White & Burrows 1991) of HST's optical telescope assembly (OTA) is *very large* compared to ground-based PSFs, both in angular and linear diameter (Fig. 1). This presents a problem even for photometry of isolated point sources, since a very careful background estimation over a large area is required — unless one totally discards the flux in the halo and restricts oneself to "core-only" photometry (Gilmozzi, priv. comm.). The PSF displays two major components: a high surface brightness *core* and a richly structured, low surface brightness *halo*. The fine structure in the halo — distracting as it is to the eye — can provide restoration algorithms with useful information about the proper origin of the recorded flux.

HST's aberrated optical transfer function — the Fourier transform of the PSF — is also very non-Gaussian; it displays three components (di Serego Alighieri & Walsh 1990; see also Beletic 1991) with almost constant power in the high frequency "shoulder" up to the diffraction limit. Fortunately the OTF shows no holes in its support, i.e. none of the spatial frequencies are missing.

The PSF is presumably the single most important auxiliary input for image restoration and obtaining a good PSF can be considered as yet another calibration step that has to be carried out alongside the science observations. Information on the PSF can be obtained in several ways: *Internal empirical PSFs* can be derived from serendipitous point source observations in the field of view — if there are any.

Since HST offers, at least in principle, a thermally controlled, stable PSF, *external empirical PSFs* can be obtained from special calibration observations. A temporally and spatially stable PSF can be expected only for the time after HST's truss structure has outgassed its H_2O content.

Finally, there is some hope that *theoretical PSFs* can ultimately be computed sufficiently well (Krist 1991) using Burrows & Hasan's telescope instrument model (TIM) software, thus avoiding the overhead, noise and sampling problems of empirical PSFs. However, a major problem for theoretical PSFs is posed by spacecraft jitter, stochastic telescope movements predominantly occurring during day-night terminator transitions. In Fine Lock mode and during quiet periods telescope jitter is as small as 7 mas RMS, but it can be as large as 20 to 40 mas RMS. In Coarse Track jitter typically is about 50 mas RMS (Blades 1991). Available engineering data allow a straightforward post-facto reconstruction of spacecraft jitter only, if data were taken in Fine Lock.

3.2. Imagery with the FOC

As the name suggests, the main purpose of the FOC is to image faint objects and its detector therefore acts as a (linear) photon counting device — but only as long as the photon flux stays well below 1 count/sec/pixel. For higher count rates, particularly when extended objects are imaged, the FOC exhibits an awkward and yet uncalibrated non-local, non-linear intensity transfer function (Greenfield, pers. comm.), limiting the usable dynamic range per pixel practically to less than 1:1000. It is thus impossible to image in a single exposure the high contrast core-halo structure of the aberrated HST-PSF with satisfactory signal-to-noise.

The aberrated FOC-PSF is practically space-invariant, but its linear and areal size presents a particular problem for the FOC with its limited field of view (FOV), despite a frame size of up to 1024×1024 pixels. For instance, a circular PSF of 4 arcsec ∅ covers about one quarter of the area of the FOC f/288 FOV (cf. Adorf 1991b). Thus edge effects will often become important: observations may be incomplete because the frame edge cuts off parts of the blurred image or, conversely, stray light from bright objects outside the proper FOV may shine in.

3.2.1. Undersampling

FOC's detector in principle undersamples the images produced by its f/48, f/96 and f/288 configurations over all, most and some of its sensitivity range, respectively (Paresce 1990). This fact, which has acquired little attention so far, restricts the choice of admissible calibration and restoration procedures. The question whether an image is really undersampled or not, is determined jointly by the *sampling density* (i.e. the distance between pixel centres) and the *signal-to-noise ratio* of the data (Adorf 1989; 1990a). Several different domains can be distinguished:

1. *Low S/N observations:* Here insufficient sampling plays little role, since the aliased high spatial frequency components are anyway buried under the noise carpet. Restoration in this case is difficult with any method.
2. *High S/N observations:* (a) When the data are *sufficiently sampled,* all calibration and restoration methods will perform reasonably well (although non-linear restoration methods can be expected to still do better). (b) When the data are *insufficiently sampled,* then the ambiguities caused by aliasing become relatively more prominent. Thus non-linear restoration methods with their potential of achieving superresolution (by disentangling aliased frequencies) are much preferable.

In principle, insufficient sampling can be compensated for by unconventional data taking strategies such as "frame dithering", i.e. fractional pixel shifts between multiple observations of the same field (Adorf 1989, 1990a). Whether pointing and tracking of HST can be sufficiently controlled to allow such data taking modes is an open question at present.

3.2.2. Calibration effects

Undersampling and the PSF-induced need for image restoration render two other FOC problems more difficult than they would be otherwise, namely *non-square frame formats* and *geometric distortions*. The FOC in its highest resolution mode records images on rectangular frames with a sampling grid of 512×1024 non-square pixels (Paresce 1990). The standard pipeline calibration takes this peculiar format into a more common square one using a simple pixel-subdivision algorithm which creates unobserved, spurious high frequencies. Where necessary, a regular sinc-interpolation scheme may be considered as a promising alternative.

The considerable geometric distortions of the FOC occur in its image intensifier stages and are quantitatively described by a set of reseaux marks (which as a side effect cause missing data). Routine pipeline calibration, which has been developed from a deterministic point of view, corrects the distortions using a bi-linear interpolation scheme for the local flux, introduces clearly visible, curved (Moiré) fringe patterns in the noise amplitudes and also generates less visible noise correlations. Again as an alternative, a sinc-interpolation scheme might be considered.

In FOC-images small-scale spatial details, crucial for image restoration, are somewhat uncertain due to yet another effect: depending on the time elapsed between instrument power-on and exposure-start, the position of the recorded image drifts on the detector (Greenfield, pers. comm.). Without information on the relative registration of the flat-field with respect to the science frames, the flat-field frames are not used as recorded, but are smoothed before application and thus do not properly calibrate small-scale pixel-to-pixel sensitivity variations, an effect known as "fixed pattern noise". Somewhat surprisingly, fixed pattern noise disturbs the restoration process much less than might be feared, and can presumably be explained as a positive effect of the large number statistics of the sizeable PSF, of which many pixels contribute to restore the flux at a given point.

3.2.3. A stochastic model for the FOC

The quantum-limited FOC-images are corrupted by signal-dependent and thus spatially non-stationary (Poisson) photon shot noise. Consequently, the stochastic model which describes the FOC imaging process must be distinct from that used in radio astronomical aperture synthesis mapping, where — apart from systematic effects — the uncertainties in the data are dominated by instrumental (detector and amplifier) noise and unobserved spatial frequencies. While for radio data the standard textbook imaging equation with a deterministic, linearly blurred signal corrupted by additive Gaussian noise is applicable, for FOC-data it is not! Restoration techniques which disregard this important fact will produce non-optimum results.

For faint enough objects, a linear imaging equation $g = H f + n$ can be considered as a good first approximation to a realistic stochastic model for the FOC. Here f, n and g denote the image or irradiance distribution to be estimated, the instrumental noise (insignificant for the FOC) and the discretely sampled data, respectively. H is a linear distortion operator describing the end-to-end effect of the telescope-instrument-detector PSF including sampling. For the spatially invariant FOC-PSF, H can be implemented via convolution. The model image f can be represented in many ways; considering it as a Poissonian stochastic field over a continuous domain allows a straightforward discussion of both the statistical and sampling aspects including bandwidth-extrapolation ("superresolution"). Most of the parameters detailing the stochastic model for the FOC can (at least initially) be taken from the FOC instrument handbook (Paresce 1990), and can be refined when more in-flight calibration data are available.

A more sophisticated stochastic model for the FOC might collect the geometric distortion, pixel-to-pixel sensitivity (flat field) variations and the intensity transfer non-linearities at high count rates into a non-local function s, leading to a non-linear imaging model (cf. eq. (1) of Sezan & Tekalp, 1990) $g = s(H f; \alpha) + n$. In view of the inherent observational uncertainties, the calibration data (flat field, dark current, bias), collectively denoted by parameter α, might here be considered as stochastic quantities.

3.3. Imagery with the WF/PC

The most prominent of WF/PC's imaging characteristics is its asymmetric, space-variant PSF (SVPSF), which varies noticeably over distances of about 50 pixels (Hanisch, priv. comm.). The PSF's spatial variability makes the task of PSF estimation daunting and the actual restoration requires high-powered compute engines. In practical restorations so far the spatial variability has therefore either been ignored or taken into account by performing restorations on overlapping tiles, on each of which the PSF was considered space-invariant, and then reassembling the restored tiles. For a promising hybrid method for iteratively estimating an empirical SVPSF using multiple point source observations on WF/PC frames see Stetson (1991).

While the WF/PC frame sizes — 800x800 pixels per CCD-chip or 1,600x1,600 pixels for the full frame — are comparable to those of the FOC, edge effects play a less dominant role here, since the PSF, despite all aberration, is relatively small compared to the larger FOV, particularly for the WFC.

Mapping the PSF with acceptable signal-to-noise is facilitated by WF/PC's excellent dynamic range (about 1:30,000 per pixel).

3.3.1. Undersampling

Almost as problematic as the WF/PC's SVPSF is its *undersampling* (see the discussions above and in Adorf 1989, 1990a), which is severe over all of the WFC's sensitivity range and affects much of the PC's usable range.

Undersampling aggravates a WF/PC specific problem, namely cosmic ray (CR) hits on its CCD-detectors, a nuisance for restoration algorithms. The contamination of WF/PC data with CRs is substantial: on a 45 min exposure any pixel is estimated to have a 10—15% chance of being affected by a CR hit (Seitzer 1990). A soft tail in the energy distribution of secondary particles and misregistrations between CR-split exposures render CRs difficult to discriminate against the insufficiently sampled pencil cores of point sources. Several CR-rejection algorithms are under investigation, including one based on a neural network (Murtagh & Adorf 1991).

3.3.2. Calibration effects

While pixel-to-pixel sensitivity variations will presumably ultimately be calibrated to a few percent or better by taking ocean flat-fields (or by some other means), they are not as well known at the moment, restricting the knowledge of fine spatial detail.

WF/PC's CCD-detector has a stable geometry which nevertheless shows some moderate, calibratable geometric distortion (Gilmozzi & White 1991). Since flux interpolation occurs on fractional pixel scales, considerations similar to those stated above for the geometric distortion correction of the FOC apply to the WF/PC, too.

3.3.3. A stochastic model for the WF/PC

A simple stochastic model for the WF/PC (Adorf 1987) looks quite similar to that for the FOC, with two exceptions: (a) detector read-out noise is significant (Griffiths 1990) and the quantity n should therefore be considered as a stochastic vector, i.e. a collection of independent Gaussian distributed stochastic variables, one for each pixel; (b) the WF/PC PSF is spatially variant and therefore the degradation operator H cannot be implemented via convolution. With this model the noise in WF/PC-data is a mixture of Poissonian (due to the quantum-limitations) and Gaussian (due to read-out).

A more refined stochastic model for the WF/PC might e.g. take into account the effects of analog-to-digital quantization.

4. PRIOR KNOWLEDGE

Prior knowledge, such as the non-negativity of physical images, is of prime importance for constraining the ill-defined stochastic inversion problem. Prior knowledge about objects can be incorporated into the prior distributions of the maximum a-posteriori (MAP) method, but can also advantageously be exploited by other restoration methods which comprise an image segmentation step.

Prior knowledge about typical data obtainable with HST's imaging instruments is abundant. This prior knowledge can naturally be decomposed into general wisdom about astronomical scenes and specific information on HST and its instruments. So far, the inclusion of prior knowledge into the restoration algorithms has just begun, but it is recognized that e.g. artifact suppression and precision improvements can be expected.

4.1. Objects

In many cases astronomical image frames contain *stellar images*, even when stars are not of immediate scientific interest, delivering PSFs for free. (Compare this with the difficulty of obtaining a PSF from an everyday scene!) The recognition of the presence of a point source — they are notoriously difficult to restore — can signal a suitable restoration method to locally relax a global smoothness constraint, thereby achieving a sharper restored point source image.

Images of objects in the other big class, *galaxies*, range in apparent size from a few pixels to as many as 10,000 or more. For these extended objects, which are easier to restore than point sources, empirical intrinsic intensity correlations may potentially aid the restoration process.

There are, of course, a variety of other sources, some of them as complicated as the Orion nebula, making up a third class of objects. For this inhomogeneous class prior knowledge is a scarce commodity and it will be difficult to arrive at highly constrained restoration recipes.

4.2. Background

Another peculiarity of astronomical imagery, especially when compared to areas outside astronomy, is the existence of a fairly homogeneous background in large parts of the sky and thus on most image frames. The background light HST sees is composed of Zodiacal light, reflected light from Earth and Moon and the galactic light attributed to a large number of distant, sometimes resolved (!), ultra-faint blue galaxies that cover the sky apparently isotropically (Tyson 1990).

The possibility of separating a typical astronomical scene into finely structured foreground objects on top of a smooth background was first explicitly noted by Frieden & Wells (1978). For an image restorer this is a peculiar situation calling for special methods to take advantage of. While the flat background makes small-scale restoration errors very obvious, this may be considered a virtue, since it allows a good check on artifacts. On a structured background these errors would still exist, but be invisible.

5. RESTORATION IN PRACTICE

Many factors influence the quality of a restoration and it is not easy to decide beforehand which are the decisive ones. Practical restoration work involves several separate steps (Fig. 2): pre-processing of the science data; pre-processing of the PSF data; the choice of a suitable restoration method; establishment of a stochastic model, if required by the method; application of the method; and potentially some post-processing of the restoration result.

Restoration experience has taught us that the restoration result depends critically on the quality of the calibration data — sometimes more than on the specific restoration method used — and that therefore the time and effort spent on data pre-processing is well invested. Most pre-processing steps are labour-intensive and involve some experimentation; only the proper restoration step itself is compute-intensive, particularly when an iterative method is being used. Most of the pre-processing steps necessary to condition the science and PSF data (Adorf 1990b, d) are fairly independent of the restoration method used.

5.1. Science data pre-processing

All restorations carried out so far have started from calibrated data. (Note that standard PODPS calibration includes *geometrical distortion correction* only for the FOC.) When a non-linear method is used which preserves or enforces the non-negativity constraint, calibration should be followed by *background-subtraction* and *zero level clipping* or some other means of removing negative excursions in the data.

A more refined treatment might include *removal of bright borders* at the frame's edge by masking, *embedding of the data frame* into a larger frame, a precautionary measure recommended in order to avoid adverse wrap-around effects of the FFTs, *apodization* of the data with some roll-off mask, a measure particularly recommended when for some reason the background differs at opposite edges, and *reconstruction of missing data* caused by reseaux marks or cosmic ray hits.

A post-processing measure could consist in *resampling* the restored image onto a finer grid, which may help the eye to pick up fine structures.

5.2. PSF data pre-processing

Restoration critically depends on the quality of the PSF and it is clearly important to use the *best possible PSF* one can obtain. Whether theoretical PSFs can be successfully used for restoration is presently an open question. In the following I therefore assume that an empirical PSF is derived from one or more observations of a point source. Obviously, such PSF images should be sharp (i.e. taken in fine lock), well exposed in order to suppress noise, and unsaturated in order to avoid non-linearities. Experience suggests that it is critically important to use a PSF acquired in the same filter as the science observation.

The same four initial pre-processing steps required for the distorted science data should also be applied to the PSF data frame. Additionally it is recommended to *centre the PSF* within the frame by

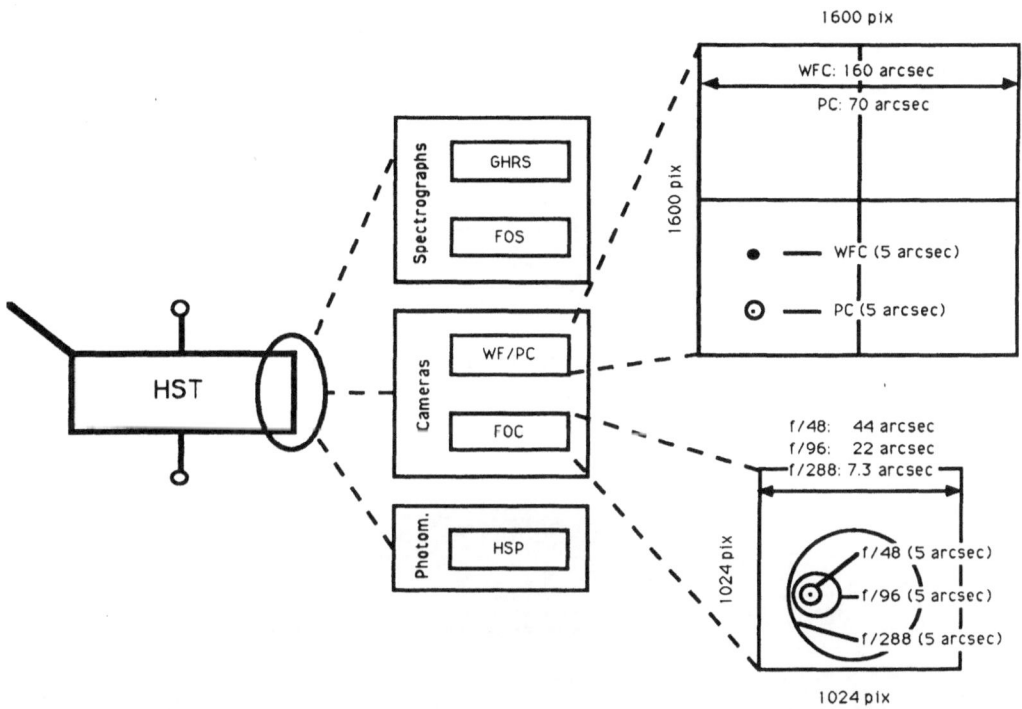

Fig. 1. Size of a 5 arcsec ⌀ circular PSF relative to the frame sizes of HST's Wide Field/Planetary Camera and its Faint Object Camera in various configurations.

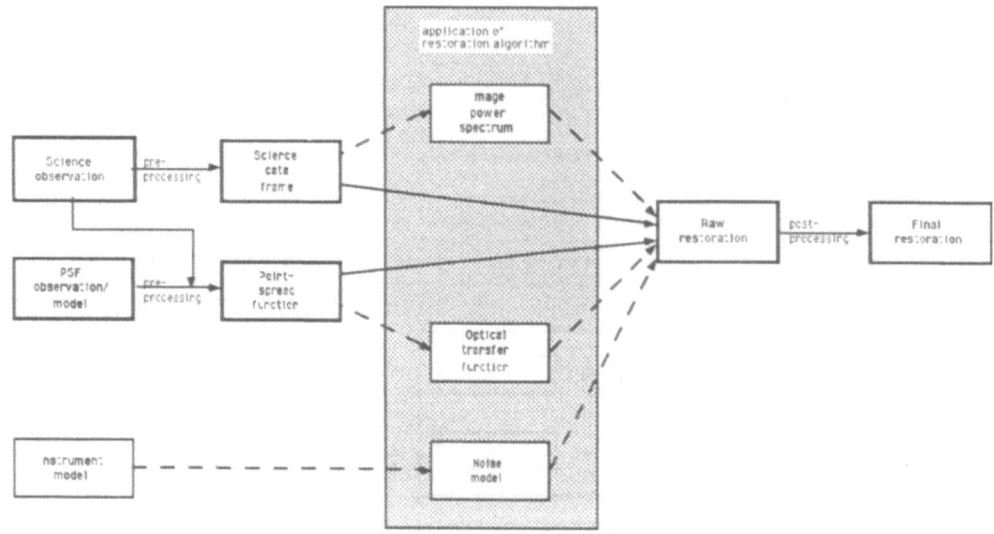

Fig. 2. Steps in practical HST image restoration.

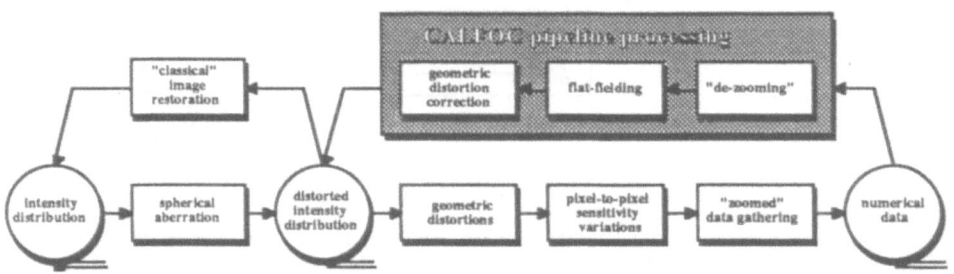

Fig. 3. Schematic view of the image formation process in HST's Faint Object Camera. STScI pipeline calibration attempts to correct for all degradations except spherical aberration, the inversion of which is left to a standard image restoration process. An envisaged "all-in-one" image restoration scheme would carry out a reconstruction of the intensity distribution at the telescope aperture directly from the raw data, using the calibration data as auxiliary input.

shifting it an integral number of pixels, although most restoration algorithms will tolerate a slightly de-centred PSF. Several restoration methods require a *normalization of the PSF* to unity and an *embedding of the PSF* into a frame of the same size as the distorted data.

Of all problems the inevitable noise in an empirical PSF seems to be the most difficult one to cope with. For some restoration methods preserving the non-negativity constraint (e.g. the Richardson-Lucy-method) both the science and the PSF data must be non-negative. Removing the background surrounding the PSF is somewhat tricky: Simply subtracting the mean background does enforce a zero mean background, but at the same time destroys non-negativity. Subsequent zero-level clipping re-introduces a non-zero mean background of a somewhat reduced amplitude at a higher level than before. In order to avoid adverse effects during restoration this remnant background should be suppressed outside the support of the PSF through *apodization* by multiplying with a roll-off mask.

The question of how to optimally combine more than one PSF observation into a single empirical PSF frame, potentially sampled on a finer grid than the individual data frames, is an interesting but presently unsolved problem.

5.3. Choosing a restoration method

The choice of an appropriate restoration method seems of prime importance, especially to newcomers to the field, but the success of a restoration attempt relies on many factors, the suitability of the method employed being only one of them. Other factors may influence the decision which algorithm to use such as availability of tested code, availability of sufficient compute power and memory, of local expertise, etc.

The choice of a restoration method is largely influenced by the data analysis goal (see above). When photometric or astrometric estimation on *restored* images is envisaged, much attention has to be paid to the quantitative aspects of the restoration methods. On the other hand, when the restored image only serves the purpose of generating an object finding list, which is subsequently fed into a standard photometry package working on the calibrated but otherwise *unrestored* data, then questions of statistical bias etc. are of lesser importance.

A variety of methods have been devised for image restoration. They can be grouped in many ways, e.g. general vs. special, linear vs. non-linear, iterative vs. non-iterative (cf. Sezan & Tekalp 1990). Another, perhaps more relevant classification scheme consists in separating deterministic inversion methods from statistical estimation methods, where methods in the latter category might be distinguished further according to the kind and generality of the stochastic model employed.

The 1990 STScI workshop on HST Image and Spectra Restoration (White & Allen 1991; see also Adorf 1990c) revealed that data analysts in the medical area have for quite some time applied restoration methods to their image reconstruction problems. Astronomical and medical imagery both work with quantum-limited signals produced by low irradiance levels (cf. Dainty *et al.* 1990), although for quite different reasons. Thus medical restoration work and particularly their methods are quite relevant to optical astronomy.

A good overview of the various restoration methods and their underlying assumptions facilitates the choice of an appropriate algorithm for a given data set and analysis goal. However, an in-depth review of all methods which are potentially applicable to HST imagery is beyond the scope of this paper. The interested reader is referred to standard textbooks on image restoration such as Andrews & Hunt's (1977), to recent general reviews such as Sezan & Tekalp's (1990) or to my own summary produced along with this review (Adorf 1991d). A comprehensive account of image restoration methods relevant to astronomy up to the early 1980's can be found in Don Wells' two excellent review articles (1980, 1983).

The most successful general methods for HST image data so far seem to be the Richardson-Lucy method and several variants from the maximum a-posteriori (MAP) family of restoration methods. The simple yet powerful, recursive Richardson-Lucy method (Richardson 1972; Lucy 1974; Shepp & Vardi 1982) and its generalizations (Snyder 1991; White 1991a) are suited to quantum-limited data and have desirable properties (Adorf 1990d, 1991d) such as robustness with respect to small-scale PSF errors.

In the versatile MAP family, which includes the maximum entropy method (Weir & Djorgovski 1991 and references therein) as a special case (Frieden & Wells 1978; Trussel 1980; Frieden 1983), the MemSys maximum entropy package (Gull & Skilling 1991) with its MEM front-end (Weir, priv. comm.) deserves special mentioning.

6. EARLY RESULTS

The Science Assessment/Early Release Observations and the first General Observer programmes have yielded quite a wealth of scientifically interesting data (cf. Hanisch 1991):

The core of R136 in 30 Doradus was imaged with WF/PC and FOC — and resolved, thus providing direct evidence that R136 is a compact star cluster and not a single supermassive star (Lindler 1991; Horne 1991; Weigelt *et al.* 1991).

FOC observations of SN1987A showed a resolved supernova core (Jakobsen *et al.* 1991), many emission knots on top of the apparent "ring", two faint "bridges" in its interior and faint outer wisps. The gain in contrast between the apparent ring and its interior is remarkable: through restoration it increased from 3:1 to 100:1 (see Panagia *et al.* 1991; Danziger *et al.* 1991). Thus the observations directly support the hypothesis that the apparent ring is a true circumstellar ring.

In the bipolar outflow of the symbiotic Mira variable R Aquarii (Paresce *et al.* 1991) various filaments and a dark obstruction could be resolved (Adorf 1991c and references therein); the quality of the FOC-data allows a detailed analysis of gas temperature and composition.

WF/PC observations of the Orion nebula revealed Herbig-Haro objects and associated shock fronts on scales smaller than 40 AU. Saturn's white spot (White 1991a) and Mars' atmospheric features (the first GO programme) have been successfully observed. On single orbit exposures the morphology of faint galaxies was determinable down to V=20 mag (WF/PC) and V=21.7 mag (FOC).

In the analysis of all these observations image restoration methods have played a decisive role. Restoration work on other observations such as the gravitational lens G2237+030 (Crane *et al.* 1991) or the Iovian moon Io (Paresce *et al.*, in preparation) is progressing.

Despite aberration, spatial detail in well-exposed parts of some of the FOC images warrants a restoration to 11 mas pixels — about a tenfold increase in spatial resolution compared to the best ground-based telescopes.

7. DIRECTIONS FOR FUTURE RESEARCH

While HST image restorers have quite a selection of working algorithms already at their disposal, much work remains to be done, before the HST restoration problem can be considered as "solved". The current goal is *quality restorations*, which allow a quantitative data evaluation including a reliable error assessment. Naturally, work in the restoration and calibration areas partially overlaps.

In the following I will concentrate on six areas deserving attention, namely PSF identification, error estimation, undersampling, calibration, adaptive restoration and high-speed computing. This list is by no means exhaustive, but indicates the kind of work either already underway or to be considered for the future. (See also the discussions in White & Burrows 1991 and in White 1991b).

7.1. PSF identification

The paramount problem of HST image restoration remains PSF identification. Here a detailed comparison of the existing library of pre-computed theoretical PSFs (Krist 1991) with empirical PSFs is urgently needed. Refined methods for PSF estimation from observations should also be investigated, particularly for the FOC, where restoration work may benefit from PSFs assembled by "data fusion" from incomplete parts or from two differently registered exposures or from two exposures taken at different levels. PSF estimation may also benefit from several recent multidimensional sampling theorems.

7.2. Error estimation

The capability of estimating accuracy and precision is mandatory for any quantitative data analysis scheme. The Wiener-filter has been favoured by some for its inherent linearity, but the Richardson-Lucy restoration method also looks amenable to a fairly straightforward error analysis, since it assembles in a non-linear fashion a linear restoration operator (Lucy 1991). The recent release 5 of the Mem-Sys package is supposed to include code for error analysis and an initial attempt to assess restoration errors for the maximum entropy method has been made (Cohen 1991).

The ultimate bounds to statistical errors are set by the Cramér-Rao minimum-variance theorem, which allows a computation of the achievable precision of any unbiased estimator (Adorf 1987 and references therein).

7.3. Undersampling and superresolution

The severe undersampling of both cameras aboard HST may require novel data analysis concepts. In aperture synthesis radio imagery, non-linear restoration methods have shown the capability of *interpolating* missing Fourier-coefficients from known ones.

Similarly, superresolution works by *extrapolating* unobserved Fourier-coefficients at frequencies above the Nyquist frequency using the observational data and the non-negativity property as constraints (but not the analyticity of the image, see Narayan & Nityananda 1986). Often a modest amount (a factor of two) of superresolution is possible and should not be treated with suspicion. Holmes (1988) has studied the band-width extrapolation behaviour of the Richardson-Lucy algorithm on simulated diffraction-limited Poisson-data of several signal-to-noise ratios and, for binary objects, obtained fairly good results even for lower SNRs.

In the case of HST the problem is not so much extrapolating spatial frequencies beyond the Nyquist limit, but unfolding the aliased spectrum, i.e. disentangling those frequencies that are aliased above and *below* the Nyquist frequency. Whether this kind of superresolution will work with the aberrated HST-data remains to be seen; first promising attempts in this direction have already been made (White, pers. comm.).

In any case it seems desirable to investigate how the high frequency and noise properties of the data transform under various resampling schemes.

7.4. Alternative calibration/restoration schemes

Routine science calibration certainly changes the high-frequency and the statistical properties of the data, which can no longer be modelled via uncorrelated noise, thus invalidating the assumptions underlying some of the statistical restoration methods. How big in practice these effects are on subsequent restorations is unclear at the moment and needs to be further investigated. In any case a better calibration of the "fixed pattern noise" seems desirable for the FOC.

In this context it might be interesting to discuss the concept of an "all-in-one" restoration scheme (Fig. 3), which would incorporate all standard calibration steps, and would work on raw data rather than calibrated frames with their distorted high frequency and noise properties. A first step in this direction has been made by Snyder (1991) and by Núñez & Llacer (1991), who introduced pixel-to-pixel sensitivity corrections into their restoration schemes.

7.5. Adaptive and multichannel restoration

In view of the non-stationary, signal-dependent quantum noise, MAP-based methods look very promising which are capable of producing restorations with spatially adaptive, S/N-dependent resolution. Methods allowing a local relaxation of a global (regularizing) smoothness constraint will be beneficial e.g. when edge restoration is important.

Multichannel restoration incorporating prior knowledge (or assumptions) about the existence of two or more classes of objects with different spatial extent or correlations look promising, too. Attempts in this direction have already been made (Weir 1991).

The more sophisticated restoration techniques such as MAP offer a "data fusion" capability, a largely unexplored possibility for data combination: two or more frames which may be differently registered, or may have been exposed to different levels, are restored into a single frame. Particularly interesting seems the possibility of joining ground-based image data containing extended, low-surface brightness structures with space-borne data, which add in fine-scale details.

7.6. High-speed computing options

HST restoration work so far has been limited more by human resources than by available CPU-power or memory. This situation may, however, soon be reversed, when more GTO and GO data have been taken and when the dust has settled about which methods to use and how to adapt them to the peculiarities of HST imagery. Past experience has made it very clear that recursive restoration methods deliver more reliable results than one-step methods — at the expense of moderately to highly increased computational costs. Using iterative methods to restore full WF/PC frames with its SVPSF is a daunting task, particularly when one simultaneously attempts to overcome undersampling (see Holmes 1988).

An initial survey of affordable high-speed computing options has been carried out (Adorf

1991a) and practical restoration work on a Cray supercomputer and a workstation with a vector-processing add-on board is envisaged by STScI and ST-ECF staff for the near future. Even more exotic architectures such as the Connection Machine are under consideration.

8. CONCLUSION

Hubble Space Telescope with its current spherical aberration encodes astronomical "scenes" in a peculiar, very non-local way, requiring unusual analysis methods for subsequent information decoding. The aberration causes a loss of sensitivity and discriminatory power, but HST's high resolution imaging capability is far less compromised than originally feared, as convincingly demonstrated by the results obtained within the Early Release and Science Assessment Observations programmes.

Given enough signal-to-noise, much of the spatial information can be recovered using statistical restoration methods suited to HST's current imaging characteristics: high-contrast signals which often contain point sources; signal-dependent, spatially non-stationary noise; a large, richly structured, wavelength-dependent, stochastic and, for the Wide Field and Planetary Camera, space-variant point-spread function; fairly sizeable image formats; geometric distortion combined with partially severe undersampling; imperfect flat-field maps; missing data points; and for the Faint Object Camera a non-linear, non-local intensity transfer function.

The development of quantitative image restoration methods which take into account all of these (and potentially other) effects is quite a challenge and, as the experience from radio astronomy shows, will take some time. However, it is certainly worth the effort, since computer restoration is bound to stay: even in post-COSTAR times it will be needed for HST archival research and also for the diffraction-limited images expected from the repaired Hubble Space Telescope.

ACKNOWLEDGEMENTS

This review has benefitted from discussions with and information from Ron Allen (STScI), Dave Baxter (STScI), Chris Burrows (STScI), Ian Evans (STScI), Bob Fosbury (ST-ECF), Richard Hook (ST-ECF), Ron Gilliland (STScI), Roberto Gilmozzi (STScI), Perry Greenfield (STScI), Robert Jedrzejewski (STScI), Hashima Hasan (STScI), Bob Hanisch (STScI), Richard Hook (ST-ECF), Keith Horne (STScI), Jon Holtzman (Lowell Obs.), Ivan King (UC-Berkeley), Leon Lucy (ST-ECF), John Mackenty (STScI), Fionn Murtagh (ST-ECF), Jorge Núñez (Barcelona), Don Snyder (Univ. Washington), Bill Sparks (STScI), Peter Stetson (DAO), Jeremy Walsh (ST-ECF), Gerd Weigelt and collaborators (MPIfR Bonn), Nick Weir (Caltech), Don Wells (NRAO) and Rick White (STScI).

REFERENCES

Adorf, H.-M.: 1987, "Theoretical limits to the precision of joint flux and position measurements with CCD-detectors", (unpublished research note)

Adorf, H.-M.: 1989, "On the HST Wide Field and Planetary Camera Undersampling Problem", in: *Proc. 1st ESO/ST-ECF Data Analysis Workshop, European Southern Observatory, Garching, 17—19 Apr 1989*, P. Grosbøl, F. Murtagh, R. Warmels (eds.), pp. 215—218

Adorf, H.-M.: 1990a, "Coping with the undersampling of WFPC images", *ST-ECF Newsl.* 12, 9—10

Adorf, H.-M.: 1990b, "How to prepare data for HST image restoration", *ST-ECF Technical Report, 6. Sep 1990 (orig.), 29. Oct 1990 (rev.)*

Adorf, H.-M.: 1990c, "Restoration of HST images and spectra — A report on the workshop held at the STScI, Baltimore on 21—22 August 1990", *ST-ECF Newsl.* 14, 15

Adorf, H.-M.: 1990d, "Restoring HST images — an ST-ECF perspective", *ST-ECF Newsl.* 14, 8—12

Adorf, H.-M.: 1991a, "Does HST image restoration need personal supercomputing?", *ST-ECF Newsl.* 15, 8—11

Adorf, H.-M.: 1991b, "Restoring HST Faint Object Camera images", in: *Proc. 3rd ESO/ST-ECF Data Analysis Workshop, Garching, 22—24 Apr 1991*, P. Grosbøl et al. (eds.), (in press)

Adorf, H.-M.: 1991c, "Richardson-Lucy restoration of the symbiotic Mira R Aquarii in the light of [OIII]", *ST-ECF Newsl.* 15, 11—13

Adorf, H.-M.: 1991d, "Image restoration methods — a brief overview", *ST-ECF Technical Report, 19. May 1991*, 8 pp.

Allen, L., Angel, R., Mangus, J.D., Rodney, G.A., Shannon, R.R., Spoelhof, C.P.: 1990, "The Hubble Space Telescope Optical Systems Failure Report", NASA, Nov 1990

Andrews, H.C., Hunt, B.R.: 1977, "Digital Image Restoration", Prentice-Hall, Inc., Englewood Cliffs, NJ 07632

Beletic, J.: 1991, "HST image processing: Determination of best focus and effects of photon noise", in: *Proc. Workshop "The Restoration of HST Images and Spectra", Space Telescope Science Institute, Baltimore, 21.—22. Aug. 1990, R.L. White, R.J. Allen (eds.)*, pp. 13—20

Blades, C.: 1991, "Letter to General Observers, 16. April 1991", Space Telescope Science Institute, Baltimore, MD 21 218

Brown, R.A., Ford, H.C. (eds.): 1991, "Report of the HST Strategy Panel: A Strategy for Recovery", Space Telescope Science Institute, Baltimore, MD 21 218

Burrows, C. *et al.*: 1991, "The imaging performance of the Hubble Space Telescope", *Astrophys. J. Lett.* **369**, L21—L25

Cohen, J.G.: 1991, "Test of the photometric accuracy of image restoration using the maximum entropy algorithm", *Astron. J.* **101**, 734—737

Crane, P. *et al.*: 1991, "First Results from the Faint Object Camera: Images of the Gravitational Lens System G2237+0305", *Astrophys. J. Lett.* **369**

Dainty, J.C., Morris, G.M., Tyson, J.A.: 1990, "Quantum-limited imaging and information processing", Introduction to the special issue on "Quantum-Limited Imaging and Information Processing", *J. Opt. Soc. America* **A 7**

Danziger, I.J., Boucher, P., Gouiffes, C., Lucy, L.B.: 1991, "SN 1987A: Observations of the Later Phases", ESO-CERN-Texas Symposium on Relativistic Astrophysics, Brighton, Dec. 1990, in: *Proc. of the New York Academy of Sciences,* (to appear)

di Serego Alighieri, S., Walsh, J.: 1990, "The HST point spread function", *ST-ECF Newsl.* **14**, 4—7

Doxsey, R., Reppert, P.: 1990, "Observatory Status", *STScI Newsl.* **7**, 11—12

Duff, M.J.B.: 1986, in: "Trends in parallel processing applications", *Proc. 2nd Internat. Workshop on Data Analysis in Astronomy, 17—30 April 1986, Erice, Sicily,* V. Di Gesu, L. Scarsi, P. Crane, J.H. Friedman, S. Levialdi (eds.), Pergamon Press, New York & London, p. 379

Fisher, R.A.: 1922, "On the mathematical foundations of theoretical statistics", *Philos. Trans. Royal Soc. London* **222**, 309

Frieden, B.R., Wells, D.C.: 1978, "Restoring with Maximum Entropy. III. Poisson Sources and Backgrounds", *J. Opt. Soc. America* **68**, 93—103

Frieden, B.R.: 1983, "Unified theory for estimating frequency-of-occurrence laws and optical objects", *J. Opt. Soc. America* **73**, 927—938

Gilmozzi, R., White, R.L.: 1991, (internal report on geometric distortion of the WF/PC), Space Telescope Science Institute, Baltimore

Gilmozzi, R.: 1991, "Spacecraft jitter: its effect on the HST PSF (and on the "breathing")", *STScI internal report*

Griffiths, R.: 1990, "Hubble Space Telescope Wide Field and Planetary Camera Instrument Handbook", Space Telescope Science Institute, Baltimore, MD 21218, May 1990

Gull, S.F., Skilling, J.: 1991, "MEMSYS 5 User's Manual", Maximum Entropy Consultants Ltd.

Hanisch, R.: 1990, "Restoration of High Spatial Resolution in HST", *STScI Newsl.* **7**, 16—18

Hanisch, R.: 1991, Monthly Report March 1991, Space Telescope Science Institute, Baltimore, MD 21 218

Hanson, K.M.: 1990, "Method of evaluation image-recovery algorithms based on task performance", *J. Opt. Soc. America* **A 7**, 1294—1304

Heasley, J.N.: 1984, "Numerical restoration of astronomical images", *Publ. Astron. Soc. Pacific* **96**, 762—772

Holmes, T.J.: 1988, "Maximum-likelihood image restoration adapted for noncoherent optical imaging", *J. Opt. Soc. America* **A 5**, 666-673

Horne, K.: 1991, "Maximum entropy deconvolution of a Wiede Field Camera image of R136", in: *Proc. Workshop "The Restoration of HST Images and Spectra", Space Telescope Science Institute, Baltimore, 21.—22. Aug. 1990, R.L. White, R.J. Allen (eds.),* pp. 132—138

Jakobsen, P. *et al.*: 1991, "First Results from the Faint Object Camera: Supernova 1987A", *Astrophys. J. Lett.* **369**

Krist, J.: 1991, "The TIM Cookbook. A Guide to making Gourmet PSFs", *STScI internal report, Feb. 1991*

Lindler, D.J.: 1991, "Block iterative restoration of astronomical images from the Hubble Space Telescope", in: *Proc. Workshop "The Restoration of HST Images and Spectra", Space Telescope Science Institute, Baltimore, 21.—22. Aug. 1990, R.L. White, R.J. Allen (eds.),* pp. 39—49

Lucy, L.B.: 1974, "An iterative technique for the rectification of observed distributions", *Astron. J.* **79**, 745—754

Lucy, L.B.: 1991, "Restoration with increased sampling — images and spectra", in: *Proc. Workshop "The Restoration of HST Images and Spectra", Space Telescope Science Institute, Baltimore, 21.—22. Aug. 1990, R.L. White, R.J. Allen (eds.),* pp. 80—87

Narayan, R., Nityananda, R.: 1986, "Maximum Entropy Image Restoration in Astronomy", *Ann. Rev. Astron. Astrophys.* **24**, 127—170

Núñez, J., Llacer, J.: 1991, "Statistically based image reconstruction", *ST-ECF Newsl.* **15**, 6—8

Panagia, N., Gilmozzi, R., Macchetto, F., Adorf, H.-M., Kirshner, R.P.: 1991, "Properties of the SN187a circumstellar ring and the distance to the Large Magellanic Cloud", *AAS spring meeting*

Paresce, F. *et al.*: 1991, First Results from the Faint Object Camera: Imaging the Core of R Aquarii", *Astrophys. J. Lett.* **369**

Paresce, F.: 1990, "Hubble Space Telescope Faint Object Camera Instrument Handbook", Space Telescope Science Institute, Baltimore, MD 21218, May 1990

Richardson, W.H.: 1972, "Bayesian-Based Iterative Method of Image Restoration", *J. Opt. Soc. America* **62**, 52—55

Seitzer, P.: 1990, "Suggestions for WF/PC use", *STScI Newsl.* **7**, 13

Sezan, M.I., Tekalp, A.M.: 1990, "Survey of recent developments in digital image restoration", *Opt. Engineering* **29** (special issue on Image Restoration and Reconstruction), 393—404

Shepp, L.A., Vardi, Y.: 1982, "Maximum likelihood reconstruction for emission tomography", *IEEE Trans. Med. Imaging* **MI-1**, 113—122

Snyder, D.L.: 1991, "Modifications of the Lucy-Richardson Iteration for Restoring Hubble Space-Telescope Imagery", in: *Proc. Workshop "The Restoration of HST Images and Spectra", Space Telescope Science Institute, Baltimore, 21.—22. Aug. 1990, R.L. White, R.J. Allen (eds.)*, pp. 56—61

Stetson, P.: 1991, "Decomposing stars in crowded fields", *Proc. 3rd ESO/ST-ECF Data Analysis Workshop, Garching*, P. Grosbøl *et al.* (eds.), (in press)

Trussel, H.J.: 1980, "The relationship between image restoration by the maximum a posteriori method and a maximum entropy method", *IEEE Trans. Acoust. Speech Sign. Proc.* **ASSP-28**, 114—117

Tyson, J.A.: 1990, "Progress in low-light-level charge-coupled device imaging in astronomy", *J. Opt. Soc. America* **A 7**, 1231—1236

Weigelt, G. et al.: 1991, *Astrophys. J. Lett.* **369**

Weir, N.: 1991, "Recent developments in maximum entropy based image restoration":, *ST-ECF Newsl.* **16**, (to appear)

Weir, N., Djorgovski, S.: 1991, "MEM: New techniques, applications, and photometry", in: *Proc. Workshop "The Restoration of HST Images and Spectra", Space Telescope Science Institute, Baltimore, 21.—22. Aug. 1990, R.L. White, R.J. Allen (eds.)*, pp. 31—38

Wells, D.C.: 1980, "Nonlinear image restoration: what we have learned", in: *Proc. SPIE "Applications of Digital Image Processing to Astronomy"*, Vol. **264**, pp. 148—156

Wells, D.C.: 1983, "Nonlinear Image Restoration", in: *Techn. Digest "Joint Topical Meeting on Information Processing in Astronomy and Optics", St. Paul, Minnesota, 23—24 Jun 1983*, pp. Th-A10 1—4

White, R., Burrows, C.: 1991, "The HST spherical aberration and its effects on images", in: *Proc. Workshop "The Restoration of HST Images and Spectra", Space Telescope Science Institute, Baltimore, 21.—22. Aug. 1990, R.L. White, R.J. Allen (eds.)*, pp. 2—6

White, R.: 1991a, "Restoration of HST images using the Lucy method with oversampling", in: *Proc. Workshop "The Restoration of HST Images and Spectra", Space Telescope Science Institute, Baltimore, 21.—22. Aug. 1990, R.L. White, R.J. Allen (eds.)*, pp. 139—143

White, R.L.: 1991b, "Restoration of images and spectra from the Hubble Space Telescope", *Proc. 25th Conf. Computer Science and Information Processing, Baltimore, March 1991*, (in press)

White, R.L., Allen, R.J. (eds.): 1991, *Proc. Workshop "The restoration of HST images and spectra"*, Space Telescope Science Institute, Baltimore, 21—22 Aug 1990

Wilson, R.N.: 1990, "'Matching Error' (Spherical Aberration) in the Hubble Space Telescope (HST): Some Technical Comments, *Messenger* **61**, 22—24

STSDAS: THE SPACE TELESCOPE

SCIENCE DATA ANALYSIS SYSTEM

Robert J. Hanisch

Space Telescope Science Institute[1]
3700 San Martin Drive
Baltimore, Maryland 21218 USA

Abstract

The Space Telescope Science Data Analysis System – STSDAS – is a portable set of applications programs for the calibration and analysis of data from the Hubble Space Telescope. STSDAS achieves system independence by being fully layered on the Image Reduction and Analysis Facility (IRAF) developed at the National Optical Astronomy Observatories. STSDAS augments the general capabilities of IRAF, and along with other IRAF-based analysis packages such as the PROS system for X-ray data analysis from the Harvard–Smithsonian Center for Astrophysics, provides a broad suite of astronomical software within a common environment.

INTRODUCTION

STSDAS is a system of some 400,000 lines of code and documentation designed for the calibration and analysis of data from the Hubble Space Telescope. At its inception in 1981, when it was known simply as 'SDAS', the system was dependent on the VAX VMS operating system and architecture and was limited in scope to HST data analysis, excluding calibration. As it was realized that astronomers would want to be able to do their own calibrations of HST data, and that it was not cost-effective, either for end-users or for the software development effort, to have a system dependent on a proprietary operating system, the scope and objectives of the system were modified. The most fundamental change was the decision to fully layer the STSDAS software

---

[1]The Space Telescope Science Institute is operated by the Association of Universities for Research in Astronomy, Inc., for the National Aeronautics and Space Administration.

Data Analysis in Astronomy IV, Edited by V. Di Gesù *et al.*
Plenum Press, New York, 1992

on the Image Reduction and Analysis Facility – IRAF – from the National Optical Astronomy Observatories (see the paper by Tody in this volume). For a review of the historical aspects of the development of STSDAS, please see Hanisch (1989).

The system now includes the capability of performing all HST data calibration using 'pipeline' type processing: all calibration options are controlled by the values of keywords in the input data file, and the software operates with essentially no user intervention. This allows the identical code to be run from within the on-line Post Observation Data Processing System (PODPS) for HST (see the paper on HST in this volume) as is used in the off-line STSDAS software. Users are free to tailor the off-line calibration and inspect intermediate data products that are not typically made available in the on-line system. STSDAS also includes a suite of calibration support software. These programs are used to analyze HST calibration data and generate the appropriate sensitivity parameters, flat fields, wavelength scales, etc., that are utilized in the pipeline calibration processing. The calibration software is organized into separate IRAF packages, one for each major scientific instrument. STSDAS also provides access to the HST Calibration Data Base via simple database queries.

STSDAS includes a number of applications packages of a more generic nature, although some of these have features that are tailored to HST data. For example, there are packages for Fourier analysis, isophote fitting, and synthetic photometry, to name a few, but at this time the synthetic photometry package only has HST filters available. In order to cope with the spherical aberration problem on HST, a new set of image restoration algorithms is being provided. Several of the most basic techniques – Fourier inverse, Wiener filter, and Lucy-Richardson algorithm – have already been implemented.

Other packages in STSDAS complement the general graphics, image display, and data manipulation packages in IRAF. For example, the STSDAS task *igi* (for interactive graphics interpreter) allows for the production of publication quality graphics using a command interpreter syntactically similar to MONGO. Other ST-specific graphics capabilities include the ability to make finder charts and overlays from the ST Guide Star Catalog, and to superimpose the HST field of view diagram on an image. STSDAS provides its own FITS I/O package in order to accommodate the idiosyncracies of the STSDAS disk data format, and to provide support for I/O of tabular data.

IMPLEMENTATION

The majority of the STSDAS software is written in SPP – the Subset Preprocessor – which is the standard programming language of IRAF. SPP is a high-level, preprocessed language that combines the familiarity of Fortran with some of the more desirable features of C (such as pointers and structures). The use of a pre-processor means that code that is properly written is almost guaranteed to be portable from one operating system to another, and experience has verified this with STSDAS.

Certain sections of STSDAS are written in Fortran 77 using an F77 interface to the IRAF Virtual Operating System – the F77VOS, or IRAF77 interface – developed at ST ScI. The F77VOS interface was implemented to ease the transition from the earlier VMS-dependent totally Fortran implementation of SDAS. A large body of existing spectral calibration software was migrated into the portable STSDAS using this interface. The F77VOS interface is also useful to scientists wishing to develop their own software within the IRAF environment, since it provides access to many of the capabilities of the VOS without the need for learning a new programming language. The access to image data is similar to the IRAF IMFORT interface, but F77VOS provides many other capabilities such as graphics I/O and table I/O. F77VOS programs operate in the IRAF environment, and are not allowed to call standard Fortran I/O routines.

The bulk of STSDAS software development is now carried out on Sun workstations and a Sun fileserver, although all software is validated on local VAX VMS systems and DECStation 3100s. Indeed, all of the HST pipeline calibration is actually run on VAX systems, whereas the majority of interactive use of STSDAS is on Unix-based workstations.

Users at ST ScI typically want to use the latest revision of the software, thus we make a weekly update to the development version of STSDAS on all fileservers at ST ScI using the source code distribution program CODA. Users also have the standard release of STSDAS available on all systems, allowing a fall-back to the previous version of a package. Major releases of STSDAS occur approximately once per year, and updates and major bug fixes are supplied to off-site users in patch kits available via anonymous ftp (or on tape, on request). The entire system can be retrieved using anonymous ftp, and is installed using IRAF-supported facilities (*rtar* to read the distribution files, and *mkpkg* to compile and link the software). An execute-only installation of the system requires about 20 MB of disk storage space; a full-source installation requires 40 MB.

Because of their independent origins, and some special requirements for HST data, the on-disk data structures for the main IRAF system and STSDAS are different. IRAF image files consist of two parts: a header file (with extension .imh) that has commonly accessed information in binary and ancillary header information in ASCII) and a binary pixel file (extension .pix). STSDAS data files also consist of two parts. The header file (with standard extension .hhh) contains ASCII text and follows the rules for construction of a FITS file header. The binary data is in an associated file (with standard extension .hhd). STSDAS data structures are based on the FITS group format (Greisen and Harten, 1981) but do not follow this standard in detail. The primary difference between IRAF and STSDAS format data files is that STSDAS files may contain a number of images in the same file.

The IRAF image I/O interface provides a subroutine library for applications programming that hides the details of the data structures from the programmer and from the user. As a result, virtually all tasks in STSDAS and IRAF can read and write data in both formats transparently to the user. The STSDAS user specifies the image subset, or image group number, as part of the filename, i.e., input.hhh[4] denotes subimage number 4. If no group number is given, the first image in the file is assumed. IRAF data files contain just one image per file.

SYSTEM CAPABILITIES

Calibration and Synthetic Photometry

The calibration software for each scientific instrument on HST is grouped together in an STSDAS package: *wfpc*, *foc*, *hrs*, *fos*, *hsp*, and *fgs*. The primary task in each package is the corresponding pipeline calibration program *calxxx* ('xxx' is the

99

instrument name, thus, *calfoc*, *calwfp*, etc.). Other tasks in these packages support the analysis of calibration data, and are used primarily by the instrument scientists at ST ScI.

A major feature of the STSDAS *wfpc* package is a modified version of the IRAF *imcombine* task. The WF/PC data is severely affected by cosmic rays, and given the sharp core of the point-spread function it can be difficult to distinguish a cosmic ray hit from s faint star. Users are advised to split their WF/PC exposures into two or more separate frames, and anti-coincidence testing can then be used to discriminate between stars and cosmic rays. The *combine* task implements this cosmic ray rejection algorithm.

An important adjunct to the instrument-specific calibration packages is the synthetic photometry package *synphot*. The *synphot* package allows users to compute the expected throughput for any HST instrument and filter combination using model spectra or observed spectra as input. The output can be given in a number of various units, and the standard astronomical filter passbands (like Johnson *UVBRI*) are supported. The package is being used in support of the absolute photometric calibration of HST.

The *synphot* package could easily be generalized to support throughput calculations for other observatories, given the appropriate filter transmission curves, allowed filter combinations, and telescope characteristics.

Analysis

STSDAS users can run all of the applications packages provided with the standard IRAF system on HST data and IRAF users can run STSDAS applications on NOAO data by virtue of the common image I/O interface. Thus, the NOAO IRAF project and ST ScI STSDAS project coordinate their applications software development.

The primary packages available for general data analysis within STSDAS are the *isophote* package for isophote fitting and surface photometry, the *fourier* package for Fourier analysis (one- and two-dimensional transforms, cross- and auto-correlation, and power spectra), and a *statistics* package containing an implementation of the non-parametric survival analysis programs developed by E. Feigelson and T. Isobe (Feigelson and Nelson, 1985, Isobe et al., 1986). We have just recently implemented several standard image restoration algorithms, and these have no particular dependencies on HST data. A general purpose curve- and surface-fitting package is available (*fitting*), and a general time series analysis package is supported.

Other Facilities

STSDAS supports a number of other analysis and data manipulation utilities, foremost of which is the STSDAS table system. STSDAS tables are binary data structures which provide for the simple manipulation of tabular information. Columns in tables are accessed by name, so that it is easy to transfer data from one program to another using a tables interface. A large variety of tools are provided for working with tables, from an interactive editor to statistics routines to database-like interfaces. Tables are used as the output format from a number of STSDAS programs, and using the table TeX/LaTeX filtering task these tables can be incorporated into TeX files very easily. Because of the general utility of the table system, it is available separately from the rest of STSDAS as a stand-alone *tables* package.

Another package that is useful not only to HST observers is the *gasp* package. This package, which is named after the Guide Star Astrometric Support Package developed at ST ScI, provides access to the Guide Star Catalog on CDROM. Users can overlay guide star positions on HST images, or on other image files if the coordinate system is at least approximately known. A plate solution task allows one to recompute the coordinate transformation parameters for an image in accordance with the Guide Star Catalog reference frame.

DOCUMENTATION

The STSDAS system has two major user's guides: the *STSDAS User's Guide* and the *STSDAS Calibration User's Guide*. The former provides an overview of the system capabilities and gives an introduction to use of the IRAF command language. The *Calibration User's Guide* gives more detailed information on the procedures and ancillary files needed to properly calibration HST data sets. Both a *Site Manager's Guide* and an *Installation Guide* are provided for remote STSDAS user sites.

For programmers we provide reference manuals for the F77VOS interface and for the table I/O subsystem. Users who need to understand the details of STSDAS data file formats, and the instrument-specific header information, may use *Interface Control Document 19: PODPS to SDAS*, which is maintained and provided by the STSDAS group.

SUMMARY

The STSDAS system, in conjunction with NOAO's IRAF system, provides the fundamental calibration and analysis needs for observers using the Hubble Space Telescope. The system continues to evolve and expand to meet the needs of HST GOs and ARs, with a current emphasis on image restoration software to deal with the HST spherical aberration problem. Additional software will be written to support the second generation scientific instruments (WF/PC II, the Near Infrared Camera, and the Space Telescope Imaging Spectrograph).

REFERENCES

Feigelson, E. D., and Nelson, P. I., 1985. *Astrophys. J.* **293**, 192.

Greisen, E. W., and Harten, R. H., 1981. *Astron. Astrophys. Suppl.* **44**, 371.

Hanisch, R. J., 1989. In *Data Analysis in Astronomy III*, Di Gesù, V. et al., eds., p. 129 (Plenum Press, New York).

Isobe, T., Feigelson, E. D., and Nelson, P. I., 1986. *Astrophys. J.* **306**, 490.

DETECTING COSMIC RAY HITS ON HST WF/PC IMAGES USING NEURAL
NETWORKS AND OTHER DISCRIMINANT ANALYSIS APPROACHES

F.D. Murtagh[1] and H.-M. Adorf

ST-ECF, Karl-Schwarzschild-Str. 2
D-8046 Garching b. München

ABSTRACT

We describe initial experiments in detecting cosmic ray hits on single Hubble Space Telescope Wide Field/Planetary Camera CCD-frames. Classifiers are trained on images with (what are considered to be) the cosmic ray hits marked. A range of classifiers are assessed. The IRAF *cosmicrays* task is found to be conservative, having a low detection rate combined with a low false alarm rate. The classical k-nearest neighbours method offers a versatile and broad-ranging alternative. With the rather homogeneous input data used, the various multilayer perceptron algorithms used did not appear to offer advantages over the variants of k-nearest neighbours.

INTRODUCTION

The Wide Field and Planetary Camera (WF/PC) detector of the Hubble Space Telescope consists of 8 Texas Instruments CCD chips, four for each of the wide-field camera and the planetary camera. Each CCD provides 800 × 800 pixel frames, allowing a 1600 × 1600 image to be constructed for the WFC or for the PC. Being essentially above the Earth's atmosphere, the likelihood of pixels being contaminated by cosmic ray hits is far greater than in the case of ground-based detectors. It has been estimated that in a 45 minute WF/PC exposure, 10 to 25% of all pixels will be hit by a cosmic ray (Holtzman and Westphal. in Adorf, 1991). The extensive contamination due to such cosmic ray hits (CRs) makes it imperative to remove them during data reduction.

For our purposes. cosmic rays can be divided into two classes. – high energy primary, and low energy secondary particles. High energy cosmic rays pass straight through the detector. On average, they cause electrons to be deposited in two to three pixels. Specifically. in one set of 87 CRs, the following numbers of contiguous pixels per CR were found:

# pixels	1	2	3	4	6	13	25
# CRs	17	6	2	2	1	1	1

80 to 100 electrons are deposited per μm travelled through a pixel. A WF/PC CCD pixel has approximate dimensions of $15 \times 15 \times 8$ μm^3 in length, breadth, and thickness. One of the Planetary Camera chips (PC8) is however notably less thick. Depending on the traversal angle, between 8 and 20 times (say) 80 electrons are deposited in a pixel. Diffusion along the path traversed by the high energy cosmic ray also takes place, but is

---

[1] Affiliated to Astrophysics Div., Space Science Dept.. European Space Agency

Data Analysis in Astronomy IV, Edited by V. Di Gesù *et al.*
Plenum Press, New York, 1992

sufficiently small to be ignored. A conversion factor of 7.5 brings the electron counts to ADU used in the image data.

Secondary particles are produced by cosmic rays hitting the surrounding spacecraft components. In particular, the detector housing and "shielding" give rise to these low energy secondaries and the number of electrons deposited by them can be very high.

It was originally foreseen and recommended (see Griffiths, 1990; Lauer, 1988) to find CRs by aligning two frames taken in "split mode", i.e. repeated observations. Thus the flux discrepancies of individual pixels could be monitored using a standard STSDAS algorithm. However, due to HST's current somewhat poor pointing characteristics, accurate alignment of two frames would require resampling of one of the frames which, due to severe undersampling, is impossible (Adorf, 1990). Other difficulties with CR-splits include additional overhead time, – read-out time, and pre-flash time – in the case of short exposures; and the additional read-out noise. Consequently, in addition to the existing approach working on frame pairs, there is a perceived need for a classifier which, given a pixel and some of its neighbourhood, reliably detects CRs on *individual* WF/PC frames and discriminates them from similar looking star images.

This is a typical pattern recognition problem and the question is how to set up a system. Two approaches seem feasible: feature- (or astronomical object-) based; and image- (or pixel-) based. For a feature-based classifier one would have to define a good set of features describing the image context in a region of interest surrounding the pixel in question. A trainable classifier would then be trained on these features together with the correct answers "CR" and "non-CR" to be provided externally. An image-based classifer works directly on the pixel data in the region of interest. Working directly on the data (the latter case) off-loads the construction of a good feature set from us to the classifier. This is the approach pursued here. Further discussion on this issue is to be found in Adorf (1991).

The aim of the present work is to develop a viable approach to single frame cosmic ray handling. In the first phase of this work. a range of WF/PC datasets which provide "ground truth" have been obtained. Secondly, the theoretical and empirical properties of various classifiers are being assessed. This assessment is reported on in this paper.

For different objectives, various questions related to classifier training and assessment have been pursued in depth in Adorf and Meurs (1988), Prusti et al. (1991) and Meurs and Adorf (1991). In choosing particular classes of objects (e.g. galaxies or young stellar objects), the superiority of automatic classifiers over hand-crafted classifiers has been demonstrated.

METHOD

The method employed was to: (1) have cosmic ray hits on images flagged by a human expert: (2) use this information to train and test various classifiers; and (3) apply the classifiers to other images.

Calibrated WFC images of the open cluster NGC 1850, with 10 seconds exposure, and without any preprocessing, where used in the experiments reported on below. In addition, other images of the faint Lynx galaxy field were made available by Ivan King (WF/PC IDT), and a marked version of part of this by Nick Weir (Caltech, Pasadena). Jon Holtzman (WF/PC Investigation Definition Team) marked the same 200 × 200 image on two different occasions, separated by several months. The second marking session was in the context of marking an 800×800 image, from which the 200×200 image was extracted. The first frame had 85 cosmic ray pixels zapped, and 102 in the second, with 74 common to both. If the first marking is taken as ground truth, this corresponds to a redetection rate of 87%, whereas if the second marking is regarded as ground truth then the redetection rate drops to 73%.

For this study, we used two variants of a neural network method, and the well established k-nearest neighbours method, which can be briefly described as follows.

1. The multilayer perceptron (MLP) is the most widely-used neural network paradigm. It may justifiably be viewed as non-linear function mapping, or as non-linear regression (Murtagh, 1991a,b). Optimization approaches to weight determination in the multilayer perceptron include backpropagation (Pao, 1989), and conjugate gradient (Barnard and Cole, 1989). Backpropagation is based on the chain rule. and often uses two user-specifiable parameters, the learning and momentum fractions. It is well-known to be a slow training procedure, but its use can favour the finding of the global optimum in weight space. Conjugate gradient provides fast convergence to a set of weights in the training phase. with the sacrifice of mapping accuracy provided.

   Irrespective of the approach taken to determining weights, the MLP architecture must be set up by the user. The three-layer network (input, hidden. and output layers) is the most common architecture, offering a good compromise beween effectiveness and parsimony. The number of hidden layer neurons must be chosen by the user with the same considerations in mind. We report below on results obtained for some of the different numbers of hidden layer neurons experimented with.

2. K-nearest neighbours discriminant analysis is a non-parametric method. the error rate of which is asymptotically no worse than twice the Bayes error rate. See Murtagh and Heck (1985). or SAS (1985).

The multilayer perceptron and k-nearest neighbour discriminant analysis have often been contrasted. They are both non-parametric, and may be viewed as providing piece-wise linear separation between classes.

In addition to the MLP and k-NNs, a number of other non-statistical approaches to handling cosmic rays on single frames have been considered. Among these are:

1. IRAF task *cosmicrays*: This is a tunable classifier based on the flux ratio between peak and local background. Preliminary results with it are presented for comparison purposes below.

2. DAOPHOT II: This well-known stellar photometry package was used to determine objects other than stars. To some extent, with an appropriate choice of PSF and detection threshold, it can be used to find cosmic rays. However, it should be stressed that seeking cosmic rays is not the aim of this package.

3. MIDAS routine *INVENTORY*: as in the case of DAOPHOT, the principal role of this object detection package does not include finding cosmic rays. It is not surprising that we found the present implementation of this package to be unfavourable for finding CRs. This was due to the default local smoothing which (by design) helped to destroy the influence of CRs.

4. In image restoration, how well an object is fitted may clearly indicate possible cosmic rays. Such an approach, based on maximum entropy restoration has been investigated by Nick Weir. A similar approach has been proposed by Leon Lucy (ST-ECF).

CLASSIFICATION TRAINING AND TESTING

The classifiers described above are based on appropriate training and test sets. For such sets, individual *pixels* were considered, rather than sets of contiguous pixels which could be characterised as *objects*. A 200×200 subframe of an NGC 1850 image was used. For computational convenience, this subframe was normalized to have maximum intensity equal to 1. Otherwise no processing was carried out on the data used.

Peaks were determined in the 200 × 200 image used. Among various possibilities experimented with, results reported below are for centre pixels greater than or equal to the local median value, where locality is defined with respect to the 3 × 3 pixel neighbourhood. This yielded 23002 peaks from a total number of 40000. The 3 × 3 subimages corresponding to all peaks were output. Among these peaks were 98 known cosmic ray pixels (CRs).

In the case of many classifiers, if one were to use a training set comprising 98 CRs and 22904 non-CRs, the results would be overwhelmingly influenced by the very different priors. This "relative frequency problem" can be avoided by selecting a small subset from the non-CRs. Approximately 100 non-CRs were sampled uniformly from the 22904 values to hand. This yielded a training set of 98 CRs, and 111 non-CRs. (In proceeding in this way, we are employing the classifiers as non-paramtric exploratory tools – at least as a first approximation – and eschewing the use a priori class probability information. See Davies, 1988.)

A test set was chosen as a different 200 × 200 frame on this NGC 1850 image. Using the same definition as previously, 23236 peaks were found. Of these. it was known that 72 were cosmics, and 23164 were non-cosmics.

CLASSIFIER ASSESSMENT

The results obtained from any classifier may be arranged in a contingency table:

	Classifier results	
	#CRs	#nonCRs
#CRs	a	b
#nonCRs	c	d

Teacher/trainer (rows: #CRs, #nonCRs)

Two indices may be used to summarize useful information in this table: *RD, rate of detection:* $a/(a+b)$; and *RFA, rate of false alarms:* $c/(c+d)$.

The performance of the classifier can be represented as follows. Note that a cosmic ray (however this is defined, – e.g. a pixel corresponding to a cosmic ray) can be (i) detected, or (ii) missed. Similarly, a star can be properly rejected as not being a cosmic ray, or can occasion a false alarm.

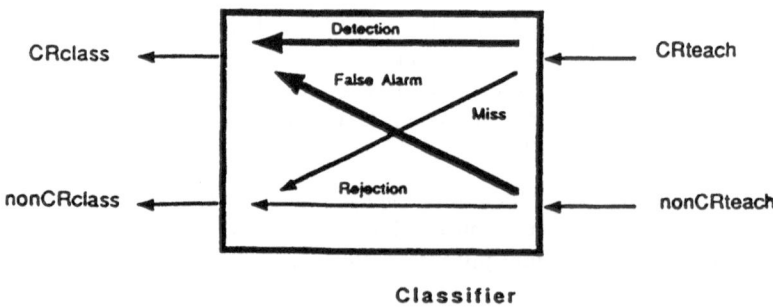

Classifier

RD. defined above, is an estimate of the *probability of detection*, $P(CR_{class} \mid CR_{teach})$. We wish to maximize RD. RFA is an estimate of the *false alarm probability*, $P(CR_{class} \mid CR_{teach})$. We wish to minimize RFA. The *OC* or *operating characteristic* diagram is the

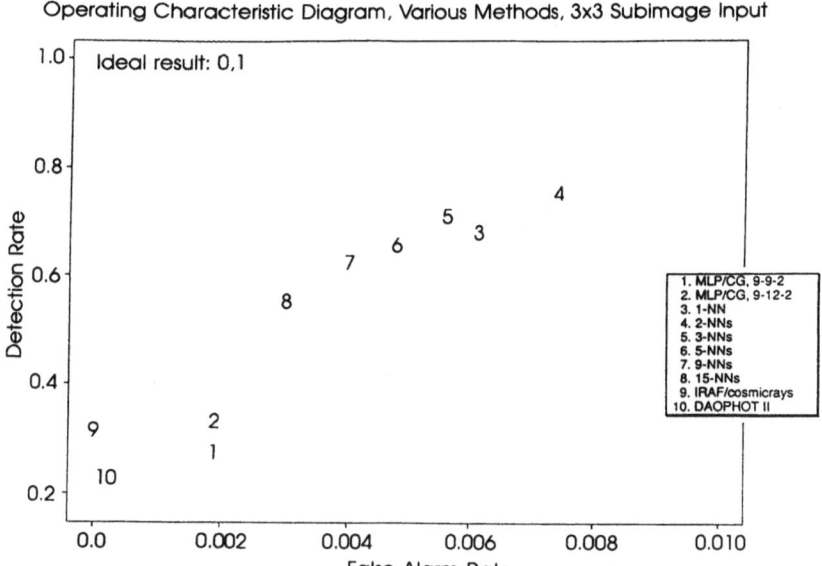

Figure 1. OC diagram for application of various trained classifiers (and two non-statistical classifiers). Note that most of the results appear on an approximate curve in (RFA, RD)-space.

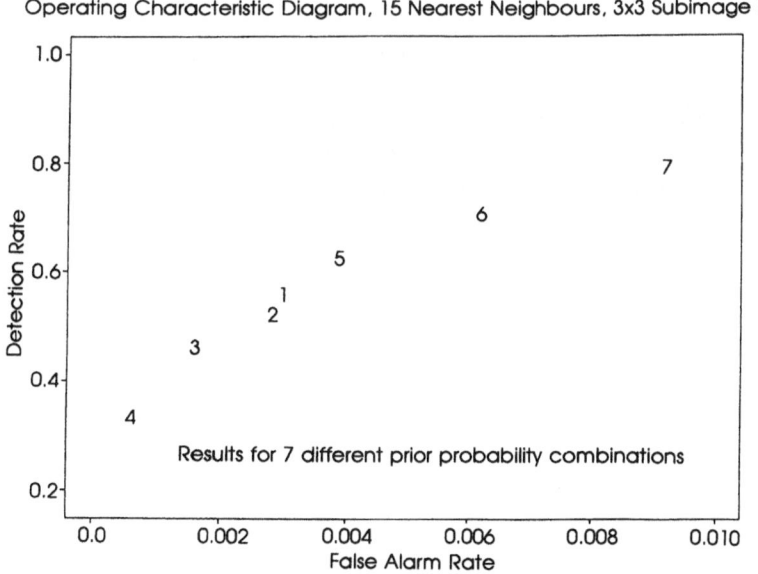

Figure 2. OC diagram for 15-NNs; note (again) the RD vs. RFA trade-off curve.

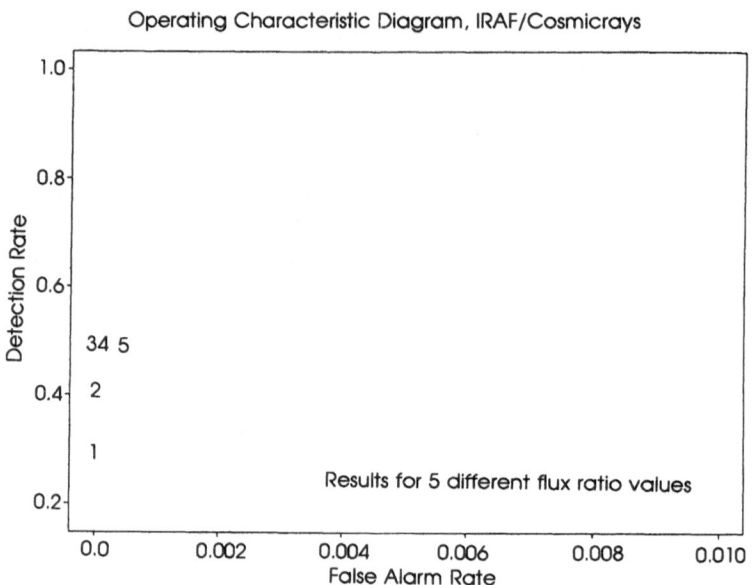

Figure 3. OC diagram for IRAF task *cosmicrays*.

plot of RD, RFA couples. Each point defines an operating point of a classifier (Melsa and Cohn, 1978). Within the basic cycle of train-test-apply, the results of training and testing a statistical classifier are represented conventionally using an operating characteristic diagram.

RESULTS

Figures 1–3 present the operating characteristic diagrams found for various methods. Figure 1 includes the results of different three-layer multilayer perceptron architectures, using the conjugate gradient optimization method. Note the curve traced out by the different methods: it reflects the trade-off inherent in maximizing RD while minimizing RFA.

Given a free parameter in any method, an OC curve can be traced for it. In Figure 2, such a diagram for the 15-nearest neighbours method is shown.

Figure 3 shows the results obtained from the IRAF task *cosmicrays* which works on a quite different basis, and does not require any learning. The sequence of values shown are a sequence of results obtained by relaxing the flux ratio threshold. Further values have the same RD value, but disimproved RFA values. This method does rather well, but is conservative in that the detection rate is small.

The OC diagrams shown here could be used to implement a cosmic ray detection strategy. For instance, the Newman-Pearson strategy would hold the RFA value fixed, at some upper acceptable limit, and find the best RD value corresponding to this; or, given that the priors convey much information, one could use a maximum a posteriori (MAP) strategy, which would minimize the overall error.

CONCLUSIONS

We have used marked WF/PC data sets for classifier training and assessment, and have assessed three methods for cosmic ray detection: about 10 variants of k-nearest neighbours; 2 variants of the multilayer perceptron; and the IRAF *cosmicrays* task.

- The k-NNs method is computationally acceptable for low parameter dimensionalities (e.g. 9, corresponding to 3×3 subimages). It is robust, and is widely available. It is a general, established method with nice asymptotic properties.

- The MLP using backpropagation is computationally demanding. Backpropagation was found to provide results comparable in accuracy to the conjugate gradient approach, but with orders of magnitude greater computational requirements (e.g. 5 hours CPU time on a SPARCstation 1 compared to 15 seconds for conjugate gradient). However improved supercomputing or otherwise parallelizable hardware capability could be of substantial benefit in this regard. Various speed-ups over basic backpropagation were not tried in this study. The results obtained from using backpropagation do not appear to be sufficiently attractive at this time.

- The multilayer perceptron using conjugate gradient was orders of magnitude faster than backpropagation. Its classification performance appeared to be equal to, or slightly worse, compared to k-nearest neighbours.

- The IRAF *cosmicrays* task has an excellent false alarm rate, but also a low detection rate. It does not require training.

In addition to allowing assessment of one classifier over another, the OC diagram can be used to implement a decision strategy for handling CRs. The best such strategy will be sought in the near future. The training set used was small: it will be increased in size. Finally, the classifiers will be applied to a greater range of images.

ACRONYMS, SOFTWARE SYSTEMS

ADU	Analog/digital unit(s)
CCD	Charge-coupled device
CR	Cosmic ray hit
DAO	Dominion Astrophysical Observatory, Victoria, British Columbia
DAOPHOT	Photometry system, produced by P.B. Stetson, DAO (q.v.)
ESO	European Southern Observatory, Garching b. München
HST	Hubble Space Telescope
IDT	Investigation Definition Team
INVENTORY	Faint object analysis package, produced by A. Kruszewski, Warsaw Observatory, available in MIDAS (q.v.)
IRAF	Image Reduction and Analysis Facility, produced at NOAO (q.v.)
MAP	Maximum a posteriori
MIDAS	Munich Image Data Analysis System, produced at ESO (q.v.)
MLP	Multilayer perceptron
NGC	New Galactic Catalog
NN	Nearest neighbour
NOAO	National Optical Astronomical Observatories, Tucson, AZ
OC	Operating characteristic
PC	Planetary camera (see WF/PC)
RD	Rate of detections
RFA	Rate of false alarms
ST-ECF	Space Telescope – European Coordinating Facility, ESO (q.v.)
STScI	Space Telescope Science Institute, Baltimore, MD
STSDAS	Space Telescope Science Data Analysis System, produced at STScI (q.v.) available in IRAF (q.v.)
WFC	Wide field camera (see WF/PC)
WF/PC	Wide Field and Planetary Camera, instrument on board HST (q.v.)

ACKNOWLEDGEMENTS

We are particularly grateful to J. Holtzman, for initial motivation and discussion; and for providing marked data. We are also grateful to: I. King, N. Weir, R. Hanisch, L. Lucy, P. Stetson, J. MacKenty, R. White, S. Ortolani and others for discussion, suggestions, and data.

REFERENCES

1. Adorf, H.-M.. "WFPC: options for overcoming undersampling", in C. Jaschek and F. Murtagh, Eds., *Errors, Bias and Uncertainties in Astronomy*, Cambridge U.P., New York, 1990, pp. 71–78.

2. Adorf, H.-M., "Cosmic ray detection on individual WF/PC frames", ST-ECF Technical Report (DA)-21, 5 pp., April 1991.

3. Adorf, H.-M. and Meurs, E.J.A., "Supervised and unsupervised classification – the case of IRAS point sources", in W.C. Seitter, H.W. Duerbeck and M. Tacke, Eds., *Large-Scale Structures in the Universe – Observational and Analytical Methods*, Springer-Verlag, Berlin, 315–322, 1988.

4. Barnard, E. and Cole, R., "A neural net training program based on conjugate-gradient", Oregon Graduate Center TR CSE 89-015, Beaverton, Oregon, 1989.

5. Davies, E.R., "Training sets and a priori probabilities with the nearest neighbour method of pattern recognition", *Pattern Recognition Letters*, **8**, 11–13, 1988.

6. Griffiths, R., *Hubble Space Telescope Wide Field and Planetary Camera Instrument Handbook*, Version 2.1, Space Telescope Science Institute, Baltimore, 1990.

7. Lauer, T.R., "The reduction of Wide Field/Planetary Camera images", Princeton Observatory preprint, Dec. 1988.

8. Melsa, J.L. and Cohn, D.L., *Decision and Estimation Theory*, McGraw-Hill, 1978.

9. Meurs, E.J.A. and Adorf, H.-M., "The selection of extragalactic source samples from the IRAS database", *Proc. IIe Rencontres de Blois: Physical Cosmology*, 1991. in press.

10. Murtagh, F., "Review of Y.-H. Pao. *Adaptive Pattern Recognition and Neural Networks*", *Journal of Classification*, 1991, in press.

11. Murtagh, F., "Multilayer perceptrons for classification and regression", *Neurocomputing*, 1991b, in press.

12. Murtagh, F., "The multilayer perceptron for discriminant analysis: two examples", *Proc. 15. Jahrestagung der Gesellschaft für Klassifikation*, H. Goebl and M. Schader (Eds.), Springer-Verlag, 1991c, submitted.

13. Murtagh, F. and Heck, A., *Multivariate Data Analysis*, Kluwer Academic, 1985.

14. Pao, Y.-H., *Adaptive Pattern Recognition and Neural Networks*, Addison-Wesley, 1989.

15. Prusti. T., Adorf, H.-M., and Meurs, E.J.A., "Young stellar objects in the IRAS Point Source Catalog", 1991, in preparation.

16. SAS, *SAS User's Guide: Statistics*. SAS Institute, Cary, NC, 1985.

THE ROSAT MISSION

THE ROSAT MISSION

H.U. Zimmermann

Max-Planck-Institut für Extraterrestrische Physik
Karl-Schwarzschild Straße 1
D-8046 Garching bei München, Fed. Rep. of Germany

INTRODUCTION

On June 1st 1990, shortly before midnight, an Delta-II rocket launched the X-ray satellite ROSAT successfully from Cape Canaveral into orbit. 15 days later the ROSAT telescope began its operation with an observation of the central parts of the Large Magellanic Cloud. This long awaited event marked the begin of a new phase in X-ray astronomy during that an extension of our present database on X-ray sources by an order of magnitude is expected.

It meant at the same time the end point of 15 years of planning, costing, developing, testing and integrating the satellite and its scientific instrumentation to build this large observatory. First ideas to perform an All-Sky-Survey with a truly imaging telescope date back to the year 1975, went through many changes and became concrete in 1982, when the German government finally decided to build the largest European research satellite and its scientific instrumentation. Important preconditions were the decision of NASA to join the project by providing one of 3 focal plane detectors as well as a launch with the shuttle, while the British complemented the payload by a smaller XUV telescope.

The development of the satellite, the instruments and the ground segment was already far evolved, when in January 1986 (somewhat more than 1 year before launch) the catastrophic explosion of the shuttle suddenly endangered the whole project. Because priorities within NASA were heavily affected by that event a launch of ROSAT with the shuttle could no longer be ensured within the next many years. This could have had disastreous consequences for the quality of the already existing instrumentation (and the budget) and therefore a decision for a change-over to a rocket flight was urgently searched and finally agreed upon. Thus 3 years after the planned shuttle transport ROSAT finally reached the planned orbit of 580 km altitude and 53 degrees inclination.

Data Analysis in Astronomy IV, Edited by V. Di Gesù *et al.*
Plenum Press, New York, 1992

WHAT CAN X-RAY ASTRONOMY DO FOR US ?

Astronomy today is no longer bound to observations in a single window of the electromagnetic spectrum but can observe celestial phenomena in almost all frequencies between the long wavelength radio band and the high energetic gamma radiation. More than 2 decades of research with always better instrumentation have made X-ray astronomy to one of the major disciplines in the investigation of our universe. From the nearest stars to the most distant quasars, allowing a glimpse on the universe still at its youth, almost all types of astronomical objects can be observed also in X-rays. X-radiation is always a sign that high energy processes are at work, processes that have an essential importance for the evolution of matter condition in the universe. The physical phenomena leading to intense X-ray emission are the extremely hot phase of matter with temperatures of more than 10^5 K originating from mass accretion or star explosions, but may also be highly relativistic electrons in magnetic or strong radiation fields. Thus there are specific object classes and phenomena, where X-rays carry most of the emission energy and therefore exhibit the prime source of information: the hot intercluster gas in clusters of galaxies; mass accretion in X-ray binaries, where one partner is a collapsed object, a neutron star or a black hole; the hot surface radiation of single neutron stars slowly cooling down; the hot gas originating from supernova explosions. For many other objects like galaxies with active nuclei, quasars or certain types of stars X-rays provide very essential complementary information.

ROSAT continues the sequence of major milestones in X-ray astronomy that began in 1962 with a rocket flight during that Giaconni and coworkers accidentally discovered the strongest X-ray source in the sky, Sco X-1. From 1970 to 1972 the UHURU satellite performed the first all-sky survey leading to the detection of about 340 X-ray sources on the whole sphere. Seven years later the HEAO-1 satellite raised this number in a second all-sky survey to about 840 sources. All these experiments used collimated, non-imaging instruments and therefore sensitivity was heavily affected by background. A new era in X-ray astronomy began with the launch of the EINSTEIN observatory (1979 - 1981), the first satellite mission using a truly imaging X-ray telescope. The EINSTEIN observatory collected an enormous amount of detailed information from about 10 % of the sky and demonstrated that almost all classes of astronomical objects emit measurable X-ray flux. In the following years the ESA mission EXOSAT (1983 - 1986) and the Japanese GINGA mission (since 1987) extended the observations towards the harder X-ray energies. In the order of 5000 X-ray objects were known at the begin of the ROSAT mission.

MISSION AIMS

After completion of an initial Calibration and Verification Phase of 6 weeks duration ROSAT has performed from August 1990 until January 1991 the first all-sky survey with an imaging telescope possessing a sensitivity about a factor of 1000 higher than that of Uhuru. During that phase ROSAT scanned the sky continously along a great circle perpendicular to the earth-sun direction. Thus sky strips of 2 degrees width (the field-of-view of the instrument) and 360 degrees length are observed during each orbit. Due to the applied scan law these sky strips lie perpendicular to the ecliptic plan and their intersection with the ecliptic plane proceeds with a rate of 1 degree per day (solar motion). In this way the ecliptic poles receive during the half year survey period up to 50 ksec exposure while sky fields near to the ecliptic plane are seen for about 500 sec. More than 50000 sources are expected from that survey providing a fantastic X-ray database for unbiased statistical investigations of most astronomical object classes. No further all-sky survey is planned in the foreseeable future, thus the ROSAT results will also strongly influence the observations of the next generation of large X-ray missions like AXAF and XMM near to the end of this decade.

The survey is followed by Pointed Observation Phases, each of a duration of half a year, where the 3-axis stabilized observatory will investigate selected targets all over the sky. Usually the observation times during these mission phases are appreciably longer than during the survey. For very deep exposures this will result in point source sensitivities up to a factor 10 higher than achieved by EINSTEIN and appreciably better spatial and spectral resolution.

The design lifetime of the ROSAT spacecraft is 1.5 years, but instrument consumables (detector gas) on-board have been foreseen for a lifetime of at least 3 years.

THE ROSAT INSTRUMENTS

ROSAT carries two telescope systems on-board oriented in parallel: a large X-ray telescope, operating in the energy range 0.1 to 2 keV, and a smaller XUV telescope sensitive in the range from about 20 to 200 eV. The X-ray telescope consists of a fourfold nested Wolter 1 mirror with an entrance diameter of 84 centimeters. Manufactured and poslished by the German company Zeiss the X-ray mirror became listed in the Guiness Book of Records as the largest X-ray mirror with the smoothest surface. Because of the short wavelength of X-radiation the resolution of X-ray mirrors is not diffraction limited but quality strongly depends on the accuracy of the shaping and the smoothness of the mirror surface. For the ROSAT telescope the mean roughness of the surface is only about 3 Angstrom that is in the order of atomic dimensions. One of 3 detectors sitting on a carousel can be moved into the focus of the telescope. Two of these are position sensitive proportional counters (PSPC) developed and built in our institute (Pfeffermann et al. 1987). To avoid polution of the counter gas a continuous stream of the argon/xenon based gas mixture flows through the multiwire proportional detectors. The 1 micrometer thin entrance window extends the sensitivity down to X-ray photons of less than 100 eV. Each of the 2 identical proportional counters has its own filter wheel in front of the window allowing to move a boron filter or radioactive calibration sources into the telescope beam. The detectors offer a positional resolution of about 20 arcsec and a moderate energy resolution corresponding to 4 spectral bands over the sensitive energy range. This detector has a 2 degree wide field-of-view and has been used during the all-sky survey. The third detector is a High Resolution Imager (HRI) channel plate detector allowing resolution of a few arcsec on a 38 arcmin field-of-view but providing no energy information. It is a rebuild of the HRI on-board the EINSTEIN observatory with higher quantum efficiency. Both the PSPC and the HRI detectors are used during the Pointed Observation Phases. Prior to launch the telescope and the focal instrumentation have undergone a full X-ray test in our 130 m long test facility at Neuried near München.

The XUV telescope or Wide Field Camera (WFC) is a three-fold Wolter-Schwarzschild telescope with 2 channel plate detectors in the focal plane (Barstow et al. 1988). The positional resolution is about 1 arcmin in the center of the 5 degrees wide field-of-view. Filters defining 4 wavelength bands are used to obtain spectral information.

BASIC DATA FLOW AND SATELLITE CONTROL

Photons entering the telescope undergo 2 reflections on the highly polished mirror surfaces before they are absorbed in the focal instrument. For each photon detected the PSPC instrument delivers 2-dimensional position information of the event in detector coordinates, in addition the pulse-height of the event as a measure of its energy and the arrival time are registered. The data of each event are immediately written to the onboard tape recorder together with auxiliary information from other satellite subsystems (housekeeping and attitude data). During the 5 to 6 contiguous contacts per day between the satellite and the groundstation - the German Space Operation Center (GSOC) at Oberpfaffenhofen/Weilheim near to München - the onboard recorded data are transmitted on a high speed link (1 Mbit/sec) to ground. About 20 % of the data are immediately passed on via a dedicated 64 kbit/sec line to the ROSAT Scientific Data Center (RSDC) for instrument health and data quality checks. In parallel to the high speed transmission of the recorded data a second satellite transmission channel (8 kbit/sec) provides data from all subsystems in real time during the typically 8 minutes of ground contact. These data are immediately analysed at GSOC and in parallel routed to the RSDC for special online data analysis. If unforeseen difficulties with the instruments are encountered this scheme guarantees during the contact cycle a quick reaction up to the next contact 95 minutes later. After the 5 to 6 contiguous contacts per day the satellite performs its operation typically for 16 hours uncontrolled. Therefore quite some intelligence has been installed on board to react on possible failure modes during this time span.

THE ROSAT SCIENTIFIC DATA CENTER (RSDC)

Located at the Max Planck Institut für Extraterrestrische Physik (MPE) at Garching near München, the RSDC is the central point for all interfaces between the observers and the German Space Operation Center responsible for the control and daily operation of the observatory. Also all interfaces to the main data centers in the UK (RAL), the USA (GSFC/NASA) and the German WFC data center (AIT, Astronomisches Institut der Universität Tübingen) are supervised by the RSDC. As part of the RSDC a separate control center for the WFC telescope has been installed at MPE.

The tasks of the RSDC include:

- full responsibility for all observing program activities providing organisational, editorial and technical support during the so-called Call for Proposals, the proposal selection and the mission timeline production phases,

- instrument control and detailed checks of all relevant instrument data in real and near-real time,

- the Standard Processing of all X-ray data to achieve a uniform calibration standard and provide observers with the results of a standard image and source analysis. From that protocol the user can then decide whether the Standard Analysis results already satisfy his needs or a more detailed analysis will be required.

- extended observer support, offering advice, documentation and visitor service, providing users with sophisticated analysis software (EXSAS) as well as working facilities at the RSDC. In addition there is the facility to retrieve data and results from the ROSAT Archives.

Almost 50 computers and workstations form the computational hardware basis of the RSDC, providing more than 80 working places with computer access. The computers are grouped into 2 DEC/VMS based workstation clusters where all mission relevant operations are performed and where also the optical disk archives reside and 1 DEC/Ultrix cluster being fully dedicated to ROSAT data analysis using EXSAS.

Up to now more than 120 years of manpower have been invested in the RSDC. About 40 persons are presently providing this service to the scientific community. Included in this number are 7 operators working in 2 shifts per day.

STATUS AND OUTVIEW

The ROSAT observatory is now in orbit since more than 10 months and has performed very successfully during that time. A few problems, mainly caused by satellite hardware, have resulted in the loss of some of the redundant hardware components on-board, but fortunately did not affect the performance of the instrumentation.

One main aim of the mission, the All-Sky Survey, is now 97 % complete. The parts of the sky for which data have already been looked at show a barely expected wealth of information in point sources and especially in extended structures. It will need many years of investigator skill to obtain a full overview on the potential of that large database. Many collaborations have been established and have already begun to identify part of the newly detected X-ray sources in the optical and radio band. This effort serves mainly to identify methods that can be used to classify the bulk of the detected X-ray sources in a statistical sense into one of the main astronomical object classes.

Besides the All-Sky Survey many pointed observations both with the PSPC and the HRI have been taken during the initial Calibration and Verification Phases and since the begin of the first Pointed Observation Phase in February 1991.

REFERENCES

Barstow M. & Willingale R. (1988), JBIS, 41, 345.
Pfeffermann E. et al. (1986), SPIE, 733, Soft X-Ray Optics and Technology, 519.
Zimmermann H.U. et al. (1986), Data Analysis in Astronomy II, 155.
Zimmermann H.U. & Harris A.W. (1991), Databases and On-line Data in Astronomy.

MISSION PLANNING WITH ROSAT

S. L. Snowden[1,2] and J. H. M. M. Schmitt[1]

[1] Max Planck Institute for Extraterrestrial Physics
[2] University of Wisconsin — Madison

ABSTRACT

The mission planning activities for the satellite bourne X-ray observatory ROSAT are discussed. Responsibility is shared between the Max Planck Institute for Extraterrestrial Physics (MPE), which provides the scientific and calibration program input, and the German Space Operations Center (GSOC), whose responsibility it is to generate a mission timeline satisfying all operational constraints. An optimum solution for the mission timeline is achieved using an efficient networking procedure.

INTRODUCTION

The ROSAT observatory is a joint German, British, and American space project. ROSAT carries two coaligned telescopes, the X-ray Telescope (XRT, Trümper 1983) covering the energy range 0.1 - 2.0 keV and the Wide Field Camera (WFC, Sims et al. 1990) covering the energy range 0.06 - 0.19 keV. It is a versatile and powerful instrument which is being used to address a multitude of problems in X-ray astronomy. ROSAT was launched with a Delta II rocket from Kennedy Space Center on 1 June 1990 into a circular 580 km orbit with an inclination of 53°. The high inclination was necessary to enable contact with the ground station at Weilheim, Germany (geographic latitude 47.5°). In normal operation, the Weilheim station provides the only communication with the satellite.

Mission planning and control of ROSAT is the responsibility of the Max Planck Institute for Extraterrestrial Physics (MPE; Garching, Germany) and the German Space Operations Center (GSOC; Oberpfaffenhofen, Germany)

OPERATIONAL CONSTRAINTS

There are a number of operational constraints, both in satellite communication and permitted satellite activities, which strongly affect ROSAT mission planning. The most important of these is the lack of real-time control. Since communication with the satellite is normally only possible through the Weilheim ground station, contact is limited to roughly eight minutes per orbit on five consecutive orbits (out of 15 orbits) per day. This forces all commanding to be preprogrammed and all commands for a day to be uploaded during the contact cycle.

Data Analysis in Astronomy IV, Edited by V. Di Gesù *et al.*
Plenum Press, New York, 1992

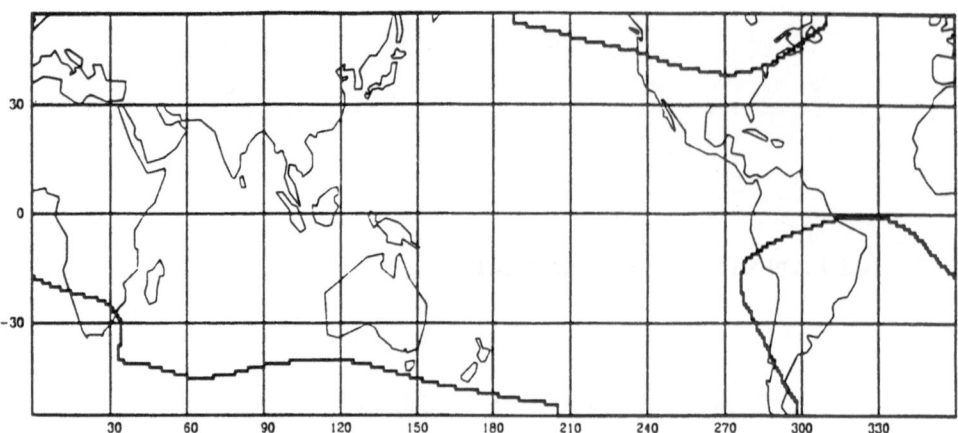

Fig. 1. Contours outlining the particle belts and SAA. The excluded regions are to the north of the northern contour and to the south of the southern contour.

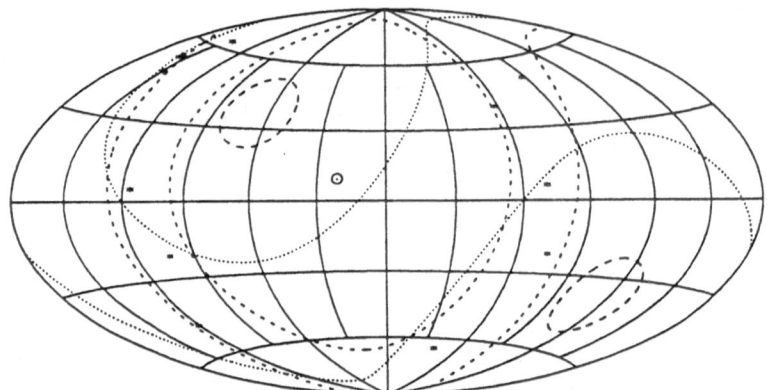

Fig. 2. Aitoff equal area projection in equatorial coordinates (zero centered, RA increases to the left) of sky visibility on 13 April 1990. Short dashed lines show the region of allowed observations, dotted lines show the region affected by the AO constraint, and the long dashed lines show the continuous viewing zones.

The following constraints affect satellite activities and must be considered when satellite commands are generated. First, the X-axis of the satellite, which is normal to the solar panels and perpendicular to the optical axis of the telescopes, must point within 15° of the sun. This limits the amount of sky accessible at any given time, however, any given target can be observed for at least thirty days twice a year. Second, the scientific instruments must be "safed" (covered by safe filters) whenever the angle between the look direction and ram direction is less than 28° to avoid damage from atomic oxygen (AO). Third, the scientific instruments must be turned off when the satellite is in the Earth's particle belts and the South Atlantic Anomaly (SAA) to avoid excessive count rates induced by high energy charged particles. Figure 1 shows the contours of these regions; the excluded areas lie to the north of the northern line and to the south of the southern line. Fourth, the scientific instruments must be safed when the observation zenith angle is greater than 97°, to avoid excessive count rates from scattered solar X-rays. Fifth, during ground station contacts with Weilheim, the angle between the satellite antenna and the ground station must be less than 120°.

Figure 2 demonstrates how the constraints affect the part of the sky which can be observed on 13 April 1991 (stars indicate the planned observations for the day). The short dashed curves outline the 30° wide swath of possible observations. The dotted curves outline the 56° wide region affected by the AO constraint. Targets lying in the viewing swath and also in the AO constraint region can be observed but only during part of the orbit. The long dashed curves outline regions where, when they lie within the viewing swath, continuous viewing observations are allowed (except for time excluded by the particle belts and SAA). The continuous viewing zones and AO constraint regions move to the right (west) at ∼ 4° per day following the precession of the satellite orbit while the viewing swath moves to the left (east) at ∼ 1° per day following the motion of the sun.

MISSION PLANNING

Overview

For operational convenience, the satellite lifetime has been divided into mission phases as shown in Table 1. The survey and pointed observation (Announcement of Opportunity or AO-n) phases each last half a year since this is the minimum length of time required to have access to the whole sky. Mission planning starts after the time allocation committee's decision of what targets should be observed has been made. The first step in the process is for MPE to receive the list of approved targets from the International Users Committee (IUC). There are three levels of priority in the target list: P1, P2, and P3. P1 proposals have the highest priority and comprise no more than 5% of the observation time. They are "time critical" observations that require special processing and include observations that are to take place at a specified time, observations that are repeated at specified intervals, and observations that are required to be as continuous as possible. P2 observations have second priority and are required to be observed during the AO-n phase but have no special constraints as to when they are scheduled. Together, P1 and P2 observations comprise roughly 65% of the available observing time. The P3 observations are not required to be observed in the AO-n phase but provide targets to fill the remaining available time.

MPE processes the target list and passes the results and other peripheral information to GSOC as a list of "requests" and data files. The formats of all MPE and GSOC exchanged files are specified and controlled by the *Project: ROSAT Mission Timeline Generator (RMTG) GSOC/MPE Interface Document* (Ashbolt and Garton 1990). GSOC then processes these requests and peripheral information to check for proper format and consistency. An acknowledgement/error file is returned to MPE with the results allowing MPE to submit corrections if necessary. GSOC next processes the accepted requests and data with the ROSAT Mission Timeline Generator (RMTG) to generate a long term

TABLE 1. ROSAT Mission Phases

Mission Phase	Dates
Satellite Check Out	1 June 1990 - 16 June 1990
Instrument Calibration	16 June 1990 - 11 July 1990
Survey Verification	11 July 1990 - 16 July 1990
Pointing Verification	16 July 1990 - 30 July 1990
All-Sky Survey	30 July 1990 - 1 February 1991
AO-1 (Pointing)	7 February 1991 - 15 August 1991
AO-2 (Pointing)	TBD - TBD
:	: :

timeline (LTL) covering the entire mission phase. This is a two-pass process with time critical observations being specified in the first pass and all other observations being placed in the second pass. The LTL is sent to MPE for approval. If the timeline is not acceptable, the cycle starts again.

The LTL is obviously based on a long term orbit prediction. While the orientation of the orbit plane is predictable with sufficient accuracy over this period of time, the uncertainty of the phase angle of the satellite is about 180° (48 minutes) by the end of a six-month period. To correct for this uncertainty, the RMTG generates a short term timeline (STL) each week for the following week's observations using updated orbit information. The process of mission timeline generation is diagrammed in Figure 3.

It is very important to realize that the orbit phase discrepancy between the LTL and STL has little effect on the scheduled observations. Since there is little uncertainty in the orientation of the orbital plane, the targets available to the satellite at a given orbit phase angle remain unchanged, i.e., there is no variation of earth blockage or AO constraint. Since the time that the satellite reaches a given orbit phase angle does change slightly, the sub-satellite geographic coordinates and therefore the belts and SAA relative to the satellite, changes also. This has the effect of changing the exact time of a possible observation but not whether the observation is possible or not (though it can affect P1 observations which require absolute timing to a greater degree). In practice, the STL typically differs only slightly from the LTL with the same

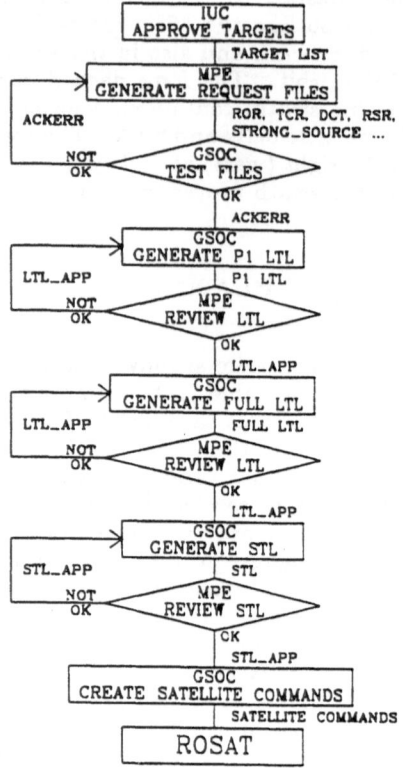

Fig. 3. Diagram of the timeline generation process.

targets being observed and only the times being slightly shifted. Figure 4 shows the total number of observations per STL (solid line) and the number of discrepancies between STL and LTL for AO-1 (dotted line). As is obvious from Figure 4, the number of discrepancies is very small and the observatory is essentially executing the originally planned LTL with a continually changing offset in time.

Communication Between MPE and GSOC

Communication between MPE and GSOC is by electronically transferred (over a dedicated line on the SPAN network) data files with specific formats. Table 2 lists the communication files and describes their purpose. The files sent from MPE to GSOC consist of satellite activity requests (e.g., ROR, RCR, TCR) which are used to command a specific action of the satellite, peripheral files such as the BELT_CONTOUR and STRONG_SOURCE, and timeline approval files. The files sent from GSOC to MPE consist of mission timelines and acknowledgement/error files.

Generation of the Mission Timeline

The heart of mission planning is the generation of the mission timeline. The mission

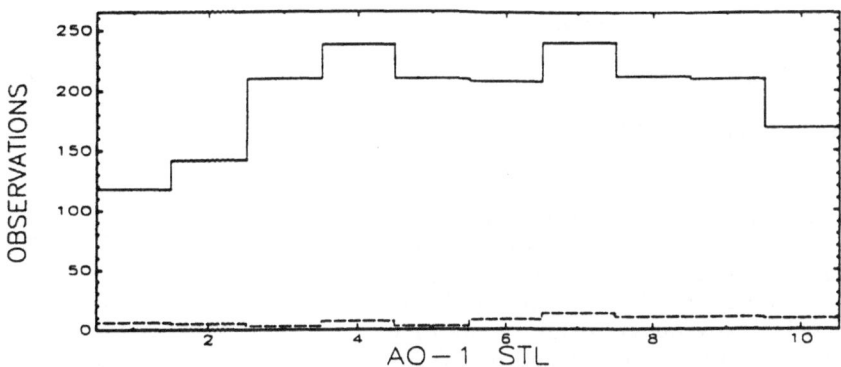

Fig. 4. Number of observations in each STL (solid line) and the number of discrepant observations between LTL and STL (dashed line).

timelines, both LTL and STL, consist of entries specifying all of the satellite activities as well as peripheral information such as the elements of the orbit. Table 3 lists some of the more important entries along with a short description. The mission timeline is used both to generate the commands to operate the satellite and for a number of housekeeping and record keeping activities at MPE.

The first step is the generation of a basic timeline. The basic timeline consists essentially of all timeline entries which are specified by the orbit geometry. Examples are the BELT, DAY, and CONTACT entries. The permissable times for observations are broken into "slots". A slot is typically the time between two belt entries which can be between 900 and 5100 seconds. The longer interbelt periods (> 2500 seconds) are broken into two slots. The second step is to schedule the highest-priority time-critical (P1) observations. The last step is to fill in the remaining slots with observations as efficiently as possible.

TABLE 2. MPE and GSOC Communication Files

File	Description
MPE	
ROR	Requests a pointed observation
TCR	Time critical modification to pointed observation
RSR	Request a survey operation
RCR	Request a scientific instrument calibration
DCT	Request a XRT detector change (HRI and PSPC)
CSR	Request to change the status of a timeline entry
TER	Request to change a timeline entry
xTL_APP	MPE approval response to GSOC mission timeline
BELT_CONTOUR	File containing the particle belt and SAA contours
STRONG_SOURCE	File containing a list of very bright sources
GSOC	
ACKERR	GSOC Acknowledgement/error response
SVIEWING	Visibility timetable for pointed observations
LTL	Long term mission timeline (binary and ASCII)
STL	Short term mission timeline (binary and ASCII)

The filling of the remaining slots is made more complicated by the imposition of additional constraints. First, as mentioned above, the remaining observations are divided into two categories with different priorities. All observations in the highest priority class (P2) must be observed; together with the P1 observations, this accounts for ~ 65% of the observation time. The remaining 35% of the observation time is filled from the lowest priority (P3) category. Second, once an observation has been started it should be completed. Many targets require more time than is available during one slot and it is desirable that few partly completed observations are left at the end of the mission phase. For example, in AO-1, the requested observing times per observation range from 1 ks to 125 ks with a mean of 7 ks. Third, the scheduled observation time when grouped by country should be divided 50% USA, 38% German, and 12% Great Britain. Fourth, the scheduling should be done with the greatest possible efficiency. To increase the observation efficiency, slews (the action of reorienting the satellite between observations of different targets) are placed to the greatest extent possible during times when the detectors can not operate, i.e., during belt periods and Earth block (when the observation zenith angle is greater than 97°).

Given the very large number of possible solutions, a more clever method of determining the optimum timeline than simple evaluation is necessary. The problem is cast as a large network where each step corresponds to a slot and each node in the step corresponds to a possible observation. The solution is a multiple decision process where a decision must be made for every slot.

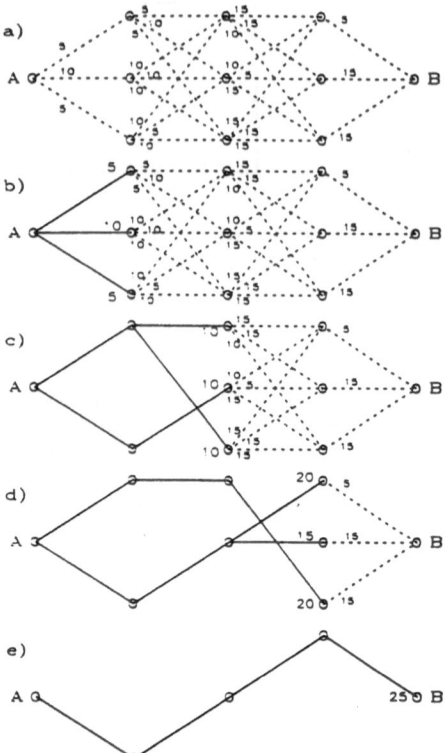

Fig. 5. Diagram of the network solution procedure, see text for description.

TABLE 3. Sample Mission Timeline Entries

MTL Entry	Description
NEWORB	Orbital elements on an orbit by orbit basis
DAY	Indicates when the satellite is sunlit
BELT	Indicates when the satellite is in a particle belt or the SAA
SLEW	Indicates when the satellite is slewing
SCAN	Gives scanning information when satellite is in survey mode
CONTACT	Indicates when the satellite is in ground station contact
POINT	Gives satellite pointing direction
XOBSER	Gives target observation information for the XRT
WOBSER	Gives target observation information for the WFC

The following is a simple example of this process. Figure 5a shows a network of paths to get from point A to point B through three intermediate steps (slots) each with three alternative nodes (possible targets). A "cost" is assigned to each intermediate segment and the desire is to get from A to B at minimum cost. In the scheduling application, the cost is related to the unused observing time in the given slot, i.e., an observation filling the entire slot has zero cost. A straightforward evaluation of the minimum cost would require 45 additions and the comparing of 26 number pairs. Figures 5b to 5e show an alternative approach. Figure 5b shows the minimum costs and paths to step 1. There are three paths to each of the the three points at step 2; finding the cheapest path to them requires nine additions and the comparing of six number pairs. Figure 5c shows the minimum cost paths and their costs to step 2. The same process is repeated to reach step 3 and the result is shown in Figure 5d. The final path is shown in Figure 5e. Note that the minimum cost path from point A to point B does not need to (and in this case does not) include the minimum cost points at all steps. The total number of additions required is 21 while 16 number pairs must be compared.

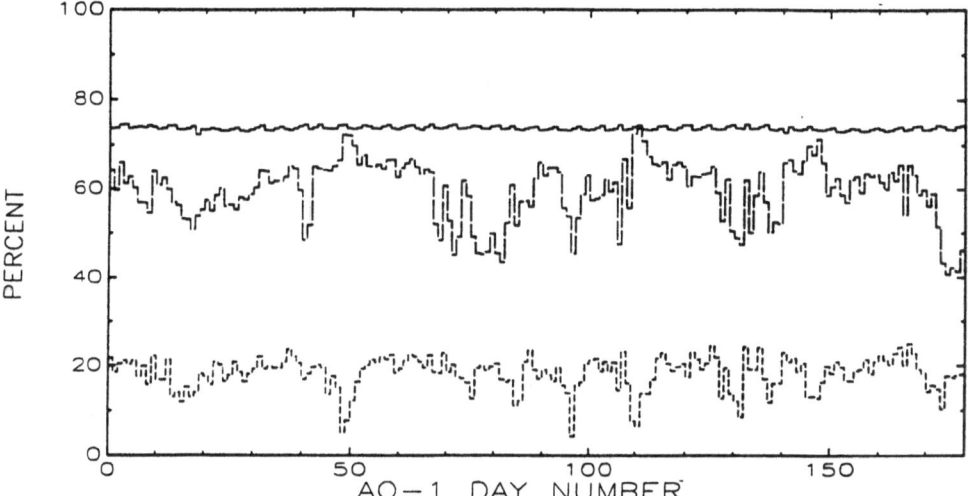

Fig. 6. RMTG efficiency. Shown are the total available time (solid, total time minus time in belts), the total observation time (long dash), and the total time in slews (short dash) per day as a day percentage for AO-1.

It is obvious that the above method can applied recursively to more (N) intermediate steps (of M nodes); the number of operations scales only as $N \times M^2$. The number of operations required for simple evaluation, on the other hand, scales as M^N. The savings in CPU time by using the network method for solving such a complex system is, in a word, astronomical. With the network method it requires roughly 24 hours of CPU time on a VAX workstation to generate the long term timeline for six months of ROSAT pointed observations. The solution is typically very efficient in using the possible observing time. Figure 6 shows the total observation time per day and the total possible observation time per day (length of day minus time in belts) on a daily basis in AO-1. The average observation time per day is 51,500 seconds for a total efficiency of 60%. The satellite spends an average of 26% of each day in the particle belts and SAA, therefore, the RMTG schedules observations for greater than 80% of the available time.

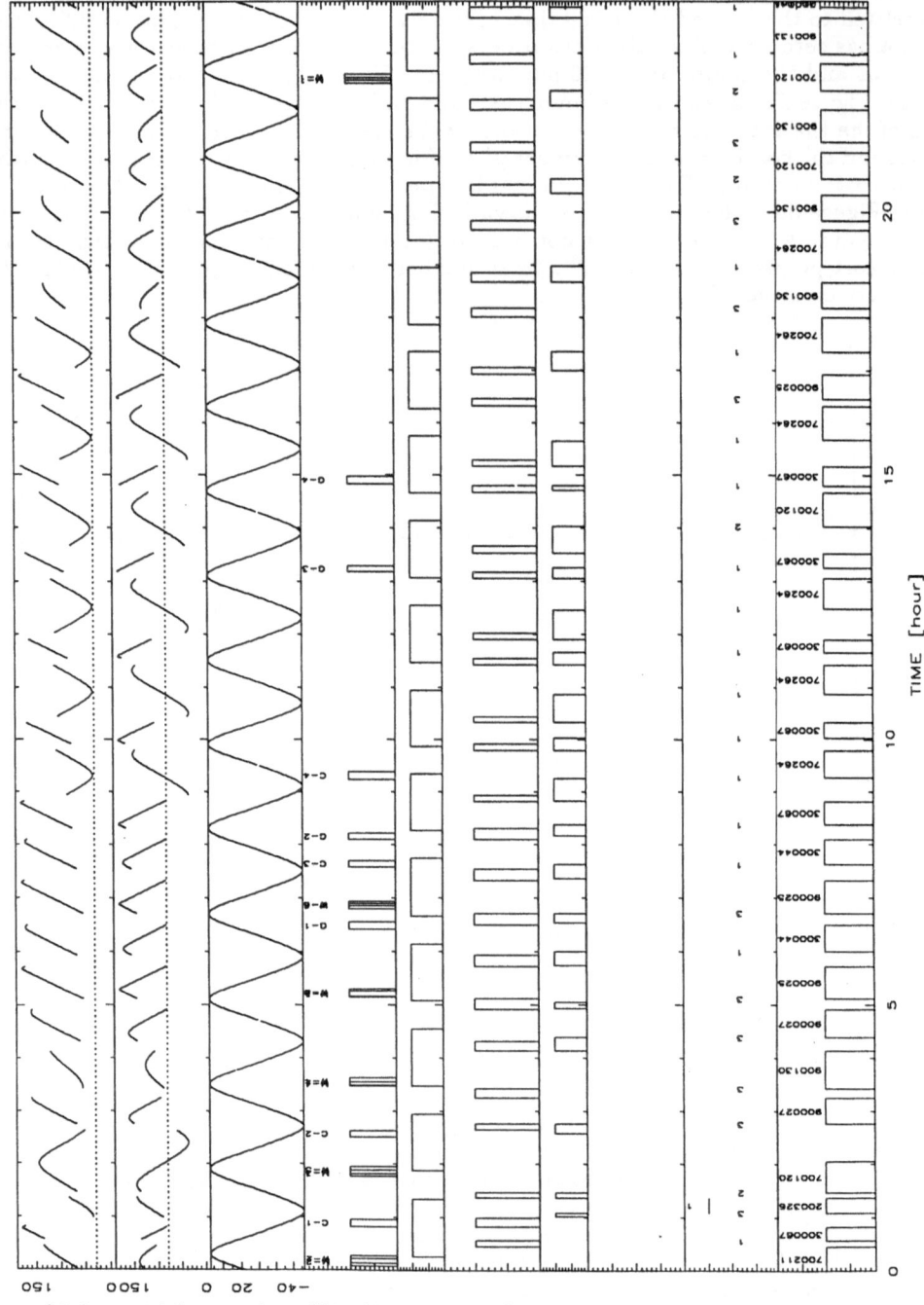

Fig. 7. Sample mission timeline plot, see text for description.

128

Figure 7 shows an example of the mission timeline for 13 April 1991. The first row (top) shows the ram angle, the angle between the look direction and the velocity vector. If this angle is less than 28° the scientific instruments must be safed. The second row shows the Earth pointing angle, the angle between Earth center and the look direction. The third row shows the geographic latitude of the sub-satellite point. The fourth row shows when ground station contacts occur. The fifth row shows when the satellite is sunlit. The sixth row shows when the satellite is slewing. The seventh row shows when the satellite is in a particle belt or the SAA. The eighth shows when a scientific instrument calibration occurs. The ninth row shows which scientific filters are being used. The tenth row shows which target is being observed.

Data Base Management

The complexity and magnitude of the ROSAT mission planning process requires a versatile and efficient data base management system. At MPE the INGRES system (Relational Technology Inc., Alameda, CA, USA) is used for nearly all data storage. All ROSAT requests (ROR, TCR, RCR, etc.) and all mission timeline entries are stored in various INGRES tables which facilitates the evaluation of the timelines as well as providing easy access to the data for other purposes.

FINAL REMARKS

The ROSAT mission planning is a joint effort between the Max Planck Institute for Extraterrestrial Physics and the German Space Operations Center. MPE has the responsibility for determining the scientific and instrument calibration program and defining the regions over the Earth where the satellite can safely operate (away from the particle belts and SAA). GSOC has the responsibility for processing the MPE input and generating the mission timelines.

Communication between MPE and GSOC is by electronically exchanged files of specified formats. The generation of the mission timeline is a several step process which includes both a long term timeline covering six-month mission phases and short term timelines which cover a one week period and are generated one week prior to execution. The solution of the mission timeline is by an efficient network procedure. Data base management at MPE uses the INGRES system.

Acknowledgements

We gratefully acknowledge the effort contributed by H. Frank, D. Garton, P. Graff, and R. Swoboda of GSOC to ensure the smooth operation of the mission timeline generation process.

This work was supported in part by the Max-Planck-Institut für Extraterrestrische Physik, the Bundesministerium für Forschung und Technologie (BMFT), and the National Aeronautics and Space Administration under grant NAG 5-1438.

References

Ashbolt, C. M. and Garton, D. 1990, *Project: ROSAT Mission Timeline Generator (RMTG) GSOC/MPE Interface Document*, GSOC Document IF-ROS-GS-470/2000.
Sims, M. R. *et al.* 1990, *Opt. Eng.*, 29, 649.
Trümper, J. 1983, *Adv. Space Res.*, 2, 241.

THE ROSAT DATA ARCHIVE AND OTHER PROJECT APPLICATIONS OF A DBMS

Joachim Paul

Max-Planck-Institut für extraterrestrische Physik

D-8046 Garching

INTRODUCTION

In 1984 it became clear that a database management system
(DBMS) would be required to handle the huge amount of data that will
be produced during standard processing (SASS) of ROSAT data. The
DBMS got used for a number of 'normal' applications like mission
planning (J.Schmid/S.Snowden) and the ROSAT proposal utility MIPS
(J.Behnke GSFC). This paper discusses three 'hybride' applications
which split the transaction into data retrieval, execution of VMS
commands and update of the database:
- the ROSAT data archival system : ROAR   (FORTRAN+EQUEL)
- the steering system for SASS   : STEER (ABF/OSL)
- a software management tool   : PORT   (FORTRAN+EQUEL)

Database Selection Criteria

INGRES was selected from several competitors as at that time it
was the only database management system DBMS that had did fulfill
our requirements. The criteria were set up to match the profile of
our users. As none of us was a database specialists, we looked for
- a relational data organization - easy to understand
- simple but powerful query language - fast access to database
  for ad-hoc queries. (QUEL,SQL)
- embedded query language with precompiler interface - to allow
  complex algorithms for updating database being written in
  FORTRAN (EQUEL,ESQL)
- sufficient tools make use of the database prior to coding -
  for preparation of project and guests. (Monitor,report writer
  QBF=query by forms)

INGRES Interfaces

A user can embed into FORTRAN nearly the same database command
that he uses to update/query the database interactively. This helps
testing complex queries as they can be developed interactively prior
to implementing them into the FORTRAN code. From FORTRAN not only
the data manipulation functions are accessible but also a forms
system to design sophisticated user interfaces.

ABF/OSL is an application generation tool that uses the INGRES language OSL (operation specification language) and allows the user to develop an application in a fraction of the time it takes to do the same from FORTRAN. This language is optimized for data transfer between tables and the forms system. It can be linked with object modules generated from other languages but the parameter handling is very poor. As long as the user only wants to move, display and update data in the database without fancy manipulations this is the best choice for the implementation.

Transactions and Recovery

As INGRES is a transaction oriented DBMS we soon ran into problems with the locking of datasets. While a transaction is open all rows that are affected by the pending query are locked. They are not accessible for other users and their transactions have to wait until the entries are released i.e. the transaction is committed or aborted. This is a problem for all programs that run for a long time while transactions are open. All of our 'hybride' applications follow the same scheme:
- retrieve processing parameters from the database
- calculate results, run programs, execute DCL commands
- update database

As the processing steps for all three applications take some minutes to complete, the database slows down processing, if transactions remain open during the second step.

The DMBS system guarantees the consistency of the database. A transaction will either be completed or rolled back. It will be rolled back in case of
- system crash
- program/process abort
- user abort of transaction.

We decided to split the transaction into half, one transaction for the retrieval and one for the database update. We now had to weigh the risk of database inconsistency against implementation of mechanisms for a transaction system outside of the DBMS.

For all three applications the second step is formed from a sequence of operations including DCL commands. We had to think of recovery strategies to handle all cases when the DBMS rolls back its transaction. The code must ensure that operations
- can be made undone (possible for COPY, RENAME commands),
- can continue/ignore step if step is not critical (creation of temporary file which can be deleted/purged after restart)
- must be able to continue if rollback is not possible (as DELETE cannot be made undone).

If the user aborts the transaction the procedure has to roll back all operations. If the system/process did abort the procedure must look for aborted actions at restart time and must roll back the previously started actions. The application program has to store the information what it is going to go and how fare it has already proceeded in a way that the information is available after system restart. This can either be the DBMS or a text file. The application has to look for pending operations and undo/complete them after restart.

In addition we had to be prepared for the worst case: a bug in the program which adds in some cases faulty data to the DBMS. No

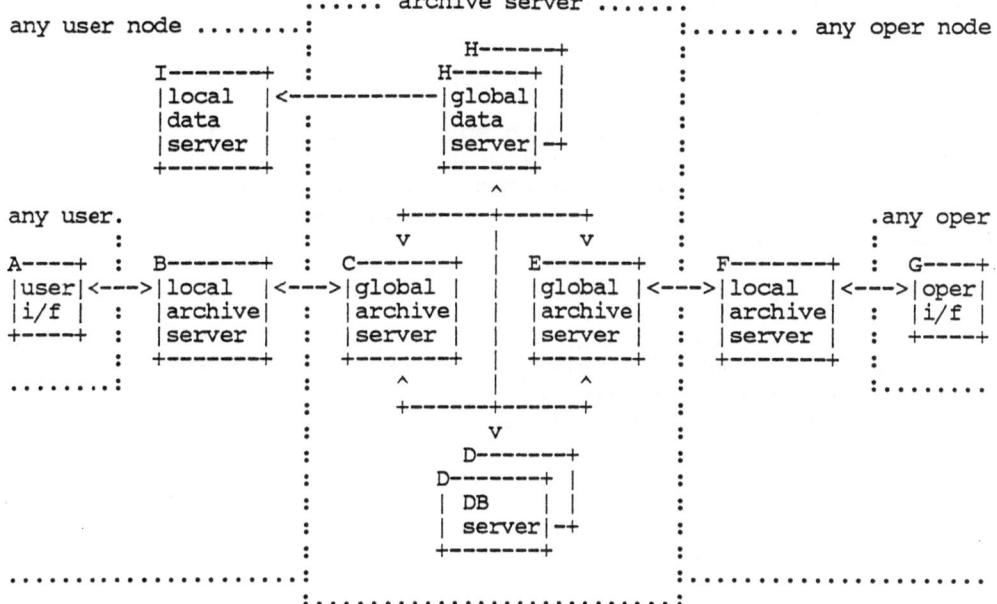

Fig. 1   Archive servers

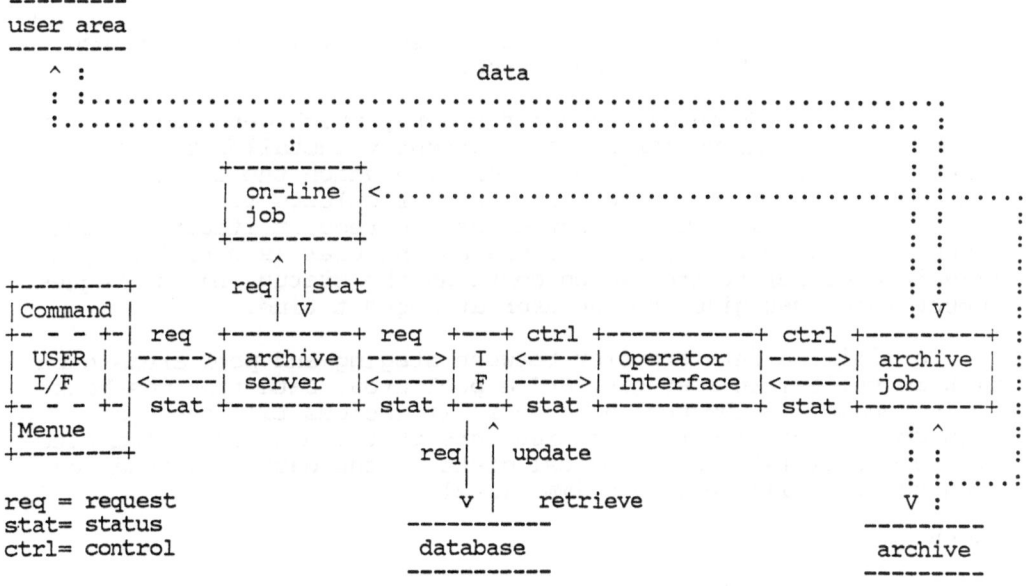

Fig. 2   Archive control and data flow

DBMS security mechanism can save you from these inconsistencies. The application designer has to think of ways to load the database with all critical information using data that is stored outside the database and can be corrected prior to loading them into the DBMS.

THE ROSAT DATA ARCHIVE (J.Paul, H.Köhler, A.Bohnet)

This package was designed in spring 90 but due to lack up time only the basic features got implemented up to now. The archive is intended to serve users on different machines and even on different operating systems. For that reason a client/server architecture was chosen (Fig. 1).

Hardware

The archive uses two RV20 optical disk drives (1GB per side) for saving and four magnetic disks (1.7 GB each) for staging the data. All data transfer to and from a user uses the staging disks. The RV20 loads its data from and to the staging disks.

Optical disks were chosen as only with this media it is possible to store and access this 200-300 GB without great efforts. The cost for the disks are relatively high (about $500) but a number of benefits make it worth while to use them:
- less space
- less work
- no refresh required

Request handling(Fig. 2)

The archive serves as an extended Backup facility which controls access rights, volumes, disk space. Any request for data must use the primary key which was passed to the archive with the data. The archive system internally uses a different key which can change when parts of the database are reconstructed.

The user sends out a request for data, which gets queued in a database table. As no servers are currently installed the user generates a request file in a common area which the archive reads and loads into the database. The operator selects requests from this table for execution. The execution of the request itself is done in batch jobs. When the request completes the user is notified by a file in a different area which contains the status and the request number which was given to the user at request time.

To minimize data transfer between staging and pool areas each disk is split into a staging and a pool area. Data that was backed up is renamed to the pool area where it resides till the space is required for other data. User requests that can be fulfilled without reload of files from optical disks as the data is already on a staging disk will be served immediately.

Database

The database is formed by four tables:
- arc_catalog   contains the primary key, and all data that is related with a single dataset
- arc_volumes   contains location of volume, and all data that is related with a single volume.

- arc_request   contains request information
- ret_catalog   logs the retrieve requests.

There will be new tables to control multiple access when the data become public:
- to keep locks for data on the pool disks
- to keep access control information (user, protection mask...)

Recovery

The request to backup/copy data is in the database till the request was completely fulfilled. If the copy from/to a user is interrupted, the target directory is cleared before the next attempt is made to copy the data. As all data transfer from and to a user is handled by batch jobs VMS ensures the proper restart of a job after a system crash.

If the system crashes while a backup is written to optical disk manually intervention is required, as the file structure on the volume might not be closed correctly.

To recover from program bugs the datasets contain all necessary information to fill the essential columns of the the archive table. This information is in the BACKUP header stored as a comment string.

Listing of save set(s)

```
Save set:          B00490003583.0001
Date:               1-MAR-1991 09:30:02.56
Command:           BACKUP/LIST=ARC_LOG:BACKUP.LOG -
                   ARC_STAGE:[ARCSTAGE.SAV.M22600...] -
                   MUA0:B00490003583.0001/LABEL=A024AA -
                   /TAPE_EXPIRATION=31-DEC-1999 00:00:00.00 -
                   /BLOCK_SIZE=16128
                   /COMMENT=PROJECT =SAV|KEYWORD =SAV_M22600_P2
                   |ORIGIN  =R$DATA_9:[SAV.M22600...]
                   |DIRECTORIES=      1|FILES   =        18
                   |SIZE    =   151723
```

```
[ARCSTAGE.SAV.M22600]P2.DIR;1                              1
[ARCSTAGE.SAV.M22600.P2]100351_SAV_M22600_2.LOG;2         30
:
```

PORT - A SOFTWARE MANAGEMENT TOOL (J.Paul)

PORT was developed at MPE in 1985 as more and more software was compiled for the Standard Analysis Software System (SASS). It keeps track for each module that was release to the common software pool.

Depending on the type of the module different actions are taken to ensure consistency among the modules in the pool. For an updated common block this does mean, that all programs that include this structure are recompiled, the libraries are updated and all programs that link the recompiled modules become relinked.

Processing Flow

When a user starts up PORT the program connects to the DBMS, opens the logfile and spawns a backend process. Depending on the

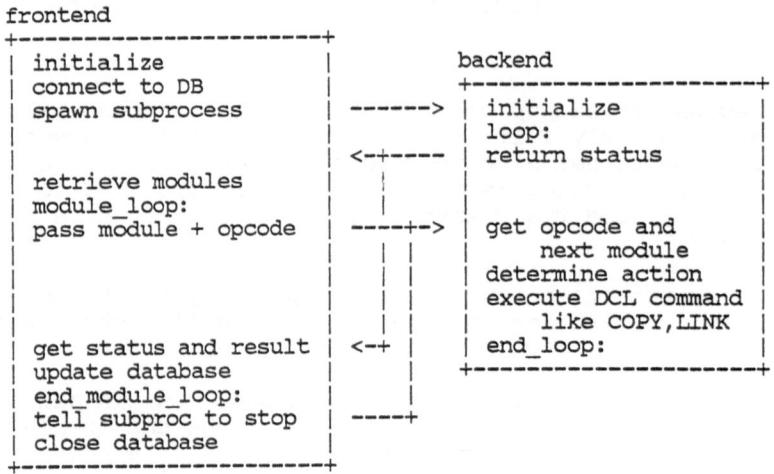

```
frontend
+-------------------------+
|   initialize            |            backend
|   connect to DB         |            +-------------------------+
|   spawn subprocess      | ------>  | initialize              |
|                         |          | loop:                   |
|                         | <-+----  | return status           |
|   retrieve modules      |   |      |                         |
|   module_loop:          |   |      |                         |
|   pass module + opcode  | ----+->  | get opcode and          |
|                         |  | |     |     next module         |
|                         |  | |     | determine action        |
|                         |  | |     | execute DCL command     |
|                         |  | |     |     like COPY,LINK       |
|   get status and result | <-+ |    | end_loop:               |
|   update database       |    |     +-------------------------+
|   end_module_loop:      |    |
|   tell subproc to stop  | ----+
|   close database        |
+-------------------------+
```

Fig. 3   Control flow for PORT frontend and backend process

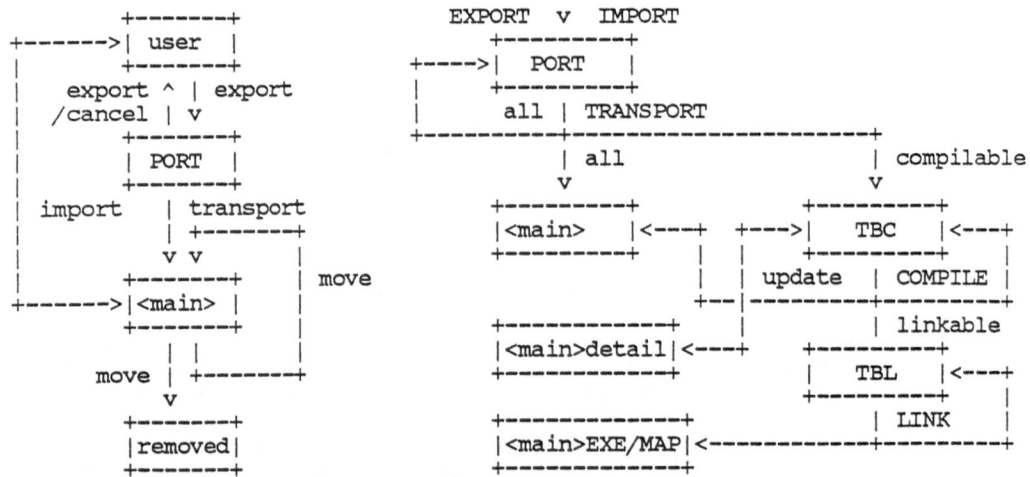

Fig. 4 PORT operations   Fig. 5 PORT database structure and flow

selected operation and input file names information is retrieved from the database for the processing steps. The backend process executes different sequences of commands for different file types. After processing of each file the database is updated. (Fig. 3)

Startup and command execution in a backend process were chosen for various reasons:
- normally a user/operator handles more than one file.
  It is faster to start up the backend process once and execute the DCL commands in the backend while the database is open than to connect to the DBMS for each file twice - first to retrieve the information and second to update the database after the execution of the commands.
- after processing of each file the "transaction" for this file is committed by updating the database.
- Relational databases return the full set of all entries that match the qualification. With the backend process we are able to process one file after the other as it is returned by the DBMS.

Types of Operations (Fig. 4)

- EXPORT to send a new module to the pool (/CANCEL for undo)
- IMPORT to retrieve a module from the pool (/CANCEL for undo)
- TRANSPORT to embed the new module
- COMPILE to (re-)compile modules
- LINK to (re-)link modules
- MOVE to move modules from one package to the other
- REMOVE to remove obsolete modules

Database design and update (Fig. 5)

All modules that are exported or imported will be stored in table PORT. This prevents different users to try to update the same module simultaneously.

When the operator "transports" a module the entry for the file is moved from the PORT table to the <main> table where it is stored using a unique internal id. In this table all entries sum up over the life cycle of the module, but only the most recent version of the module is the "current" one. All code modules that must be compiled add an entry to table TBC (= to be compiled).

The operator can now select COMPILE to compile all modules that are stored in TBC. The file is compiled and some tables are updated with information which is taken from the header of the file. For FORTRAN modules this header contains
- the names of all include files,
- all subroutine names that are called and
- all extensions to the standard.
The detail tables are used to resolve all dependencies among the modules. If the module was a main program its name is added to TBL (= to be linked) else the names of all programs that link the new/ updated module are added to TBL.

The last option the operator has to select is LINK to link all programs that are stored in TBL.

Recovery

For the recovery we had to balance the effort of coding an ex-

ternal transaction handling against the case that two users would try to address the same module at nearly the same time:
- normally users work only with modules they have written.
- only one operator should install new modules, compile or link as these steps can affect other users if they are running programs while the upgrade is in progress (which normally implies that there are inconsistencies or a short time).
- there are only two points that must be undone for recovery. This is when modules are copied, the original is deleted and the new module is not jet inserted in the database.
- the program is only used interactively. If the program/system did crash the user/operator can easily undo the last step manually.

When the database became inconsistent due to a programming bug we reloaded the database from the contents of the directories, the file headers and the map files. We lost information which is not critical for the system like history of the files, the exporting user and the export date.

STEERING SYSTEM FOR STANDARD ANALYSIS OF ROSAT DATA (R.Gruber)

The steering system is used to keep track and to control all processing of ROSAT data for the standard analysis that is applied to all data.

Processing Flow

The operator selects and prepares a job for execution. The job itself is submitted to a bach queue and the operator can prepare the next job. The job now executes a sequence of application programs. When done the job updates the database and checks for datasets which can be processed at the next higher level. If some are found they are prepared and scheduled for execution in the same way as the operator had initiated the processing. (Fig. 6)

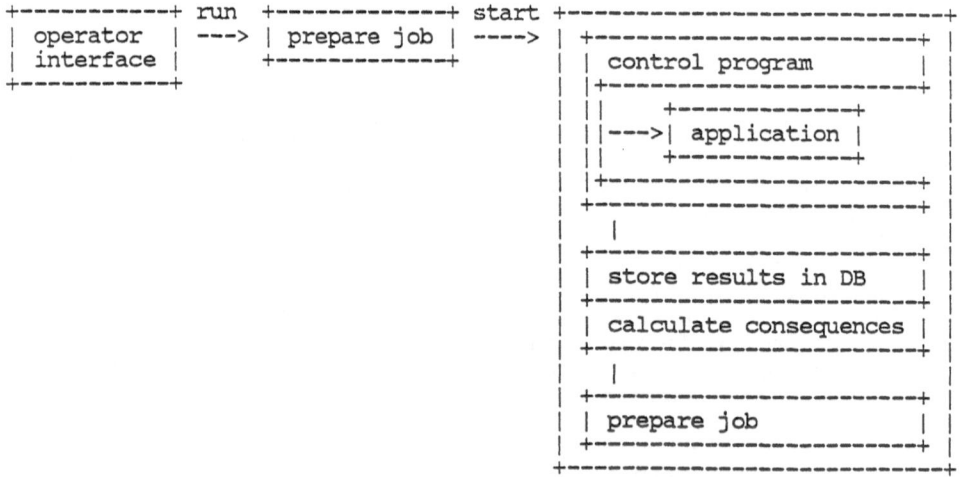

Fig. 6 STEER control flow

Preparation of a job does mean that all relevant information is extracted from the database and inserted into a parameter file. The control programs and the application programs use this file as input to determine their behaviour. The application programs write back to the parameter file all results that are required by the next program or that are to be loaded into the database.

This structure was chosen to
- decouple the application programs from the database. This
  * speeds up the program as it does not have to start the DBMS
  * makes testing easier as the parameter file can be edited
  * makes testing independent from other users in contrast to using a common database.
  * permits start of the job later than preparation
  * makes it possible to schedule the job for execution on machines that do not run the DBMS
- centralizes DBMS access. Only one program must be tested and programs can be developed even when no DBMS is available.

This was possible as the processing parameters are 'static' and the different jobs need not to write back their results as soon as possible. Only the operator interface must response in a reasonable time.

The operator interface was designed for only one user per processing level per database. This is a severe restriction, but there is no real need to have several operators working on the same data. On the other hand we had not to think of uniuely flagging each entry to keep track of an operator's work.

Recovery

The application does take care of aborted processings. For each aborted/incomplete processing a new version is generated and the job is restarted with the new version.

All database updates are transactions. Program/system crashes while updating the database are handled by the DBMS. The update step can even be executed twice as the database does accept duplicate entries but it does not really append them to the tables.

Recovery from bugs in programs that update the database is very difficult. Reloading all database updates from the parameter files could rebuild the database. But as there are several hundred runs per day, which get backed up to free space it is nearly impossible to recover the database in that way as all data must be reloaded from the archive to disk. The only positive aspect is that data is never deleted or updated and, therefore, it cannot happen that data, that once was entered correctly, gets damaged.

INTERACTIVE ANALYSIS WITH EXSAS

H.U. Zimmermann, T. Belloni, G. Boese,
C. Izzo, P. Kahabka, O. Schwentker

Max-Planck-Institut für Extraterrestrische Physik
Karl-Schwarzschild Straße 1
D-8046 Garching bei München, Fed. Rep. of Germany

INTRODUCTION

Data from observations with the ROSAT X-ray telescope will routinely undergo a Standard Processing in the ROSAT Scientific Data Center (RSDC) at the Max-Planck Institut für Extraterrestrische Physik (MPE, Garching). As has been elaborated in a previous contribution the Standard Processing performs two major tasks: in a first step data are normalized and calibrated and then a Standard Image and Source Analysis is performed. The Principal Investigator of the observation will eventually receive the data and an analysis protocol containing the results of the Standard Analysis. Often a user may not be fully satisfied with the findings of the Standard Analysis and therefore wish to proceed with a more detailed data evaluation on his own. It is for this kind of investigation that we have developed the Extended Scientific Analysis System (EXSAS).

THE EXSAS ENVIRONMENT

EXSAS has been designed and developed by the RSDC for analysis of X-ray data in general, with particular emphasis on the analysis of data from the ROSAT X-ray and XUV instruments. EXSAS comprises a large collection of application modules as typically used by X-ray astronomers in the data analysis of this wavelength regime. Therefore the user can perform all of the analysis already done in the Standard Analysis also in EXSAS with the choice of using different selection criteria. Additionally many other and more sophisticated analysis methods are available. EXSAS will normally be run interactively on workstations, but may also be run in batch mode.

The package is embedded in the well known astronomical image processing system ESO-MIDAS, developed, maintained and distributed by the European Southern Observatory (ESO). ESO is only a hundred meters off from our institute. Therefore in the past we often succeeded to include our specific needs in the ongoing development of the MIDAS system and always found very good support.

In the following we list some of the basic features and design criteria of the EXSAS/MIDAS environment that should enable observers to easily work with the system for a deeper investigation of their data.

- EXSAS/MIDAS is highly portable; in our institute we run the package on both VAX/VMS and two different UNIX systems (ULTRIX and HP-UNIX).

- MIDAS has a fairly wide distribution and acceptance within the astronomical community in Europe, but can also be found in other continents. The 'Deutsches Astronetz' has selected it as the standard image processing system some years ago. Thus EXSAS can be installed with minimum effort at most of the European astronomical sites.

- The system uses only standard file formats (tables, images and in a few cases ASCII files) and strongly standardized sets of interface routines for all internal and external input/output.

- EXSAS applications are all written in standard Fortran 77, a language that every astronomer knows well. This enables the user to easily understand existing software modules and if necessary change them according to his specific needs.

- It is easy to implement own user software into the EXSAS/MIDAS environment.

- MIDAS contains already a number of packages that otherwise would have had to be developed for EXSAS like extended image and graphic presentation/manipulation as well as general fitting and statistical packages.

- All EXSAS applications maintain and update a Standard EXSAS Header for general information transfer and history information on performed actions.

- All main application packages have instrument independent structures. Therefore the implementation of further instruments is fairly easy.

- Besides on-line help and tutorials an EXSAS Users Guide and an EXSAS Cookbook are available.

TRANSFERRING DATA INTO EXSAS

In order to make data distribution independent from the specifics of different Operating Systems, all ROSAT data distributed by both the RSDC and the German XUV Data Center at Tübingen (AIT) are delivered as a so-called ROSAT Observation Dataset (ROD) in FITS format (Flexible Image Transport System: a data transfer standard that is operating system independent and has been widely accepted by the astronomical community). A ROD consists of more than a dozen tables and images containing all primary and auxiliary data needed for further detailed analysis: photon event data, instrument housekeeping and quality parameters, attitude and orbit files, calibration data and also selected results from the Standard Analysis performed at the Data Centers.

EXSAS APPLICATION PACKAGES

For better overview EXSAS applications have been grouped into 4 application packages performing the following tasks:
- Data Preparation (bringing data to the proper form for subsequent specific analysis work)
- Spatial analysis
- Spectral analysis
- Timing analysis

In the following a short overview on the main functionality of the different packages will be given.

Data Preparation

The package comprises:

- Selection and binning of photon event data with the following properties:

  • Selection with respect to time, pulseheight and a basic spatial constraints (box, ring, sector; cursor selectable)

  • Selection according to the actual count rate

  • Selection history is kept in file header

  • Counting of selected events

  • Binning in time (lightcurves), in pulseheight (spectra) or spatially (profiles, images)

  • Any combination of all above

- Instrument Correction of

  • Spectra: dead time, vignetting, filter transmission, point spread function dependencies are used in either attitude or photon position mode to calculate a correction vector for subsequent spectral fitting

  • Lightcurves

- Spectrum manipulations allowing

  • Spectral rebinning (by explicit definition or using constant signal to noise ratio in each bin)

  • Background offset corrections

  • Algebraic operations, rate and error calculations

Spatial Analysis

- A sophisticated package for detection of pointlike and moderately extended sources in images or photon event data has been implemented:

  • Source detection in image data using a local background

- Source detection in image data using a smoothed background
- A maximum likelihood method to evaluate best estimates for the position, intensity, extent and corresponding errors using the information of all individual photon events involved

- Ring and box integrated profiles from image and photon event data

- Utilities for coordinate handling and presentation

Spectral Analysis

Functionality of this instrument independent analysis package includes

- Construction of a wide range of model spectra (by free combination of standard and/or user defined models) and comparison or fit to an observed spectrum

- Calculation of corresponding errors using either the covariance matrix or a chi-square grid search method

- Determination of photon and energy fluxes, luminosities etc.

- Standard plotting of results on fits, error ellipses, chi-square contours

Timing Analysis

The package contains

- Power spectra calculation including automatic peak detection

- Auto-Correlation and Cross-Correlation methods

- Period folding with barycentric correction of photon arrival times

- Different statistical tests on source variability

- Standard plotting and display facilities

GENERAL SUPPORT FOR EXSAS

The EXSAS software system is on request available from the RSDC. Only prerequuisite is the access to a MIDAS installation (MIDAS will be distributed by ESO). The ROSAT Data Center will provide full support and error services for all main EXSAS packages for the foreseeable future, updating and extending the functionality in regular intervals. An EXSAS Users Guide and an EXSAS Cookbook complement the internal documentation (help faciltities, tutorials). For ROSAT observers that intend to evaluate their data at the RSDC a number of image processing workstations and a limited visitor service is available at MPE.

PROS: DATA ANALYSIS FOR ROSAT

D.M. Worrall, M. Conroy, J. DePonte, F.R. Harnden, Jr., E. Mandel, S.S. Murray,
G. Trinchieri[†], M. VanHilst, and B.J. Wilkes

Harvard-Smithsonian Center for Astrophysics
60 Garden Street, Cambridge, MA, 02138, U.S.A.

INTRODUCTION.

The Post-Reduction Off-line Software (PROS) is a portable X-ray analysis software system that is integrated into the Image Reduction and Analysis System (IRAF), which was developed at the National Optical Astronomy Observatories (NOAO; see Tody, this volume). The PROS consists of a suite of applications tasks and an interactive image display program, SAOimage, which runs under the X-windows system. The PROS is being developed on a network of SUN workstations by our group at the Smithsonian Astrophysical Observatory (SAO) in coordination with the IRAF development group at NOAO and the Space Telescope Science Data Analysis System (STSDAS) group at the Space Telescope Science Institute (STScI). The PROS has been written for the analysis of pointed data from the ROSAT X-ray telescope, and developed and tested using *Einstein* Observatory data. Goddard Space Flight Center (GSFC) has the primary responsibility for testing the PROS as part of the joint GSFC and SAO activities in support of the U.S. ROSAT Science Data Center (USRSDC). In addition, a number of sites in the community received the β-test version of the PROS for evaluation in late 1990.

A design consideration for the PROS project has been to provide software that can be extended to other X-ray missions. Penn State University is currently adapting the software for the analysis of data from the forthcoming *Astro-D* mission, and we have begun discussions with the newly formed High Energy Astrophysics Archive (HEASARC) group at GSFC concerning the extension of the PROS to support the analysis of X-ray archive data.

DESIGN

The PROS is software which ROSAT guest observers may use to complete the scientific analysis of their data. As early as ~1984, when SAO first proposed to NASA to provide scientific software for the USRSDC, it was apparent that a system involving distributed data and software would be the most satisfactory for the user community. Thus, the issue of portability was very important in the conception and design of the PROS, as was developing software which would run in a modern workstation environment, preferably an environment where many of the tools required to analyze X-ray data were already provided. STScI's decision in 1984 to use the IRAF environment for STSDAS assured us that by also choosing IRAF we would be using an environment already familiar to many users of space-based instruments. Only in this last point has the situation not met our expectations in that the spread of IRAF through the X-ray community has been slow up until now (a situation possibly exacerbated by Hubble Space Telescope delays). In consequence a larger than expected amount of the work of initiating X-ray users into the rich IRAF analysis environment is beginning to fall on the shoulders of the USRSDC, and feedback from scientific users of the PROS development code has not been as extensive as desired.

[†] Also, Osservatorio Astrofisico di Arcetri, Firenze, Italy

Data Analysis in Astronomy IV, Edited by V. Di Gesù *et al.*
Plenum Press, New York, 1992

We recognized early on the enormous advantages which would be derived from a data structure that represents X-ray data as a list of photons with associated attributes such as position, time and pulse-height bin (energy), as is natural for high-energy astrophysics data, but which could be accessed as an image, with optional filters for non-spatial quantities. We discussed these ideas with the NOAO group which agreed to provide such a data structure and subsequently made available the 'Quick [*i.e.*, prototype] Position-Ordered Event' (QPOE) data structure in IRAF release v2.8 in mid 1989. Among the beneficial consequences of using the QPOE data structure are:

- Non proliferation of (normally sparse) array files for a user who wishes to run various spatial analysis tasks for different selections of energy, time, *etc*, since the user specifies the filters as part of the QPOE file-name specification rather than first writing out a new file.
- All IRAF image processing tasks provided by NOAO, STScI, SAO and others, are available for immediate use on data stored in a QPOE file, as well as data stored in other IRAF-supported structures such as arrays.

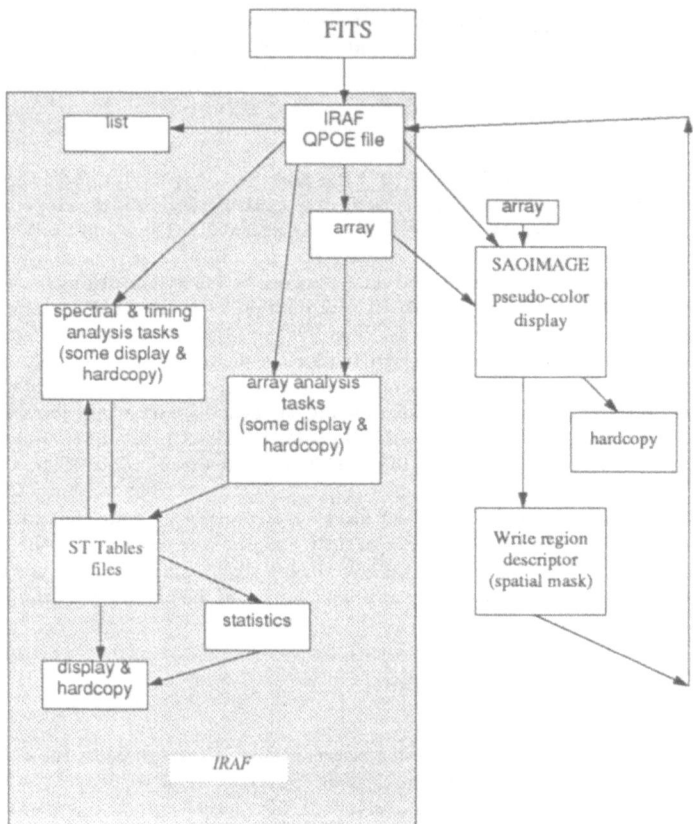

Fig. 1. PROS/IRAF simplified overall design.

The fact that so many of the image-processing analysis tools required to analyze X-ray data already exist or are planned by others, has allowed us to concentrate our development on tools which either (a) begin by requiring single-photon access to a QPOE file, such as timing and spectral analysis tasks, and/or, (b) are peculiar/unique to the needs of high-energy data, in particular because much of the data involves the regime where Poisson statistics apply.

FEATURES

The PROS/IRAF software development is coordinated with new releases of IRAF from NOAO. SAO distributes the PROS. After correct installation, the PROS appears to the user as a package called 'xray' in the IRAF top-level menu, the same menu level as the NOAO and STSDAS packages. Figure 1 is a simplified diagram showing some of the key elements of the PROS. Some of the features are listed below.

File Formats

File formats used within the PROS are restricted to QPOE, IRAF image (array), ST table, and ASCII. Conversion routines to and from the Flexible Image Transport System (FITS) data format, using the binary extension for QPOE files, are available and allow for machine-independent transfer of the data among different installations of IRAF and other astronomical data-analysis systems.

Display and Hardcopy

Some display tasks exist within the 'xray' package and its subpackages. Output to ST tables allows more extensive display and statistical analysis.

Error Arrays

PROS spatial-analysis tasks allow for generation and manipulation of associated error arrays, assuming symmetrical ($\sqrt{N}$) uncertainties. Error arrays store the square of the uncertainties for computation efficiency.

History

PROS tasks add history text strings to the headers of data files, where appropriate.

Spatial Regions

A syntax for specifying spatial regions (circles, ellipses, annuli, boxes, polygons) has been developed for spatial filtering of a QPOE or array file, and filters may be saved in ASCII region-descriptor files or as binary masks. Regions may be specified in sky coordinates if the data file has a header specifying its coordinates according to the IRAF World Coordinate System interface.

IMAGE DISPLAY

As part of the PROS, an interactive image-display program, SAOimage, was developed to provide image display with or without the use of IRAF under the X11 window system. SAOimage uses the button menu and sub-menus as shown in Figure 2, and the mouse-controlled pointer is used for such additional functions as changing the color or grey-scale mapping and drawing cursors (regions). An example of SAOimage output is shown in the center picture of Figure 3. Some of the features of SAOimage are listed below:

- Conforms to X11 specifications, and already ported to a variety of hardware/software systems.
- May be used with or without IRAF, and supports several input array file formats.
- Contains button-menu control for:
    - Change of zoom/pan (with guiding sub-window)
    - Change of color mapping (auto or user-selected scalings)
    - Display of cursors (standard shapes or user-drawn polygon)
- Tracks the pointer coordinates
- Supports bitmap display on monochrome workstations
- Prints pixel values
- Produces hardcopy on PostScript-compatible printers
- Writes cursor regions into ASCII region-descriptor file(s).

Further details of SAOimage are given in the technical report by VanHilst (1990).

DATA INTERFACES

ROSAT data are all processed by a standard analysis software system (**SASS**; see paper by Voges, this volume). This processing produces many output files in formats compatible with the **VAX/VMS** architecture on which the standard processing software runs. The output files include the photon event list, *i.e.*, the list of photons with associated positions, energy, time *etc*, and many auxiliary files such as lists of sources detected in the field. The **USRSDC** converts all relevant data into **FITS** format for distribution to users.

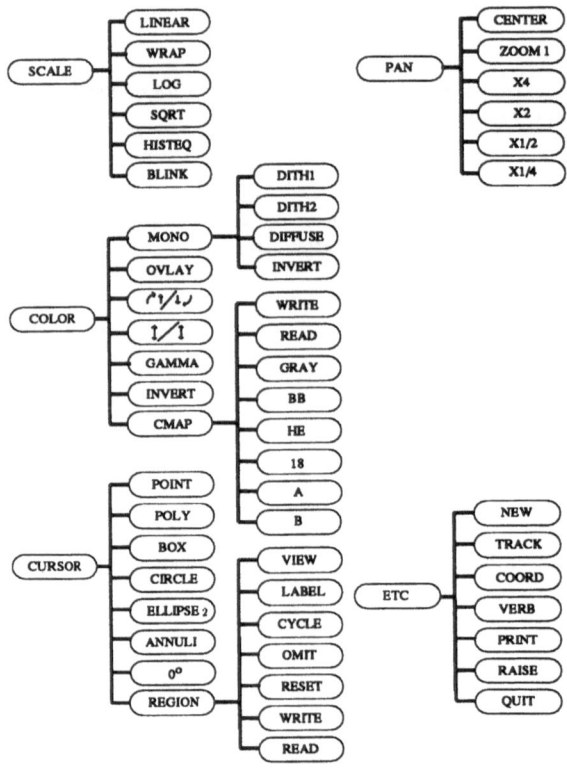

Fig. 2. Button menu and sub-menu items of **SAOimage**

The **PROS QPOE** file, as implemented for both *Einstein* and **ROSAT** data, includes both the photon event list and additional information. A standard header includes data which are required by many existing **IRAF** tasks and/or **PROS** tasks. Data required by **PROS** tasks for selection of events by single-photon access, such as information pertaining to good time intervals, are also included in the **PROS QPOE** file. As part of the **PROS** project we therefore developed tasks which would gather the data needed for a **PROS QPOE** file from the **ROSAT** standard-processing output and convert these data to **FITS** using the binary table extension. This gather-and-conversion procedure runs along with the other **FITS** conversions at the end of standard processing at the **USRSDC**. Files with position information only (integrated over energy and time) are written to **FITS** image files for subsequent conversion inside **IRAF** to an **IRAF** image file. Data such as distributions of counts in pulse-height bins for regions centered on strong sources are written as **FITS** tables for subsequent conversion inside **IRAF** to **ST** Tables files.

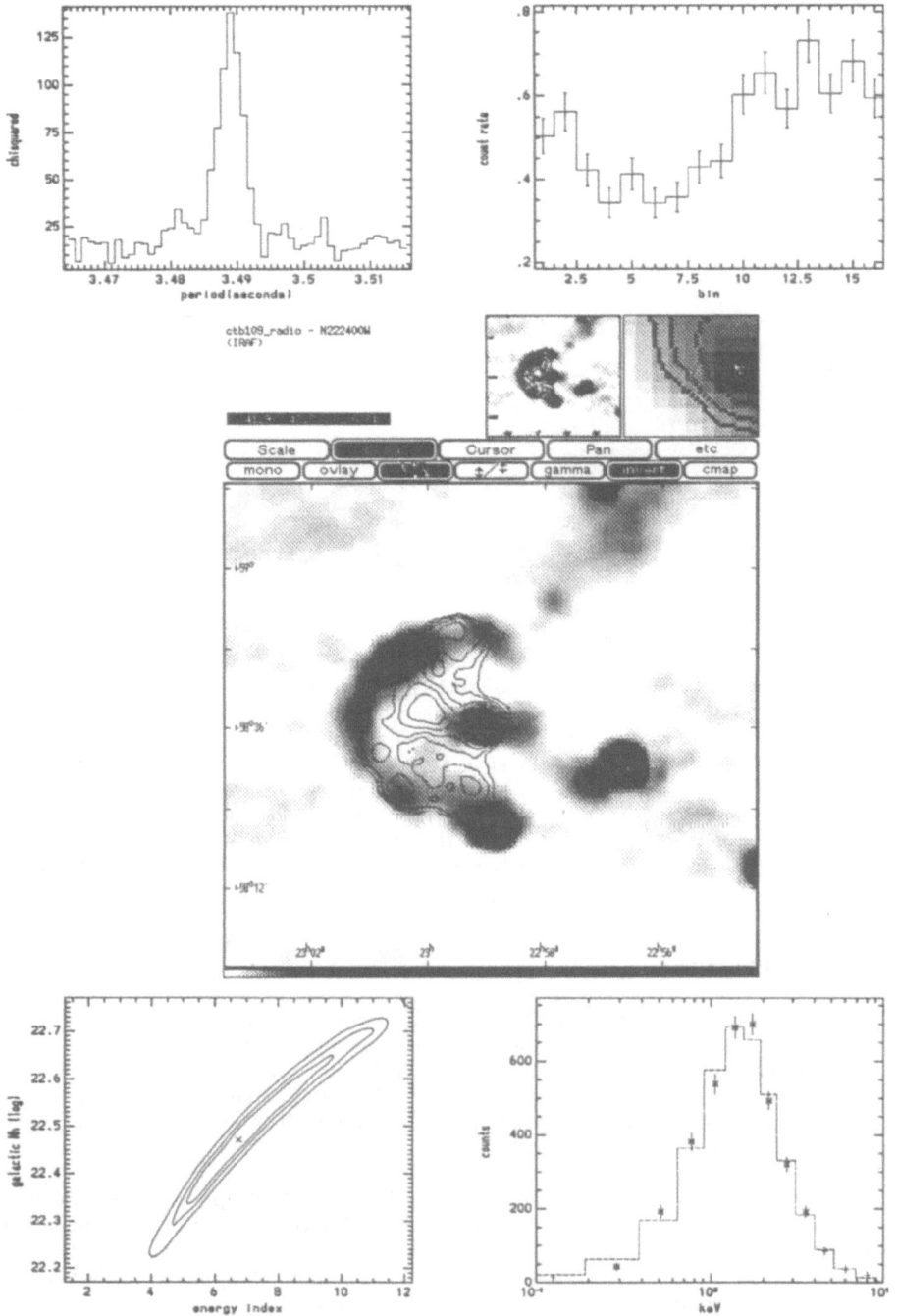

Fig. 3. Examples of PROS/IRAF analysis of the *Einstein* IPC X-ray data of the supernova remnant CTB 109 and its pulsar. *Center*: A contour map of the X-ray image, labelled in celestial coordinates, is superimposed on the 5 GHz radio image from the 300 ft survey (Condon, Broderick and Seilstad 1989). *Top*: X-ray timing analysis of the pulsar, the brightest X-ray source in the field. Tested against a constant source, the period is 3.489s (left-hand plot); folded light-curve at right. *Bottom*: X-ray spectral analysis of the pulsar. The best-fit spectrum is a power law of energy index 6.76 and line-of-sight galactic $N_H = 2.95 \times 10^{22}$ atoms cm^{-2}. The right-hand plot compares this best fit model with the data points; points used in the fit are marked with an asterisk. The left-hand plot shows the 68%, 90%, and 99% confidence contours for N_H and energy index.

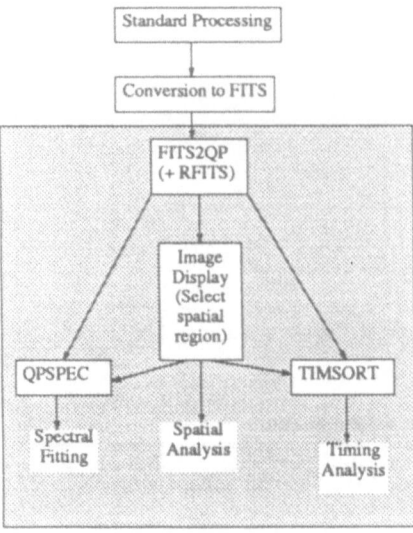

Fig. 4. Basic tasks which convert and select data inside the PROS/IRAF

Figure 4 shows by name some of the tasks which convert and select data inside the PROS/IRAF system. The PROS task 'fits2qp' creates a QPOE file with all its necessary components, and FITS conversion tasks provided by NOAO and STScI convert all auxiliary files to their IRAF resident formats. The QPOE file can be displayed and spatial regions around sources of interest selected using SAOimage. The 'qpspec' task will create a pulse-height distribution of counts within a specified region, with optional temporal filtering supplied as part of the file name specification. The 'qpoe' file is spatially sorted by y-position (with x secondary). For timing analysis the 'timsort' program will time-sort the data within a specified region (if required) and with optional energy filtering supplied as part of the file name specification.

SCIENTIFIC APPLICATION TASKS

Figure 5 summarizes the contents of the IRAF 'xray' package and its subpackages. FITS conversion routines and other file utilities appear at the 'xray' package level. In addition to the FITS conversion, the 'rosat' and 'einstein' packages contain routines which will convert files directly from standard-processing formats. These were particularly useful for us in testing the file conversions, but are likely to be superseded for most users by the use of FITS format for data distribution. The 'nimages' package contains routines which supplement the IRAF 'images' package.

Figure 3 illustrates some of the analysis capabilities of the PROS spectral, spatial, and timing packages. The test data are an *Einstein* IPC observation which reveals extended X-ray emission from a supernova remnant and periodic emission from a pulsar. Because this field is useful for testing many of the PROS tasks, we include it as test data with the PROS distribution. ROSAT test data (short calibration observations of a field containing AR Lac) are also distributed with the PROS.

The PROS spectral package allows the user to fit combinations of model forms to multiple data sets. Best-fit parameter values can be found, and a χ^2 contour plot can be drawn projected onto the plane of any two model parameters, with the user allowed the choice of a variable or fixed value for each remaining parameter. Because the package comprises a series of tools, and because IRAF provides the facility for writing scripts (macros), it is suitable for many projects. For example, it has been used in a maximum likelihood analysis to find the mean and dispersion of the spectral parameters for a population of objects for which 3,782 separate spectral fits had to be performed (Worrall and Wilkes 1990).

The timing analysis package is still rudimentary, but an overall design is in place which should

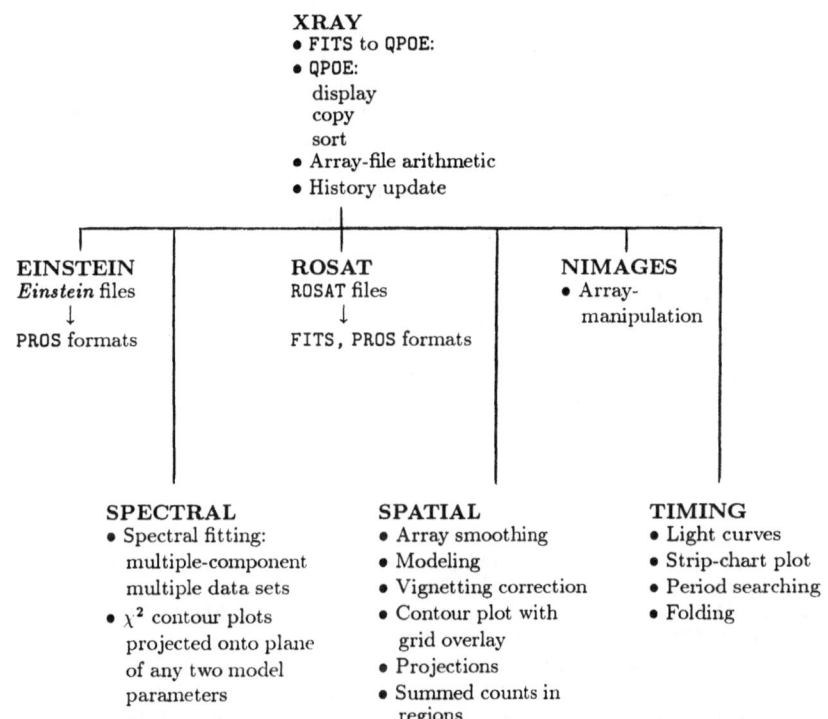

XRAY
- FITS to QPOE:
- QPOE:
  display
  copy
  sort
- Array-file arithmetic
- History update

EINSTEIN
Einstein files
↓
PROS formats

ROSAT
ROSAT files
↓
FITS, PROS formats

NIMAGES
- Array-
  manipulation

SPECTRAL
- Spectral fitting:
  multiple-component
  multiple data sets
- χ^2 contour plots
  projected onto plane
  of any two model
  parameters

SPATIAL
- Array smoothing
- Modeling
- Vignetting correction
- Contour plot with
  grid overlay
- Projections
- Summed counts in
  regions

TIMING
- Light curves
- Strip-chart plot
- Period searching
- Folding

Fig. 5. Summary of the current contents of the 'xray' package and its subpackages

permit easy addition of tasks. Figure 3 shows results of a χ^2 period search on the X-ray data from the pulsar in CTB 109, and the folded light curve at the best-fit period is presented.

Spatial analysis is illustrated by the central picture in Figure 3. A background image (from long exposures where sources have been removed) has been subtracted from the X-ray image, and then corrections for telescope vignetting have been applied. The resulting X-ray image has been smoothed and contoured and is shown overlayed on a radio image of the same part of the sky. Many image processing capabilities are provided by the IRAF system together with the many applications tasks written by various groups.

STATUS AND PLANS

At the time of writing, April 1991, we have just entered the operational and release stages of the PROS project. SAOimage has been available since 1989 from SAO, and IRAF users can request it from NOAO. We have planned a phased release for the PROS/IRAF software; the first release, which occurred in March 1991, was to sites with SUN hardware compatible with the platforms and operating systems on which the PROS was developed and tested. We intend this to be followed by versions for other platforms. During this early operational stage there will be a concerted effort to fix bugs discovered by the expanded user community, and to meet short-term goals for increased functionality and user friendliness based on the feedback we receive. A large percentage of our current effort is going into data conversion as updates to the ROSAT standard processing software continue to be received. The PROS/IRAF distribution (not including SAOimage) currently contains about 68 Mbytes of source code and executables per platform/compiler, and includes more than 50,000 lines of source code (excluding comments).

In parallel with the operational work, we plan to develop more advanced algorithms within the PROS system with a greater multi-mission emphasis. To this end, we will work with other X-ray groups such as those at Penn State University and the HEASARC. SAO's plan for the AXAF Science Center takes the concept of distributed software further and calls for the distribution of standard processing functions (which for *Einstein* and ROSAT are at a centralized location) with the post-reduction software tools, making both fully available to a remote user.

ACKNOWLEDGMENTS

We are very grateful to the systems and programming staffs of the High Energy Astrophysics Division at CfA for their contributions to this effort. The PROS was developed at SAO as part of the U.S. ROSAT Science Data Center (USRSDC) project and is supported at SAO by NASA through contract NAS5-30934. The USRSDC is jointly operated by GSFC and SAO, and a great number of scientists at both institutions have also contributed to the specification and testing of algorithms. We wish to thank those scientists for their contributions and also acknowledge the help of scientists at the Marshall Space Flight Center. Last, but not least, we would like to acknowledge the crucial help and encouragement of our colleagues of the NOAO IRAF and STScI STSDAS teams.

REFERENCES

VanHilst, M. 1990, *BAAS*, **22**, 935.
Worrall, D.M. and Wilkes, B.J. 1990, *Ap. J.*, **360**, 396.

THE STANDARD AUTOMATIC ANALYSIS SOFTWARE SYSTEM

(SASS)

Rainer Gruber

MPI fuer Extraterrestrische Physik
Giessenbachstr.2
D 8046 Garching

INTRODUCTION

As you all know, the basic categories of elementary particle physics are TIME, SPACE, CAUSALITY and every theory has to describe how these basic categories interfere with each other.

I claim that the same is true for the development of a software package that is designed to have as an input the continuous data flow from a satellite and as an automatically produced output all detectable sources with their characteristic properties.

We started in 1984 with the development of SASS and indeed if we look back on the last seven years, we must say that TIME, SPACE and CAUSALITY were the basic categories we had to fight with and which modeled our progress.

SASS has to process the POINTING as well as the SURVEY phase and the PSPC as well as the HRI. So we had a close collaboration with the Harvard Centre for Astrophysics (SAO) and, during the last year, with the Goddard Space Flight Centre (GSFC). Thus TIME SCHEDULES often imposed very hard constraints on our work.

Because of the requirement that the analysis of the data of one ROSAT day should take us less than one day, cpu and elapsed TIME PERFORMANCE governed our life.

SPACE: everybody here knows that I don't speak about rooms where people get packed more and more when the project evolves. It is DISK SPACE which imposes very severe restrictions on the processing when mass data have to be handled in a continuous way, and everybody who is involved in this job knows that this may have a very high impact on the overall peformance.

And CAUSALITY: an automatic processing requires a highly elaborate configuration management, we called it Steering System, to keep track of versions of programs and parameter sets in use, to keep track of a respectable number of files that are interconnected in various ways even when they were produced weeks apart.

Data Analysis in Astronomy IV, Edited by V. Di Gesù *et al.*
Plenum Press, New York, 1992

It requires the ability to automatically identify which data belong to which observation and have to be disentangled from or merged with other data. It requires an easy trace back of occurring errors as well as of the configuration change of data files induced by these errors. Thus mainaining CAUSALITY had to be one of our main goals.

I should mention that two more categories are necessary to describe software development completely. They interfere strongly with the foregoing ones: MAN POWER or better LACK of MANPOWER is the one and MONEY is the other category.

But before going into detail how these five categories interfered - which is an experience which seems to me to be the most interesting I can tell you - let me start with a short over-view what is the scientific aim of and what we are doing with SASS.

SCIENTIFIC AIM

Pointing

- deliver photon events that observers may use in Level 2 processing, i.e. with
  - corrected times
  - detector corrections included
  - screening for quality performed
  - boresight corrected aspect applied
  and with all appropriate housekeeping information and information on accepted times and screening parameters

- as an additional service perform a standard analysis which
  - detects sources
  - determines best positions
  - determines spectral properties, time variabilities and
  - crosschecks the found sources with sources from the catalogues available up to now

  It is clear that this kind of standard analysis is an optimized system for some averaged situation only, leaving the rest to the observer.

Survey

- deliver a full sky map, based on the analysis described above

These are the requirements which SASS is supposed to fullfill and it actually does.

BOUNDARY CONDITIONS

- processing time should be less than observing time to keep up with data production
- for the POINTING phase parallel processing should take place at GSFC under identical conditions

This last condition forbids patchwork solutions and enforced us to organize all software in a strictly release oriented form.

DATA FLOW AND OPERATIONS

Let me skip the details how we fullfilled the needs mentioned above. Let me just mention that after getting the EXABYTE with the dump data from GSOC, we split the data into streams that are related to HRI and PSPC processing and (in case of the PSPC) to POINTING and SURVEY processing and which are clearly separated from each other to insure overall integrity. The datafiles are split according to header and mission planning information. This processing step is called SAV processing.

The further processing is split up into a sequence of independend steps, which I will describe mainly for the PSPC:

In TEL processing we handle the time sorted data stream, extract events with position and energy, apply time and detector corrections and identify the observation intervals.

OBI processing handles each individual observation interval. To each event it applies the boresight corrected aspect, determines accepted times and screens the data according to predefined quality limits and calculates an exposure map.

In SEQ processing the events of observation intervals belonging to one observation get merged to build the observation master data set OMAS. The processing now mainly deals with maps referring to a definite sky position. Images are produced on which detect algorithms determine sources, extract the respective events, on which all operations mentioned above are performed.

Source detection occurs in several steps. A sliding window technique uses the surplus of events over local background. Punching out some circle around the sources gives a background map, which allows to determine another set of sources based on this smoothed background. This is performed for different energies and with different cell sizes resulting in a set of source lists which then get merged and matched to one list. A maximum likelihood algorithm takes the events from predefined circles around these source positions and determines exact position and flux. The original events around these positions are got from the OMAS file to form the base for the determination of the source characteristics.

The sources get identified with the help of our SIMBAD catalogue, which is an ectract from the Strasbourg catalogue especially produced for ROSAT. It contains 640 000 entries, specifying also supplementary information on spectral class, orientation of galaxies etc.

Whereas details may be different the overall scheme is about the same for the HRI.

SURVEY

Survey processing on the one hand shows a strong similarity with PSPC pointing processing. Nevertheless the different kind of satellite operation enforced several deviations, which in effect required an unexpectedly big additional amount of manpower. Instead of maps representing the detector field of view we represented data in strips containing a 2 degrees by 360 degrees area, adjacent on the equator and with increasing overlap in direction of and total overlap on the poles. For each event it has to be decided to which strip it belongs. Since the field of view is 2 degrees and it moves by 1 degree per day, 6 days data formally contribute to one strip. Since the maps become too big, we had to divide the strips up into 4 quadrants, resulting in an additional step for the Steering system, called SOP processing, where these quadrants are processed independently. Additional special handling is required by the poles, as may be seen from the fact, that formally the poles could only be processed after the whole survey phase, since every orbit is contributing. Every event belonging to the pole region is registered twice, once in the strip and in the pole area. This pole area encompasses a 17 degree diameter circle and is handled like a strip, getting divided into 25 quadrants.

As an additional service for MPE scientists in case of survey processing, we identify sources also with more than ninety user specific catalogues containing compiled sets of e.g. supernova remnants or galaxy clusters or e.g. a homemade set of all X-ray sources known up to now. Upper limits are calculated for those catalogue positions that have no counterpart.

SOFTWARE

To give you some first impression on the scope: HRI processing runs about 60 programs creating about 130 files, triggered by about 600 parameters. This is about the same for PSPC POINTING and SURVEY processing.

For the PSPC processing, MPE wrote 260000 lines code, 230000 lines being pure code, the rest being program documentation. The Steering System uses 90000 (50000) lines code. For subroutines that may be used by PSPC and HRI processing or which are of general use, 80000 (33000) lines have been written. In total these are 430000 (196000) code lines.

We built our own software configuration management system which controls all changes done by several programmers and which keeps track of versions and insures consistency. It is an elaborate system which proves to be extremely usefull. Joachim Paul has spent an enormous amount of energy to built and keep up this system. Whereas at the beginning of our S/W development these kind of systems were quite expensive, in fact too expensive for our budget, we are quite shure that today we would option for a commercial system. On the other hand the system as it is, is best adapted to our needs and especially when delivering a new release it proves very helpfull.

FOLLOW-UP PROCESSING

After processing, some temporary files get deleted, the rest will be put to archive. The Steering System manages a complete interface to the archive system to store the data on optical disk and to get from the archive what is needed.

In an offline process a selected set of files will be converted to FITS format, to be used by the observers. The Steering

System provides the necessary information to be able to easily get the required files from the archive.

DATABASES

We use the relational DBMS INGRES to
- log all processing information, especially on the interrelation between data file versions
- supply all information necessary to steer the automatic processing flow
- control processing parameters, restart conditions, consistency and completeness of the processing

More than 70 tables are used to fullfil this task. No data are stored in the database, except a few selected results as e.g. total accepted times.

Most of the external information needed by SASS is delivered via a separate database CALMIS. It contains the latest Mission Planning (MPL) data, corrected by daily protocols of the Attitude Monitoring and Control System (AMCS). It contains the radioactive calibration results, delivered by some detached process called Near Real Time Analysis (NRTA), which investigates housekeeping data and selected scientific data directly after dump. And it contains additional information like the results of boresighting or the continuation flags defining whether an observation has been closed or some observation slots are still missing, or information concerning the principal investigators.

A set of interface routines guarantees that the weekly update of these informations will be done in an identical manner at GSFC and MPE thus ensuring identical results.

DISK SPACE

Let me give you some figures relating to SURVEY processing to have a rough orientation.

One Master Data Record (MDR) got from GSOC contains about 70 - 90 Mbyte. Getting one MDR per day, these data are blown up in the course of processing to

MDR			70 - 90	Mbyte/day	
SAV			70 - 90	Mbyte/day	
TEL			70 - 90	Mbyte/day	
OBI	15 x 4	=	60	Mbyte/day	
SEQ			20	Mbyte/ 2 days, since 1 strip represents 2 days	
SOP	4 x 15	=	60	Mbyte/ 2 days	
TOTAL			340	Mbyte/day	

This corresponds to a blow up factor of 4 - 5.

HARDWARE

2 x	VS 3100 (16 MB)	each with	
		1 x	600 MB SASS local
		2 x	1000 MB SASS local

```
1 x      VS 3200    (16 MB)          with
                                  2  RV20  optical disks with
                                  2 x 1 GB
                                  4  800 MB disks global
                                  6  1,7 GB disks
                                     = 750 MB SASS
                                        750 MB FITSCON
                                          5 GB Archive
                                  1  Exabyte
                                  1  TK50
                                  1  Streamer Tape    6250 (BPI)


1 x      VS 3100    (24 MB)       INGRES SERVER
                                  FITSCON
        Network: Ethernet, LAT, LAVC
```

planned in addition:

```
    1  VS 3600 (Server)
    1  VS 3600/76  (fast processing)
    2  1.7 GB disks
```

On the 5 GB archive disks about a dozen days data may be stored for online use by FITSCON to convert the files to FITS format. It is clear that each time delay of FITSCON may have impact on the processing capability of SASS. Making SASS able to hold about 7 days on local disk was the option choosen to keep this potential source of trouble small.

PERFORMANCE

Figures are given for survey since most of our experience now is based on it:

```
    SAV            1.5  h/day
    TEL            1.5  h/day
    OBI        15 x 0.25 h/day
    SEQ            2.0  h/ 2days  since 1 strip represents 2 days
    SOP         4 x 3.0  h/ 2days
```

Thus to process one ROSAT day with one machine we need about 14 h/day.

EXPERIENCES

As time scheduling is concerned we must say: in most cases we had great difficulties to meet our goals. One experience seems to be quite clear: implement preliminary solutions as a short time help and it causes you a lot of work in addition.

Essentially there is a strong interference with lack of man power. Monetary reality forces you to stay always below what would be necessary. This seems to be the iron rule. The more evaluated the project the more the lack of manpower appears on the design and not on the programmers side. From a certain point of no return you are no more able to compensate, since you need more time to train a new person on the design level than to solve the problems yourselve.

On the other hand, such a big project has a lot of dark corners, where problems arise that were not foreseen by none of us and that became difficult to avoid. Performance difficulties forced us to change our processing environment from the VAX 7600 to a cluster of microvaxes. Getting rid of the problems of cluster network took an enormous amount of time of the personal involved directly as well as of testing people. Leaving the S/W development environment on the 7600 brought in the difficulty of having no adequate debug facility on the cluster.

A lot of internal requirements emerged in the course of processing, while testing and analyzing the outcome. Strong attaques on internal modularity and standard structured coding resulted which under the prevailing and ever lasting conditions of lack of manpower were not easy to fend off. But there is one big experience: you pay twice and more if you hurt the principles of structured and transparent programming for reasons of short time economy. Nevertheless there are situations where you know you have to do it and to pay.

Changing INGRES to release 6 made us feel like an INGRES Debug Department, destroying for some six weeks our possibility to deliver a new SASS release and wasting time with looking for workarounds. This is contrasted with the fact that otherwise we are quite happy with INGRES .

And still another point: Requirements seem to be small hydras, especially when they concern the ability to do something automatically. You just implemented the capability to archive the data of an observation intervall by pressing just only one menu item.In the meantime it is realized by somebody, that one observation in survey processing comprehends more than seventy observation intervals leaving you with the task to spare the operators seventy times to call the archive screen and to press this menu item. Thus new requirements evolve as soon as you fullfill the old ones.

SCIENTIFIC RESULTS

From the processing of 20% of the survey data we estimate to end up with more than 50000 sources, compared to about 5000 X-ray sources known today. This clearly depends on the threshold defined for source detection, which we took to be 4 sigma.

Up to now there are still some corrections of the aspect solution to be done, which we expect to finally allow for a source position accuracy of about 30 arcsec for the survey and about 10 arcsec for the pointing mode.

THE XUV WIDE FIELD CAMERA ON-BOARD ROSAT:

GROUND SYSTEM AND DATA PROCESSING

Alan W. Harris[1] and George S. Pankiewicz[2]

[1]Rutherford Appleton Laboratory, U.K.
[2]University of Leicester, U.K.

(Both currently at the WFC Quick-Look Facility, MPE, Garching)

INTRODUCTION

The Wide Field Camera (WFC) is a British XUV imaging telescope, rigidly fixed to the main *ROSAT* X-ray telescope and aligned parallel to it, which extends the wavelength coverage of the *ROSAT* mission to around 700Å. WFC data from the 6-month all-sky survey phase of the mission, which finished in February 1991, are still being analysed but it is already clear that the WFC has increased the number of known XUV sources by a factor of about 200.

The WFC survey generated around 50 Mbytes of data per day. Here we describe the WFC mission with emphasis on the ground system and the data-handling techniques employed. Some of the initial discoveries made with the WFC are briefly discussed.

THE WFC INSTRUMENT

The WFC is a joint project of the author's institutes, the University of Birmingham, the Mullard Space Science Laboratory and Imperial College of Science, Technology and Medicine, London. The telescope has an unobscured geometrical collecting area of 456 cm^2 which focuses radiation onto one of two identical curved microchannel plate (MCP) detectors. The resolution is 3 arcmin FWHM averaged over a 5° field of view. XUV filters provide continuous wavelength coverage from 60 – 220Å, and a further band around 600Å (Table 1). The survey was carried out in two bands, around 100Å and 140Å, with a field of view of 5°. The WFC has its own startracker which allows the celestial attitude to be derived to an accuracy of around 30 arcsec independently of the two startrackers carried by the German instrument. For a detailed description of the WFC instrument see Sims *et al.* (1990).

Data Analysis in Astronomy IV, Edited by V. Di Gesù *et al.*
Plenum Press, New York, 1992

Table 1. WFC XUV Bands.

Filter	F.o.V[1] [deg.]	Bandpass[2] [Å]
S1[3] (B_4C/C/Lexan)	5	60 – 140
S2[3] (Be/Lexan)	5	115 – 180
P1 (Al/Lexan)	2.5	150 – 220
P2 (Sn/Al)	2.5	540 – 720

[1] Non-vignetted.

[2] Bandpass limits correspond to 10% of peak transmission.

[3] These two filters were used on alternate days during the all-sky survey (two further science filters are carried on-board which serve as a redundant pair for S1 and S2).

THE GROUND SYSTEM

Data received from the WFC are recorded on tape at the German Space Operations Center (GSOC) at Oberpfaffenhofen, near Munich and, together with attitude data and WFC-relevant spacecraft data, are despatched by courier on a weekly basis to the U.K. Data Centre (UKDC) at the Rutherford Appleton Laboratory. The UKDC is responsible for checking the integrity of the data, attitude determination and initial calibration processing of the image data. The task of searching the survey data for sources was split between all five U.K. institutes involved. Each institute received regular batches of pre-processed data from the UKDC and had responsibility for source searching a particular area of the sky. The lists of sources from each institute are being combined to produce an all-sky EUV source catalogue.

During the guest-observer or 'pointed' phase, which follows the 6-month survey, the UKDC is responsible for distributing data products to guest observers. It is planned to include a basic standard analysis of the data, produced by the U.K. Survey Centre (UKSC) at Leicester University. The UKDC also provides tapes of pre-processed WFC pointed observation data to the German XUV Data Centre at Tübingen, from where WFC data will be distributed to German guest observers.

In addition to the five institutes in the U.K., the WFC ground system includes a 'Quick-Look Facility' (QLF), located at the Max-Planck-Institut für Extraterrestrische Physik (MPE), Garching. This data analysis facility receives WFC data from GSOC, via a fast link, shortly after reception and is responsible for monitoring the health and scientific data return of the WFC and compiling a first-cut EUV source catalogue. The QLF, manned by WFC project personnel, serves as the interface between the German and U.K. ground systems and has responsibility for issuing tele-commands for uplinking to the WFC via GSOC. For further details of the QLF see Harris (1988).

A schematic representation of the WFC ground system is given in Fig. 1 (see also Pye, 1988).

DATA PROCESSING

Photon events registered by the active WFC microchannel plate detector are recorded on the on-board spacecraft tape recorder in terms of their pulse height, arrival time and x, y detector coordinates. In addition, WFC housekeeping data and attitude data from the WFC startracker are written continuously to tape. The tape recorder data are read out to the GSOC (Weilheim) ground

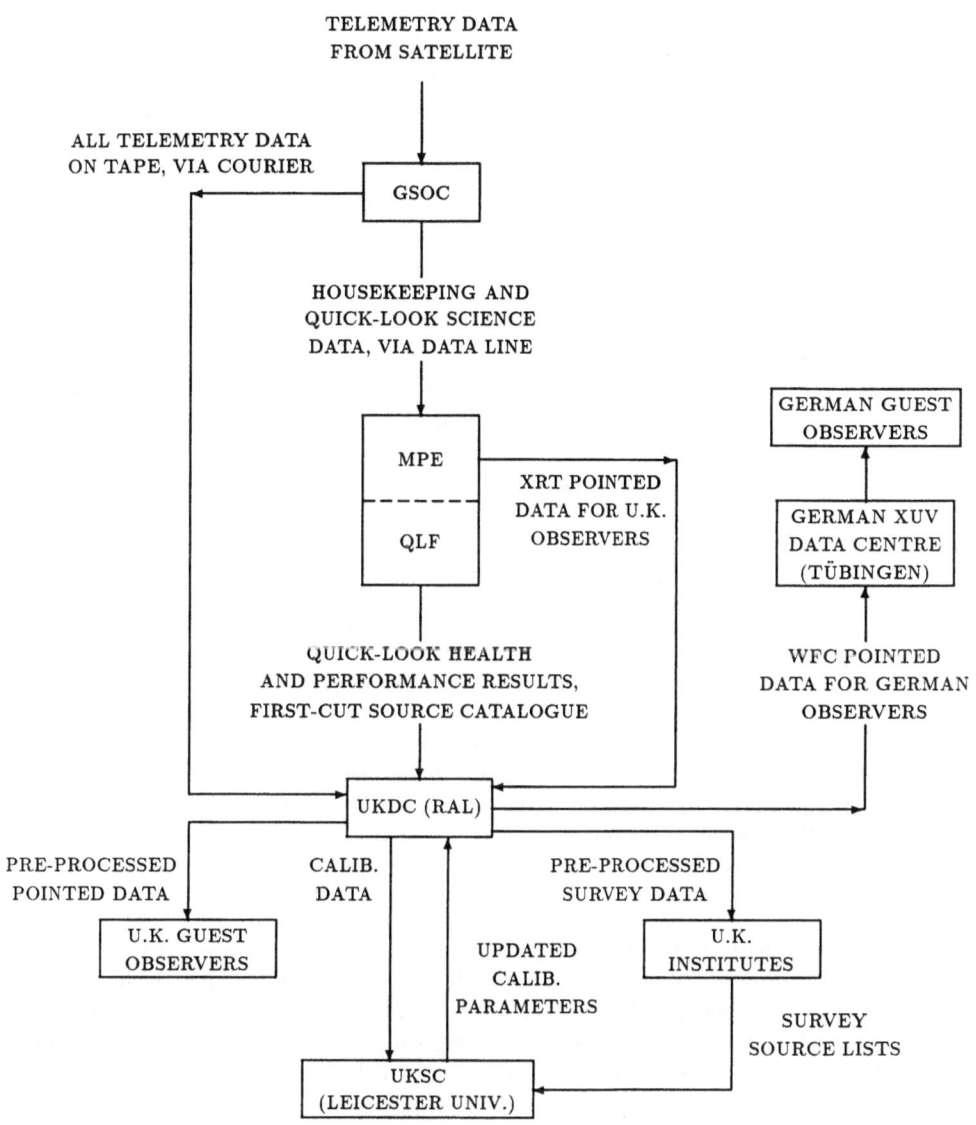

TELEMETRY DATA
FROM SATELLITE

ALL TELEMETRY DATA
ON TAPE, VIA COURIER

GSOC

HOUSEKEEPING AND
QUICK-LOOK SCIENCE
DATA, VIA DATA LINE

MPE

QLF

XRT POINTED
DATA FOR U.K.
OBSERVERS

GERMAN GUEST
OBSERVERS

GERMAN XUV
DATA CENTRE
(TÜBINGEN)

QUICK-LOOK HEALTH
AND PERFORMANCE RESULTS,
FIRST-CUT SOURCE CATALOGUE

WFC POINTED
DATA FOR GERMAN
OBSERVERS

UKDC (RAL)

PRE-PROCESSED
POINTED DATA

CALIB.
DATA

PRE-PROCESSED
SURVEY DATA

U.K. GUEST
OBSERVERS

UPDATED
CALIB.
PARAMETERS

U.K.
INSTITUTES

SURVEY
SOURCE LISTS

UKSC
(LEICESTER UNIV.)

Fig. 1. WFC ground system data flow.

station during the daily cycle of 5 or 6 contacts with the satellite. All the WFC housekeeping data and 20% of the science data (3 orbit's worth) are immediately forwarded to the QLF at MPE.

Processing of both WFC housekeeping and science data takes place under the control of dedicated 'pipeline' command procedures, with a minimum of human intervention being required. The pipelines used at the QLF and those used for processing the full WFC data return at the UKDC and U.K. institutes are very similar. The raw data (science, attitude, calibration and housekeeping files) received from GSOC each day are copied into a dedicated processing area and checked for consistency and integrity. Erroneous or corrupted data are flagged and the different types of data are re-formatted to facilitate further processing.

The Survey Phase

Survey science data processing, which is more complex than pointed observation processing, was carried out in the stages described below and shown in Fig. 2 in the form of a flow diagram. The QLF carried out all the processing stages each day as soon as the 3-orbit 'quick-look' data had been received from GSOC. In addition, orbital strip maps with celestial coordinates superimposed were generated each day at the QLF to enable an initial check for source detections to be made as part of the quick-look analysis. In the U.K., stage 1 was carried out at the UKDC and the following stages were performed at the 5 institutes involved, after receipt of pre-processed data from the UKDC.

1. An attitude solution for each event was calculated based on a detector linearization matrix, the time of arrival of the event at the detector, the positions of stars in the WFC startracker field of view at that time and the known alignment offset between the startracker and the WFC telescope. At the QLF, spacecraft attitude data are used to fill in gaps when WFC startracker data are not available, for example due to scattered light in the startracker field of view from the bright earth or moon. The UKDC makes use of the X-ray telescope attitude solution which is not available on the timescale required by the QLF.

2. After initial attitude and calibration processing, the survey data were added to a data reservoir held on disk which at any time contained events from a 360° band of sky being covered by the 5°-wide WFC scan path at that time (at the U.K. institutes, however, there was a delay of several weeks between receipt of the data at the groundstation and pipeline processing). As the scan path precessed, at the rate of 1° per day, regions of the sky for which coverage became complete were closed off and corresponding 2°x 2° image data-sets (so-called 'small-maps') were extracted from the reservoir and transferred to a separate processing area for analysis.

3. Since the scan path always crossed the ecliptic poles, the depth of coverage of a particular point on the sky depended on its ecliptic latitude. A source near the ecliptic equator would be scanned about 15 times per day (i.e. once per orbit) for a period of 5 days, after which the 5°-wide WFC scan path would have precessed beyond the source. At higher ecliptic latitudes small-maps remained active for longer. Small areas around the ecliptic poles were covered daily throughout the survey. In order to keep the active data reservoir to a manageable size, the polar regions with ecliptic latitude greater than 75° were artificially closed off and processed about once per week. The storage of event data from these regions was then started afresh.

4. The closed off small-maps were then processed to subtract the background and subjected to a point-source search routine. The maps were also displayed on a work-station to manually check for sources and anomalies. The algorithms used are part of an astronomical image display and analysis package called ASTERIX which is distributed by the U.K. STARLINK project. In addition, specialist WFC event sorting software allowed flexible manipulation of survey data in the reservoir so that images could be generated in detector coordinates, or for a given sky area in celestial coordinates, for any desired time period.

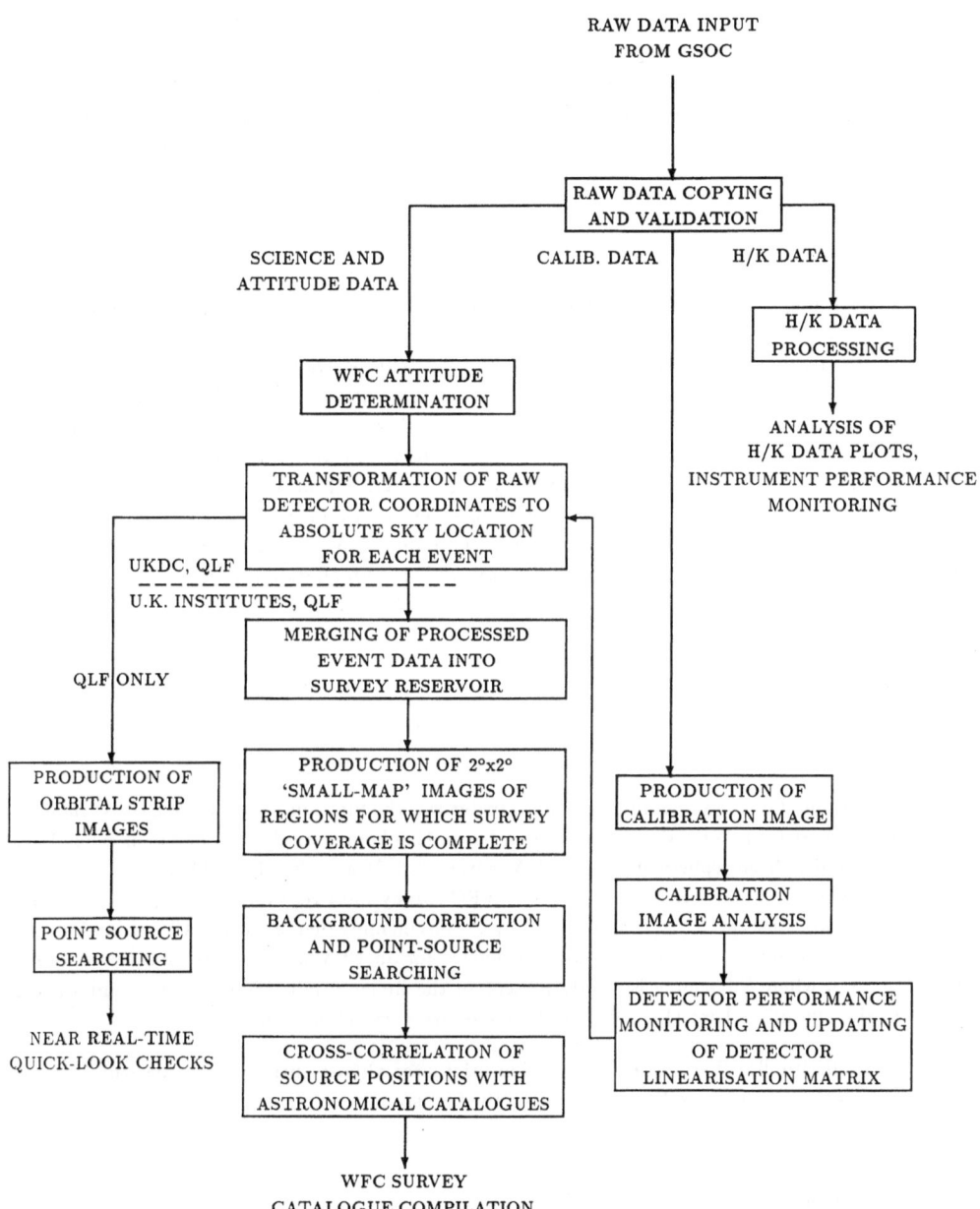

Fig. 2. WFC survey pipeline data processing.

5. The final stage of the survey pipeline processing was an automatic cross-correlation with a set of astronomical catalogues to check for positional coincidences with known astronomical sources.

The final survey EUV source count at the QLF, based on only 20% of the total WFC science data, is 138. Around 70% of these sources have been identified to date. At the time of writing, processing of the complete survey data return is nearing completion at the 5 U.K. institutes involved. The source lists produced are being merged by the UKSC into a single WFC all-sky survey catalogue. A uniform re-analysis of the survey data is planned in the longer term, in order to take advantage of calibration and software improvements which have been taking place continuously during the first catalogue compilation. It is anticipated that the final WFC survey catalogue will contain between 1000 and 2000 sources.

The Pointed Phase

The small-map scheme, which is similar to that used by the *Einstein* and *EXOSAT* projects, is also used in the pointed phase. The pointed observation small-maps are compatible with those used in the survey. During the pointed phase, guest observers will receive a standard data package from the UKDC/UKSC, consisting of pre-processed data together with an observation summary and possibly some basic scientific results as a guide to further analysis. Software for more detailed analysis is available to guest observers in the U.K. via the STARLINK astronomy network. In addition, a Guest Observer Centre, located at the University of Leicester, will provide dedicated data analysis facilities and experienced staff to support guest-observer work and provide advice on instrument characteristics, etc. In the future, once data enter the public domain, it is likely that an archive of *ROSAT* data will be set up, probably based on optical disks. Small data-sets may be accessible via the STARLINK network, while larger amounts of data could be requested on tape.

SOME INITIAL WFC RESULTS FROM THE QLF

V471 Tau

This eclipsing binary system in the Hyades cluster was the first source for which EUV variability was demonstrated from WFC data. The system has a 12.5 hour period and consists of a hot DA2 white dwarf and a chromospherically active K2V companion. It was already well known as a variable source of strong hard and soft X-ray emission. WFC survey data received at the QLF on August 9, 1990 from two consecutive scans across the object convincingly demonstrate EUV variability. An image produced from the first scan shows no evidence for an EUV source at the location of V471 Tau (Fig. 3a), whereas a source is clearly present in the image produced from the second scan, 1 *ROSAT* orbit (96 min.) later (Fig. 3b). Both images are from S2 filter data.

This result implies that the EUV radiation originates almost entirely from the white dwarf, which was in eclipse behind the K2 companion on the first scan (as confirmed by the orbital ephemeris of the V471 Tau system).

The Vela Supernova Remnant

The detection of this 15,000 year-old supernova remnant (SNR) with the WFC is the first ever detection of a SNR in the EUV. Extended EUV emission is apparent (Fig. 4) from a large arc of the 5° diameter SNR shell surrounding the Vela pulsar (marked by a cross in the figure). The EUV emission originates in gas at a temperature of a few million degrees, heated by the blast-wave from the supernova explosion. The emission has a patchy appearance, due presumably to the uneven distribution of ambient interstellar gas around the original supernova: the expanding

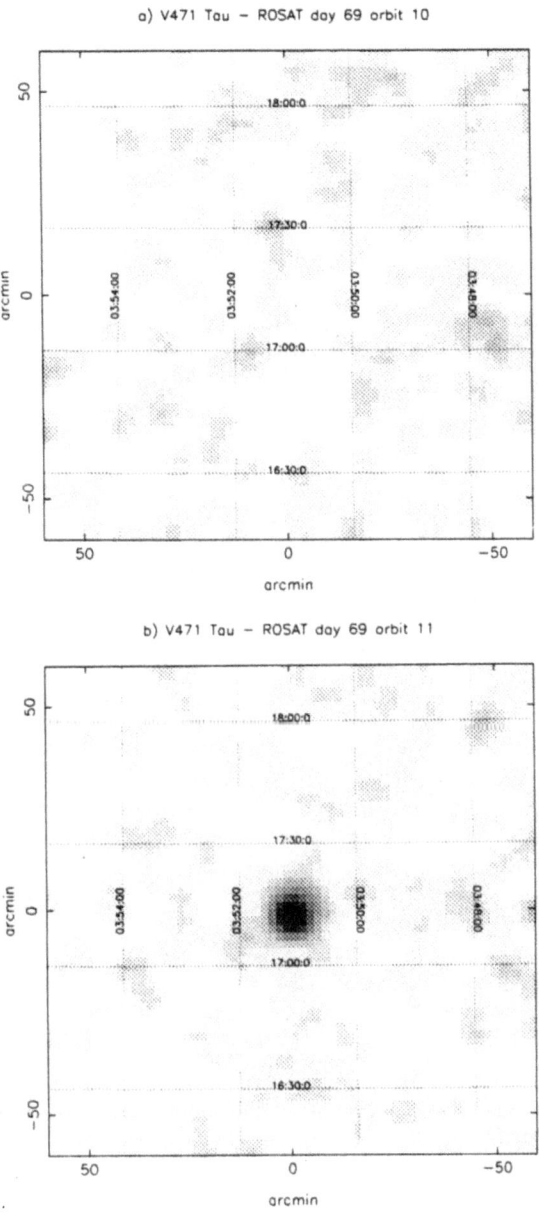

Fig. 3. Images obtained at the QLF of the eclipsing binary V471 Tau from two successive orbits during the all-sky survey. The white dwarf, which is in eclipse in (a), is visible in (b) one *ROSAT* orbit later. Coordinates are equinox 2000.

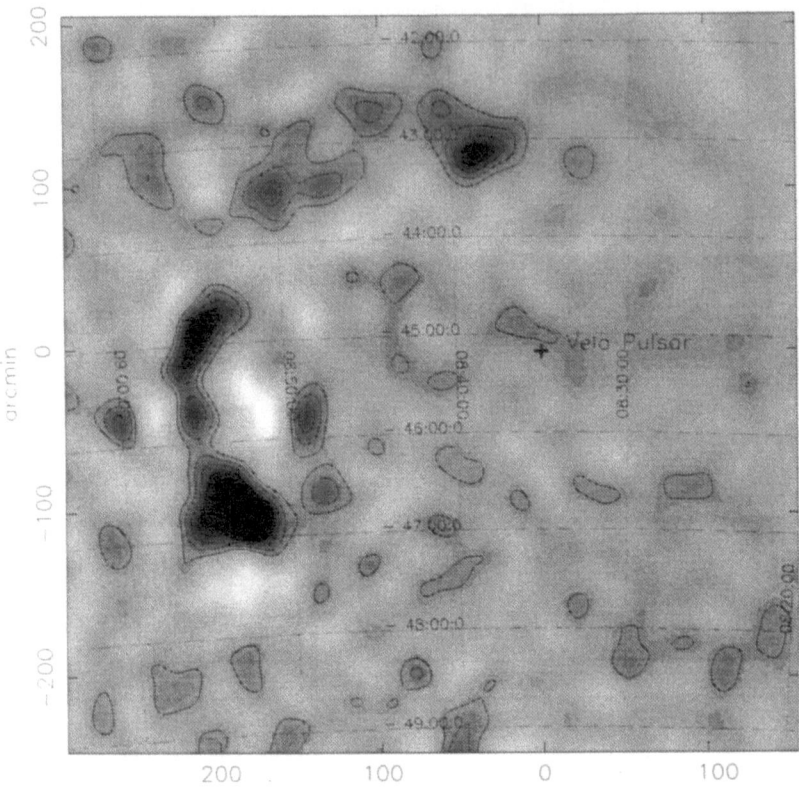

Fig. 4. A smoothed WFC image of the Vela supernova remnant from QLF survey data. Patchy EUV emission is apparent extending over about 120° of the shell to the east of the Vela pulsar.

envelope may have encountered more interstellar gas clouds in some directions than in others, leading to an irregular pattern of densities and temperatures in the shell we observe now.

By combining the WFC data with data from X-ray, optical and radio observations it should be possible to build up a clearer picture of the temperature and density distribution of gas in the Vela remnant and hence learn more about how expanding supernova remnants interact with the surrounding interstellar medium.

BY Draconis

This system is a non-eclipsing close binary (M0Ve+M0Ve), showing both flaring and a classical photometric wave. The mean photometric period (star rotation) of 3.84 days is shorter than the orbital one of 5.98 days, indicating that the system is relatively young since synchronization has not been attained. Large X-ray luminosities have been measured from the system (Golub, 1983).

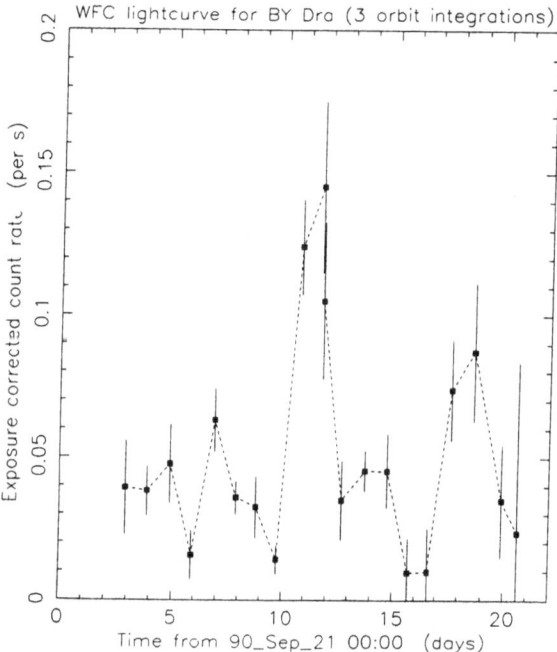

Fig. 5. A plot of exposure corrected count rates versus time for the BY Draconis system between September 23 and October 11, 1990 (S1 and S2 filter data combined). Flares, recorded simultaneously with IUE, are apparent on October 2 and 9.

In an attempt to gain multi-wavelength data of the system's flares, IUE observations were scheduled to be simultaneous with coverage of the object in the *ROSAT* survey, as part of the *ROSAT* IUE all-sky survey programme (RIASS). The WFC scanned BY Dra from Sept. 23 to Oct. 11, 1990, providing a total of 2.7 kilosec. of event data at the QLF alone. A timeline of the exposure corrected count rate (Fig. 5) clearly shows flaring activity on Oct. 2 and 9, which has since been correlated to flaring observed at UV wavelengths with IUE (M. Barstow and G. Bromage, personal communication). These results vividly demonstrate the need for simultaneous multi-wavelength observations of objects of this kind as a prerequisite to a clear understanding of the phenomena involved.

ACKNOWLEDGEMENTS

The success of the WFC project is due to the expertise and close cooperation of many scientists and engineers at the 5 participating U.K. institutes. On their behalf, we would like to express our gratitude to the *ROSAT* project staff at MPE and GSOC for their help in establishing the efficient operations and data interfaces between the U.K. and German sides of the *ROSAT* mission.

REFERENCES

Golub, L., 1983 <u>in</u>: 'Activity in Red-Dwarf Stars', P. B. Byrne, M. Rodono, eds., D. Reidel, Dordrecht.

Harris, A. W., 1988, <u>Journal of the Brit. Interplan. Soc.</u>, 41:353.

Pye, J. P., <u>ibid</u>, 337.

Sims, M. R., Barstow, M. A., Pye, J. P., Wells, A. A., Willingale, R., Courtier, G. M., Kent, B. J., Reading, D., Richards, A. G., Cole, R. E., Goodall, C. V., Sumner, T. J., Rochester, G. K., 1990, <u>Optical Engineering</u>, 29:649.

AN XUV SURVEY WITH THE WIDE FIELD CAMERA ON ROSAT

A.E. Sansom, M. Denby, R. Willingale, C.G. Page

Physics and Astronomy department,
University of Leicester, University Road, Leicester, UK

1 Introduction

The wide field camera (WFC) on board ROSAT has carried out an all sky survey in the previously unexplored XUV waveband (67 - 200 Å), in parallel with the x-ray survey (6 - 120 Å) carried out with the main telescope on ROSAT. This paper describes the processing and analysis of calibration and sky survey data (carried out at five consortium institutes in the UK), plus an overview of survey results. The XUV survey was carried out in two filters, S1A (67 - 137 Å) and S2A (112 - 200 Å), by interchanging filters once a day. The WFC has a 5° field-of-view (fov). In survey mode a strip of 360° in ecliptic latitude by 5° wide is scanned once every satellite orbit round the earth ($\sim$ 96 minutes), and this strip moves $\sim$ 1° per day around the ecliptic. Thus the interchange of filters at a rate of once per day is sufficiently rapid as to produce survey sky coverage, without gaps, in each filter. The WFC is a grazing incidence telescope, with a microchannel plate imaging detector. For descriptions of the instruments see e.g. Sims et al 1990, Barstow and Sansom 1990.

The ROSAT science analysis system in the UK is based on Starlink utilities. In this way the software is readily available to nearly all astronomers within the UK community. ASTERIX, the core of programs at the centre of the system, is an instrument-independent application suite with most programs operating on data stored in a flexible format in a readily accessed manner using the Starlink Hierarchical Data System (HDS). For ROSAT purposes interfaces into the raw satellite telemetry and instrument characteristics of the WFC are provided via event processing and calibration software packages.

The WFC survey telemetry data from the satellite were sent from Germany to Rutherford Appleton Laboratory (RAL) where the data were processed into lists of photon event positions, time tagged to 30 msec accuracy. An aspect solution was applied to convert these event positions to celestial coordinates, using data primarily from the WFC startracker, and from the main telescope (XRT) startrackers when no WFC startracker information was available. Section 2 gives an overview of the calibration data analysis, with particular reference to the detector point response as used during source searching. Section 3 describes the processing of WFC survey data at the consortium institutes in the UK. Section 4 describes some overall survey results. Section 5 gives examples of individual cases of the survey data detections, and conclusions are given in section 6.

Data Analysis in Astronomy IV, Edited by V. Di Gesù et al.
Plenum Press, New York, 1992

2 WFC Calibration Data Analysis Overview

2.1 Data Products

To get the most information from the WFC results we need accurate knowledge of how the instrument is functioning. Some in-orbit WFC calibration observations were acquired early in the mission, prior to the survey. These observations consisted mainly of a series of pointings close to the hot, hydrogen rich white dwarf HZ 43, which is one of the few sources with a measured spectrum, from EXOSAT observations. The following instrument features were calibrated using ground based data, and\or with analysis of in-orbit data.

- The detector readout is not linear. Therefore, from ground based measurements a look-up table was derived to convert detector electronic coordinates to event incidence angles relative to the instrument axis. The accuracy of this look-up table is checked at regular intervals in-orbit, using a pattern of known spot positions, illuminated by a UV lamp, and has been updated since launch.

- On-axis efficiency was measured as a function of source energy, and results from ground tests at monochromatic energies are being compared with results from in-orbit observations of sources with previously measured XUV spectra.

- Overall instrument vignetting was derived from ground based measurements on individual instrument components (detector and mirrors) as well as the integrated WFC. In-orbit source flux measurements at several off-axis positions show reasonable agreement with the smooth function derived from ground based data.

- Response of the instrument to point sources, as a function of source energy and off-axis position was derived from ground based data, and checked with in-orbit observations. This analysis is discussed in section 2.2.

- Boresighting offsets and aspect solution for the WFC pointing as a function of time are obtained with the use of the WFC and XRT startrackers. Identifications of XUV detections with catalogued sources have been used to determine and improve the accuracy of the aspect solution (see section 3.3).

- Deadtime corrections were derived as a function of imaged event rate into the instrument, for low (< 200 c/s) and high (< 400 c/s) telemetry data acquisition rates.

- Background analysis of survey data indicates that most of the background is from particles, which are unaffected by instrument vignetting. The intrinsic detector background is very low, and the in-orbit observed background can fall to values lower than were typically expected before launch. The contributions to the background vary with time, as well as with position on the sky. A separate analysis to understand the contributions to the WFC background is currently in progress.

Data products from the above calibration analyses (apart from the background analysis) are stored, as tables of values and coefficients, in a Starlink HDS file (the master calibration file or MCF). In this file allowance is made for changing values as a function of time, by indexing calibration data products on modified Julian date. This allows for real instrument changes. Changes in our understanding of the instrument calibration can also occur, and in this case any improved calibration parameters are written over the previous versions. These data products are accessed through a suite of subroutines, which provide the user with a reasonable interpretation of the calibration data (eg. interpolated, as a function of time) for use in analysis of WFC images, fluxes, time series etc.

2.2 Analysis of the WFC PSF

An analytic form was found, to describe the response of the WFC to point sources. This analytic description has the advantage that such a point spread function (PSF) can be evaluated with any centre with respect to the pixel boundaries, at a range of energies and off-axis source positions. This is useful for the WFC where the peak of the point response is undersampled, and the response changes with off-axis position due to the optical aberrations inherent in the telescope optics. The overall response of the WFC to point sources has a sharp central peak and broad wings. Source contours are circular near the fov centre, and become broader, elongated and distorted off-axis. The analytical form consists of a Moffat profile (Moffat 1969) (similar to the surface brightness profile characteristic of elliptical galaxies). An ellipticity is allowed for off-axis source positions, and an exponential component is added to the profile providing some further improvement to the description of the point response, on-axis:

$$z = \frac{1}{(1+Q)} \left[\frac{(I-1)}{\pi f a_\alpha^2} \frac{1}{(1+(r/r_\alpha)^2)^I} + Q \frac{exp(-r/r_\beta)}{2\pi f a_\beta^2} \right] \tag{1}$$

Where: $r/r_\alpha = \frac{\sqrt{x^2+y^2/f^2}}{a_\alpha}$, similarly for r/r_β.

x,y = coordinate system aligned along the PSF major and minor axes.

f = ellipse axial ratio (minor/major) for off-axis PSF.

a_α = scale length for Moffat contribution.

I = power index of Moffat contribution.

Q = relative normalisation (fraction of counts in exponential contribution: 0 off-axis).

A 2-dimensional fitting routine was used to derive best estimates for the coefficients (x,y centre of PSF, a_α, a_β, I f and Q) at a range of monochromatic energies and off-axis positions, from ground based data. The coefficients were found to be a weak function of energy and a stronger function of off-axis spot position. The PSF is stored in the MCF as a table of coefficients. MCF access routines allow the user to obtain values of the PSF (as a function of source energy and off-axis position) at a single point from the PSF centre or as a 2-d array of values.

Early in-orbit calibration observations of HZ 43, have been used to check the ground based PSF. The agreement is good, especially for data on-axis (as shown in figure 1 - corresponding to $\sim$ 124 eV). Similarly the expected PSF for the survey is plotted in figure 2. The survey data plotted in this figure is an average of 14 point source survey observations. The expected PSF agreed well with the narrowest of these, indicating possible additional broadening of some of the sources by residual aspect solution errors, or incomplete detector sampling. Survey sources consist of a superposition of source events from different off-axis pointings across the detector fov. Therefore sources which do not sample the whole detector field are likely to be broader than expected. Figures 1 and 2 also show, for comparison, a Gaussian, normalised at the 68% energy width (dotted curves). This illustrates the non-Gaussian shape of the WFC PSF.

3 Survey Data Processing at the UK Consortium Institutes

3.1 Data Distribution and Initial Processing

During the six month ROSAT survey, a processing 'pipeline' was in operation. This started at the UKDC, where Rosat WFC Observation Datasets (RWODs) were generated on a daily

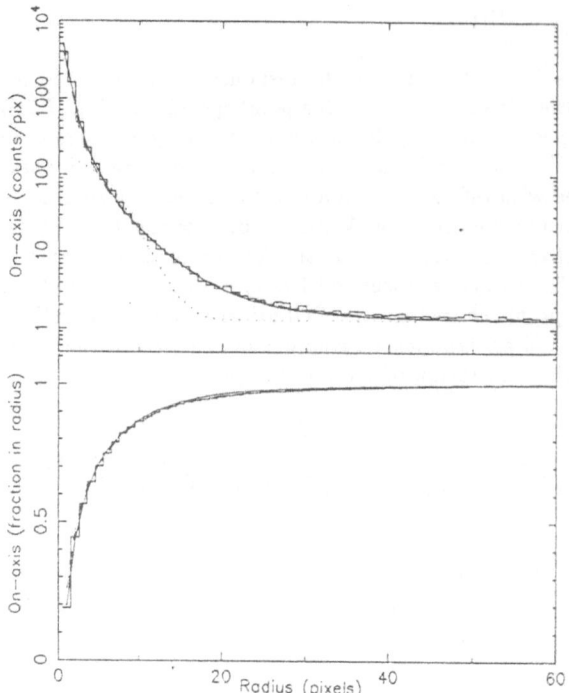

Figure 1. On-axis differential and cumulative point response curves, showing the good agreement between data (histogram) and PSF models from ground calibration (dashed curve) and in-orbit observation of HZ 43 (solid curve). A Gaussian normalised to the 68% energy width is shown for comparison (dotted curve). The pixels are 36 arcseconds for these pointed data.

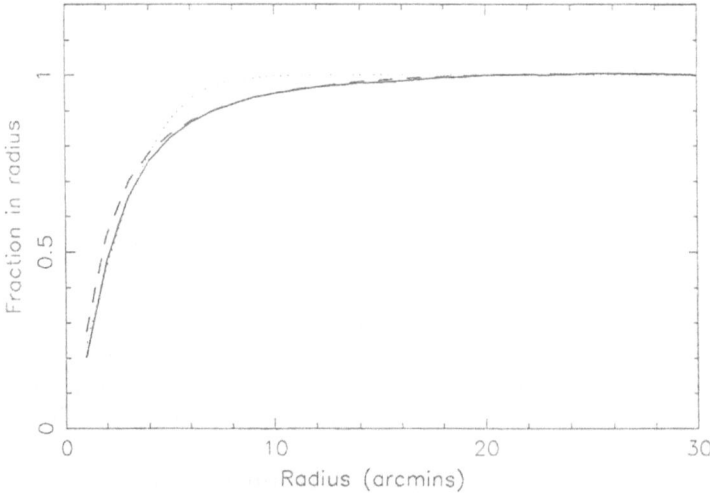

Figure 2. Survey average cumulative point response. Here a background has been subtracted from the model (dashed curve) in the same annulus as for the data (solid curve) (radii between 20 and 30 arcmins), in order to compare like with like. There is about 2% of the PSF in this annulus, for both data and model. A Gaussian normalised to the 68% energy width is shown for comparison (dotted curve). The pixel size is 1 arcminute.

basis from the raw telemetry stream. The RWODs contain pre-processed (unpacked, linearized and aspect corrected) event files together with the associated aspect and housekeeping files. These RWODs were distributed, as Backup save sets on tapes, to the five consortium institutes jointly responsible for the WFC construction, operation and survey data analysis in the UK. These are: Mullard Space Science Laboratory (MSSL), Leicester University (LU), Birmingham University (BU), Imperial College London (IC), and RAL. The sky is divided for the purpose of analysis at these institutes. This distributes the task of processing the survey data, and aims for a uniform analysis of the sky by making the same processing software available to each institute.

Starlink HDS files, used for storing processed ROSAT data in the UK, allow deeply nested, self describing structures, and a wide range of data types. The HDS datasets which ASTERIX uses are of two major types. The event dataset is essentially a photon list with position and timing information about each photon detected by the instrument. Images and time-series datasets can be generated using an event sorting program. The WFC event sorting software is interfaced with ASTERIX and fulfills two roles. For pointed mode data it creates ASTERIX datasets from the raw telemetry stream and allows exposure correction of images and time series from raw counts into fluxes normalised to the on-axis response of the WFC. For the survey mode it provides, in addition, a database management facility which ensures that the same areas of sky are merged together to give maximum coverage at any point and that the same regions of sky are removed from the system once their scan coverage is completed.

At each consortium institute the RWOD data were initially sorted into a reservoir file of events, in which overlay of events from a given sky region was achieved. Images of 2 by 2 degrees and 1 arcminute resolution were sorted in ecliptic coordinates for sky areas with completed survey coverage (in ecliptic mid-latitude regions), or at regular intervals for sky regions exposed for a large fraction of the survey (near the ecliptic poles). These ASTERIX images, and the event lists which formed them are referred to as 'small maps'. The images were searched for sources and checked for anomalies. At the end of processing for each RWOD the reservoir was updated to remove the analysed areas of sky and the whole process repeated for the next RWOD in sequence. Typical count rates for the survey were ~ 20 to 100 c/s in S1A, and ~ 40 to 120 c/s in S2A, with higher count rates close to the south atlantic anomaly, and other regions of high particle fluxes (where the detector gets switched off). Most ($\sim 2 \times 10^8$) of the imaged events come from particles, which cannot all be screened from the detector. Some ($\sim 5 \times 10^7$ c/s) are likely to be from the XUV background, and some ($\sim 3 \times 10^5$ counts) arise from sources. This is different from other wavebands, where most of the events are from discrete, astrophysical sources. There is about 1 source per 10 small maps in the WFC survey.

3.2 Source Searching

Source searching was carried out by two programs using entirely different techniques. These are described below:

    i) using a matched filter for point sources (PSS).
    ii) using a sliding box with a threshold set by Poisson statistics (NIPS).

i) PSS - this program uses a *matched filter* technique, which for point sources corresponds to cross-correlating the background-subtracted data with the telescope PSF. The first step is to estimate and remove the background, globally from each image. This is complicated by the presence of bright sources, by gradients in exposure time across the image, and changes in particle rate around the orbit. The way in which overlapping images are formed from the events reservoir also produced blank strips along the edges of some images. After allowing for sources and blank areas, the background is fitted to a set of cubic splines. The cross-correlation is computed at unit pixel spacing, producing a significance map, and then repeated with a finer grid spacing, around significance peaks (> 5 sigma), to locate sources. The exact source positions are determined using the Cash statistic (Cash 1976). Source fluxes are then calculated from χ^2 fits to find the normalisation of the PSF for each source. PSS uses the PSF from the MCF in the form of PSF fractions under a 2-d array of pixels. PSS outputs a source position, flux, associated errors and a significance. The matched filter technique has the advantage of maximizing the sensitivity of the source searching algorithm, for well defined source shapes. However, it works best for signals of known shape in Gaussian noise, whereas the background levels in the WFC survey images were typically in the range of 0.5 to 1.5 counts/pixel, rather firmly in the Poisson regime.

ii) NIPS - An alternative *sliding box* method was also developed which sums the events within a circular box of radius 2.5 arcminutes around each pixel, and then computes the Poisson probability of getting this total, given the local background level. This program computes the background level by first summing counts over the whole image in blocks of 8 by 8 pixels (one arcminute/pixel) and then applying a median filter to sets of 3 by 3 of these block means. When significant flux is detected the background is estimated again using an annulus of 6 to 18 arcminutes radius around the position, and the significance is re-computed. Once located roughly, the exact source position is determined by maximum likelihood using the Cash statistic. A circular Gaussian function, with variable width, is fitted to find source locations and widths. The source counts are then summed from a circular box of radius 6 arcmins then corrected for the WFC PSF fraction lying outside the box. The source significance is expressed as the negative logarithm of the probability (from Poisson statistics) of exceeding that level with random noise. A significance threshold of 3 plus a count rate limit of > 10 counts was applied for a source detection. The source detection threshold corresponded to a theoretical Poisson false-alarm rate of 1 in 1000 per image, assuming a uniform background. Source positions, fluxes, associated errors and a significance are output. NIPS searches for sources in the S1A and S2A survey images, then in the sum of these two images (bringing to light a small number of sources which would otherwise have been missed). The sliding box technique has the advantages of speed and robustness, plus it works with sums of unweighted events, governed by Poisson statistics, appropriate for WFC images.

The results from both programs were converted into files suitable for use with SCAR (the Starlink Catalogue Access and Reporting package). Upper limits were computed for sources only detected in one filter.

All images were also inspected by eye to check on general data quality, to ensure that no obvious sources were missed, and to find extended sources (such as supernova remnants) to which both automatic methods were less sensitive. All detected sources were then classified in three categories:

- Spurious - any source clearly affected by a data or background-subtraction defect or on the very edge of an image. The Moon also appeared in the field every 14 days and was treated as a spurious source.

- Doubtful - if the data look fine, but the source detected by one of the source searching routines is not evident to the human eye in either filter.

- OK - clearly detectable by eye and with no obvious defect. Such detections are noted as strong, medium or weak, to give a rough estimate of the source brightness.

3.3 Point Source Positional Accuracy

There are several contributions to the positional uncertainty of point sources. Residual aspect solution errors are now at the level of 14 arcsecs root-mean-square (rms), after improvements to the aspect solution made throughout the survey. Linearization of detector event positions is accurate to $\sim$ 12 arcsecs rms (with the maximum errors at the detector edges). The detector resolution is $\sim$ 20 arcsecs rms. Event positions are stored to a positional accuracy of 7 arcsecs (in order to blur out any Moire fringing which would otherwise appear in the images). Positional uncertainties from the point source search algorithms, due to finite source and background counts, and a finite point response width, contribute between $\sim$ 3 and 100 arcseconds (90% confidence error circle).

The above uncertainties add in quadrature to give an overall uncertainty of between 40 and 108 arcseconds (at the 90% confidence level). Offsets between identified XUV source optical positions (e.g. from the Hubble guide star catalogue), and XUV positions, were measured using the best estimates of the XUV positions. These show an rms scatter of 41 arcsec about the XUV mean, and a 90% radius of 50 arcsecs, in agreement with the expected errors from the individual contributions summed in quadrature.

3.4 Cross-Correlation with Other Catalogues

The detected XUV source positions are cross-correlated with a number of existing catalogues, in optical and other wavebands, stored in SCAR format. From these cross-correlations, potential identifications are obtained, within a certain positional offset from the XUV detection position. Information about the XUV sources and any potential identifications are stored in a SCAR catalogue. This is known as the WFC 'S2' survey catalogue. (The 'S1' survey was carried out on $\sim$ 1/5th of the survey data, sampled at the WFC Quick Look Facility in Munich). The survey processing will be redone as the 'S3' survey, with the improved knowledge obtained during the S2 analysis (e.g. in aspect solution, timing, deadtime corrections, point source searching, source quality flagging, instrument efficiency calibration etc). The S2 survey catalogue of XUV detections, available to the consortium institute members, has been used for the summary of survey results described in section 4.

4 Survey Results

4.1 Overview of sources detected

So far $\sim$ 90% of the 6 month survey has been processed, searched for sources with NIPS, and catalogued. Source searching with PSS is also underway. A summary of sources found so far with NIPS is given below (excluding detections noted as spurious).

Number of XUV detections (total)	=	1171	
(with detection significance > 4.0)	=	413	(in S1A or S2A)
"	=	327	(in S1A)
"	=	285	(in S2A)
"	=	197	(in S1A and S2A)

Information on potential identification with objects in existing catalogues will be analysed and manipulated with an algorithm then checked by hand to produce a subset of identified XUV sources. The criteria which will be applied to obtain probable identifications with the algorithm are as follows:-

- The XUV detection is not noted as spurious.

- The catalogued position of the potential identification is within 1.5 arcmins of the XUV source position (for XUV positions derived from the improved aspect solution).

- The potential identification is with one of:

  - A hot (DA,DO,or DB) white dwarf.
  - A catalogued hot sub-dwarf.
  - A cataclysmic variable.
  - A coronally active star or star system (RSCVn, contact binary).
  - A BY Dra type flare star system or dwarf flare star.
  - A known nearby star (within ~ 25 pc), or high proper motion source.
  - A star brighter than $m_V = 9$th magnitude.
  - A supernova remnant.
  - An x-ray source.
  - A pulsar.
  - An active galaxy or quasar.
  - A variable star.

The probability of chance coincidence with any of the above sources is generally negligible. The most populous category by far is that of stars brighter than 9th magnitude, with which there is less than $\sim 0.6\%$ probability of chance coincidence, within a circle of radius 1.5 arcmins. Identifications with other categories of sources, which are not expected to be detected with the WFC are checked for manually. In particular there are several cross-correlations with normal galaxies of magnitude ~ 16. However, the handful of such cases is consistent with chance coincidence of catalogued galaxies. Therefore there is no clear evidence so far that any normal galaxies have been detected with the WFC. Three examples of identified point sources are shown in figure 3.

Identifications of sources from existing catalogues is underway. Here we show the distribution of source types, in figure 4, for the nearest potential identifications, within 2.0 arcmins (prior to the application of the optimum aspect solutions to all survey sources). All catalogued source classes are generally found to be well represented by this nearest neighbour criterion, (with the exceptions of the ~ 11 identifications with normal galaxies, and about 10 of the ~ 200 main-sequence stars, which are likely to be chance coincidences).

Roughly a third of the XUV detections have a probable identification with an already catalogued source. Most of these are late-type stars or star systems (including RSCVns, interacting binaries, cataclysmic variables, coronally active stars, flare stars etc.) about 32 are associated with stars in the late stages of stellar evolution, and a handful are close to extragalactic sources. A further ~ 191 cases have less likely identifications at a distance of between 2 and 5 arcminutes from the XUV detection. A plot of XUV flux ratio (S1A/S2A) against S2A flux is given in figure 5 for four different categories of detected sources. This figure shows an overall correlation and some segregation of source types with XUV flux. Main sequence stars and star systems tend to have lower XUV fluxes than the identified white dwarfs. These white dwarfs tend to have high count rates particularly in S2A, since they are generally nearby, and therefore less affected by interstellar absorption which tends to cut down more flux in the waveband covered by S2A. Conversely, the identified active galaxies tend to have insignificant S2A count rates.

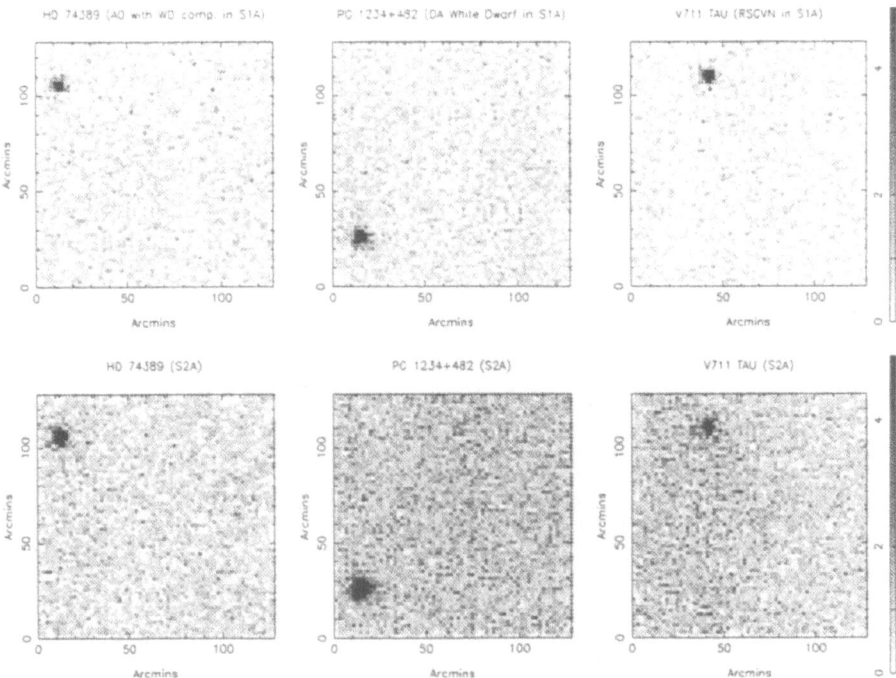

Figure 3. Examples of 6 WFC survey small maps showing 3 detected XUV sources in both S1A (upper panels) and S2A (lower panels) filters. An A star found to have a white dwarf companion is identified with the first source. The second source was classified as a hydrogen rich sub-dwarf (sdB), but optical follow-up spectroscopy has since shown it to be a hot hydrogen rich (DA) white dwarf. The third source is identified with a catalogued RSCVn.

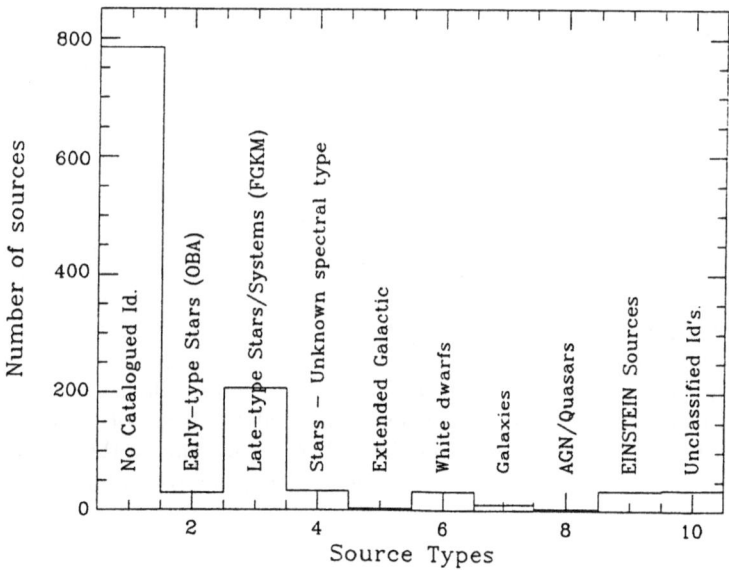

Figure 4. Histogram of nearest catalogued sources, within 2 arcmins from XUV detections showing different categories of sources.

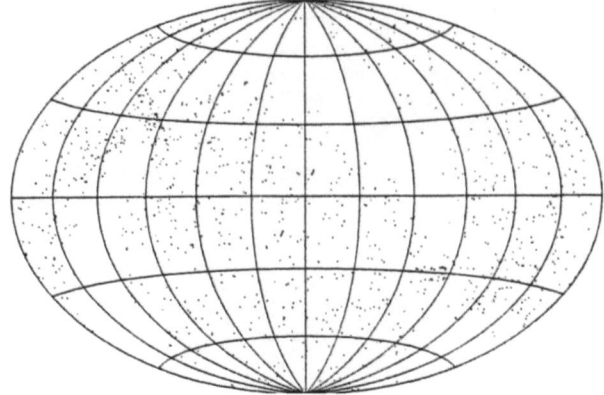

Figure 6. Positions of XUV detections, in Galactic coordinates. Blank regions are unprocessed or unobserved.

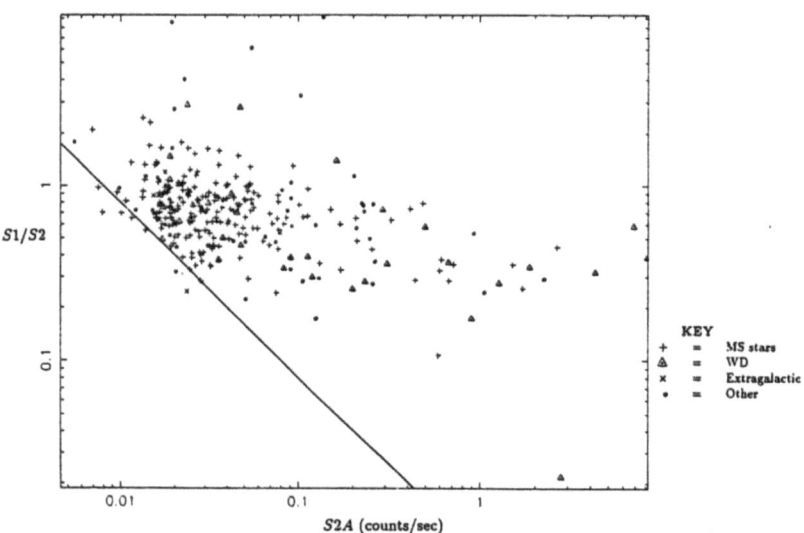

Figure 5. XUV filter ratio (S1A/S2A) plotted against S2A filter flux, for several categories of sources. This plot shows some segregation of source types, and overall trends. A lower envelope will exist in this plot due to source detectability in each filter. The solid diagonal line shows an estimate of this limit, for the expected source detectability levels of ~ 0.016c/s in S2A and ~ 0.008c/s in S1A, relevant to a typical mid-latitude exposure time in the survey of about 3000 seconds (~ 38 satellite orbits with a source in the fov for about 1.25 minutes per orbit).

Source temperatures will also affect the distribution of sources in this plot. The distribution of XUV sources on the sky is shown, in Galactic coordinates, in figure 6. This figure shows the sky areas analysed at the time of writing, with maximum exposure (and peaks in source numbers) at the ecliptic poles. Detections of the moon are evident in short lines along the ecliptic equator; these will not appear in the final point source catalogue. Increased background tended to occur at intervals during the survey, especially near the ecliptic poles, therefore the peaks in source numbers here may reflect peaks in the particle background variations rather than a real increase in XUV source detections.

4.2 Expectations for Unidentified Sources and Optical follow-up

Many of the unidentified XUV sources will turn out to be hot white dwarfs, with hydrogen rich atmospheres. This is expected since the catalogues of known white dwarfs only cover certain areas of the sky. Primarily the Palomar Green survey of UV excess objects (Green et al. 1986), which identified many hot white dwarfs, only covers ∼ 25% of the sky. XUV sources identified with some main sequence stars may turn out to be binary companions, including white dwarfs. In particular the identifications with isolated A stars are likely to be caused by an otherwise undetected binary companion. New interacting binaries, flare stars and coronally active star and star systems and a few active galaxies are expected to be amongst the types of sources discovered.

Identifications with sources, other than those made from existing catalogues, is underway. A series of optical spectroscopic observations, on the Isaac Newton Telescope, is being carried out. Using finding charts made from Schmidt plates, sources within the XUV error circle are observed at low resolution initially, and then at higher (∼ 2 Å) resolution if required for confirmation or additional information. The results are assessed in near real time, at the telescope, in order to minimize the observation time needed to make an identification. In addition individuals with telescope time on other telescopes are taking part in identifying or confirming likely optical counterparts to XUV sources.

5 Individual Cases

5.1 XUV Detections of Misclassified Sub-dwarfs

Hot sub-dwarfs (∼ 30,000 K (sdB) to 50,000 K (sdO)) are stars in the late stages of stellar evolution, which have not yet become completely degenerate. Therefore their surface gravities are lower than those of the white dwarfs which they will evolve into. The evolution of hot sub-dwarfs is not well understood (eg. Vauclair and Liebert 1987). Better constrains on the physical parameters of these stars would lead to a better understanding of their evolution. We expect sub-dwarfs to emit XUV radiation, and so we might look for examples with the WFC, and, with the application of model atmosphere codes, constrain their physical parameters (atmospheric composition, temperature and surface gravity). However, their space density is low, so the expected likelyhood of detecting sub-dwarfs with the WFC is low. So far, about five cases of catalogued sub-dwarfs have been identified with XUV detections. Optical spectra obtained for three of these clearly show them to be hot DA white dwarfs. Figure 3 shows one bright XUV example (PG 1234+482). The widths of the hydrogen absorption lines in the optical spectra are too broad for them to be sub-dwarfs. The other two cases are still under investigation. It seems that the WFC is effective at picking out hot white dwarfs, misclassified as sub-dwarfs in existing catalogues (in particular, the PG catalogue). These identifications will add to the sample of white dwarfs, and thus improve our understanding of the white dwarf luminosity function, which is important to understanding when stars in the local neighbourhood formed.

5.2 XUV Detections of Companions to A stars

The WFC survey has revealed several cases of XUV sources associated with catalogued early A stars. This was unexpected since these stars were some of the few classes of stars undetected in x-rays, by the Einstein satellite. In addition, WFC calibration observations of bright A stars showed no sign of any XUV emission. Therefore it is probable that these XUV emitters are otherwise undetected companions to the A stars. It is likely that these companions are white dwarfs (like Sirius A/B, which is one of the strongest XUV sources detected in the WFC survey). An example (HD 74389) is shown in figure 3.

6 Conclusions

The first ever all sky survey in the XUV waveband is now virtually complete. Analysis of the survey data is well underway, and will be improved with knowledge gained. Many hundreds of new XUV sources have been discovered, and some have now been optically identified, turning up previously uncatalogued white dwarfs, RSCVn and active stars, emission line galaxies etc. The survey will lead to a catalogue of XUV sources, a map of the XUV sky, and improved understanding of many individual objects and categories of sources.

Acknowledgements

There are many people who have contributed to the success of the WFC on ROSAT. We would like to acknowledge the contributions of the WFC hardware and software teams at the five UK consortium institutes. We would like to thank J. Holberg for obtaining optical spectra of three white dwarfs.

References

Barstow, M.A., Sansom, A.E., 1990, In "EUV, X-ray and Gamma-ray Instrumentation for Astronomy", SPIE Vol.134A

Cash, W., 1976, Astr. Astrophys., 52, 307

Green, R.F., Schmidt, M., Liebert, J.W., 1986, Astrophys. J. Suppl. Ser. 61, 305

Moffat, A.F.J. 1969, Astr. Astrophys, 3, 455

Sims, M.R., Barstow, M.A., Pye, J.P., Wells, A., Willingale, R., Courtier, G.M., Kent, B.J., Reading, D., Richards, A.G., Cole, R.E., Goodall, C.V., Sumner, T.J., Rochester, G.K., 1990, Optical Engineering, 29, 649

Vauclair, G., Liebert, J., 1987, In 'Scientific Accomplishments of the IUE', Y. Kondo (ed.) 355

THE GRO MISSION

THE GEO MISSION

THE GRO - COMPTEL MISSION :

INSTRUMENT DESCRIPTION AND SCIENTIFIC OBJECTIVES

V. Schönfelder[1], K. Bennett[4], W. Collmar[1], A. Connors[3], A. Deerenberg[2],
R. Diehl[1], J.W. den Herder[2], W. Hermsen[2], G.G. Lichti[1], J.A. Lockwood[3],
J. Macri[3], M. McConnell[3], D. Morris[3], A. Connors[3], J. Ryan[3], G. Simpson[3],
H. Steinle[1], A. Strong[1], B.N. Swanenburg[2], B.G. Taylor[4], M. Varendorff[1],
C. de Vries[2], C. Winkler[4]

[1]Max-Planck-Institut für extraterrestrische Physik, D-8046 Garching, FRG
[2]Laboratory for Space Research, Leiden, The Netherlands
[3]University of New Hampshire, Durham N.H., USA
[4]Space Science Department of ESA, Noordwijk, The Netherlands

ABSTRACT

The imaging Compton telescope COMPTEL is one of the four instruments onboard NASA's Gamma Ray Observatory GRO which is to be launched in April 1991 by the Space Shuttle Atlantis. COMPTEL will explore the 1 to 30 MeV energy range with an angular resolution of a few degrees within a large field-of-view of about 1 steradian. Its medium energy resolution (8.8 % FWHM at 1.27 MeV) in addition makes it to a powerful gamma-ray line spectrometer. Within a 2-weeks observation period COMPTEL will be able to detect sources which are about 20-times weaker than the Crab. With these properties COMPTEL is well suited to perform the first complete sky survey at MeV-energies. Targets of special interest are galactic gamma ray sources (like radio pulsars, X-ray binaries, the Galactic Center, the unidentified COS-B sources, supernova remnants and molecular clouds), external galaxies (especially the nuclei of active galaxies), gamma-ray line sources (e.g. the distribution of the 1.8 MeV line emissivity throughout the Galaxy), the diffuse gamma-ray emission from interstellar space, the cosmic gamma-ray background, cosmic gamma-ray bursts, and gamma-ray and neutron emission during solar flares.

INTRODUCTION

COMPTEL covers an energy range (1 to 30 MeV) which was once classified "the impossible range" in astronomy. In spite of tremendous difficulties in exploring this range, quite some progress has been achieved during the last few years. The few existing results have already demonstrated that the scientific return of MeV-gamma ray astronomy is very high.

The imaging Compton telescope COMPTEL was proposed to NASA in 1978 in response to the Announcement of Opportunity for instruments onboard the Gamma-Ray Observatory GRO by an international collaboration consisting of the Max-Planck-Institut für extraterrestrische Physik in Garching, FRG, the Laboratory for Space Research in Leiden, The Netherlands, the University of New Hampshire in Durham, USA, and the Space Science Department of ESA in Noordwijk, The Netherlands. In 1981 - at the end of a three years lasting definition phase - COMPTEL was selected by NASA as one of the four GRO instruments. The other instruments on the Observatory are BATSE (Burst and Transient Source Experiment), OSSE (Oriented Scintillation Spectrometer Experiment), and EGRET (Energetic Gamma-Ray Experiment Telescope).

Fig. 1.   Schematic View of GRO. OSSE is on the left, EGRET on the right side, COMPTEL is in the middle. BATSE consists of 8 detectors, two at each corner of the observatory.

A schematic View of the Gamma-Ray Observatory is shown in Figure 1. COMPTEL occupies the middle position of the spacecraft, EGRET is to the right, OSSE to the left. BATSE actually consists of 8 single detectors, two at each corner of the platform.

GRO will be a free flying satellite which is to be launched by the Shuttle Atlantis according to the present NASA schedule in April 1991. GRO is three-axis stabilized. The pointing accuracy is $\pm$ 0.5 degree, however, the pointing direction will be known at any time to an accuracy of 2 arc minutes. Absolute time will be accurate to 0.1 msec. GRO will have a circular orbit of 450 km and 28.5 degree inclination. This orbit guarantees a mission life time of at least 2 years and at the same time provides a low background environment. On the other side, about 50 % of the observation time will be lost due to occultation of the fields-of-view of the instruments by the Earth. The nominal observation time of GRO will be 2 weeks per viewing direction. During the first 15 months of the mission a complete sky survey will be made. During later phases of the mission detailed observations of selected objects are foreseen.

INSTRUMENT DESCRIPTION OF COMPTEL

COMPTEL has been otimized to perform the first very sensitive survey of continuum and line emission in the 1 to 30 MeV range. For this purpose it combines a wide field-of-view with imaging properties within that field. The high sensitivity is achieved by minimizing its response to undesired background events /1/.

COMPTEL consists of two detector arrays, an upper one of low Z material, (liquid scintillator NE 213A), and a lower one of high Z material (NaI (Tl) scintillator). In the upper detector (D1) an infalling gamma ray is first Compton scattered and then the scattered gamma ray makes a second interaction in the lower detector (D2). The sequence is confirmed by a time-of-flight measurement. The locations and energy losses of both interactions are measured. For completely absorbed events the arrival direction of the gamma ray is known to lie on the cone mantle of half opening angle φ around the direction of the scattered γ-ray (see Fig. 2), where

$$(1) \qquad \cos \overline{\varphi} = 1 - \frac{m_0 \, c^2}{E_2} + \frac{m_0 \, c^2}{E_1 + E_2}$$

$$(2) \qquad E_\gamma = E_1 + E_2$$

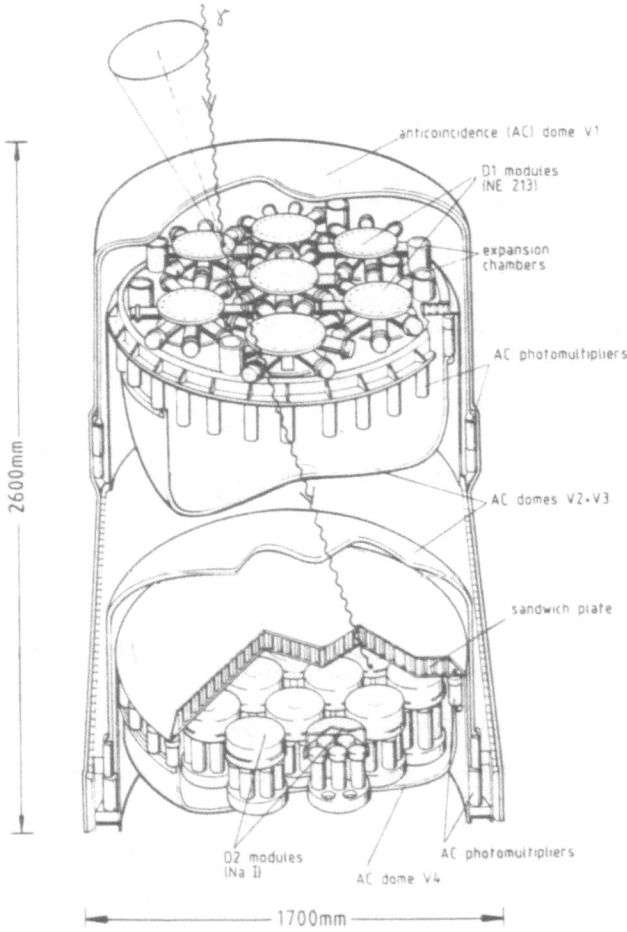

Fig. 2. Schematic View of COMPTEL

A celestial source therefore can be located from the cones of different source gamma rays. Incompletely absorbed events produce cones which do not contain the source position. The angular resolution of the telescope depends on the accuracy to which $\overline{\varphi}$ and the direction of the scattered gamma ray are determined. In some ways COMPTEL is similar to an optical camera: the first detector, which can be compared with the camera's lens, directs the light into the second detector, comparable to the film, in which the scattered photon is absorbed. Although the photons are not focussed as in case of the optical camera, COMPTEL is able of reconstructing sky images over a wide field-of-view with a resolution of a few degrees.

A schematic drawing of the telescope is shown in Figure 2. The two detectors are separated by a distance of 1.5 m. Each detector is entirely surrounded by a thin anticoincidence shield of plastic scintillator which rejects charged particles. Off to the sides between both detectors are two small plastic scintillation detectors containing weak Co^{60} sources; these are used as gamma-ray calibration sources.

The Upper Detector (D1) consists of 7 cylindrical modules of liquid scintillator NE 213. Each module is approximately 28 cm in diameter and 8.5 cm thick and viewed by 8 photomultiplier tubes. The total area of the upper detector is 4310 cm^2. The Lower Detector (D2) consists of 14 cylindrical NaI (Tl) blocks of 7.5 cm thickness and 28 cm diameter, which are mounted on a baseplate with a diameter of 1.50 m. Each NaI-block is seen from below by seven 7.5 cm photomultipliers. The total geometrical area of the lower detector is 8620 cm^2. Each anticoincidence shield consists of two 1.5 cm thick domes of plastic scintillator Ne 110. A dome is viewed by 24 photomultipliers. Each calibration source consists of a cylindrical piece of Co^{60}-doped plastic scintillator of 3 mm thickness and 1.2 cm diameter which is viewed by two 1.9 cm photomultipliers. Except for the front-end electronics of the photomultipliers, all main electronics are mounted on a platform outside the detector assembly.

A gamma ray is electronically identified by a delayed coincidence between the upper and the lower detector, combined with the absence of a veto signal from all charged particle shields and from the calibration detectors. The quantities measured for each event are as follows:

1. the energy of the Compton electron in the upper detector ($E_1 > 50$ keV);
2. the location of the interaction in the upper detector;
3. the pulse shape of the scintillation pulse in the upper detector;
4. the energy loss E_2 in the lower detector ($E_2 > 500$ keV);
5. the location of the interaction in the lower detector;
6. the time-of-flight of the scattered gamma ray from the upper to the lower detector;
7. the absolute time of the event.

The locations of the interactions are derived from the relative pulse heights of all phtomultipliers viewing one detector module. The pulse shape measurements and the time-of-flight measurements are performed in order to reject background events. A gamma-ray event is initially selected onboard according to coarse criteria established on the pulse heights of signals from D1 and D2, time-of-flight, the absence of a veto signal from all anticoincidence domes, and provided no preceding interaction has deposited a large amount of energy in the triggered cell (overloads). If the initial event selection logic is satisfied, a series of actions are initiated:

1. pulse height analysis of the 15 photomultiplier tubes of the identified modules of D1 and D2, plus the sum of the eight D1 and seven D2 PMT's;

2. time-to-digital conversion of the time-of-flight and pulse-shape discriminator circuit outputs;

3. the triggering of the digital electronics, which time-tags the event.

In addition to this normal double scattering measuring mode, two of the NaI-crystals of the telescope will be used to measure the energy spectra of cosmic gamma-ray bursts and solar flares. In case of a solar flare, COMPTEL receives a solar flare trigger indirectly from BATSE. COMPTEL then goes into an alternate event selection mode, which also allows to measure solar neutrons in addition to gamma rays.

The digital outputs of the pulse height analysis and the time-of-flight conversions are further checked against upper and lower limits, loaded via serial telecommands. The data are stored in the event buffer, with those events satisfying the final selection criteria having first priority and those failing one or more criteria having second priority for transmission through the telemetry. Calibration data are transmitted regularly either as events or as spectra. Estimates of the orbital trigger rate and the need to transmit an adequate sample of calibration events are consistent with an event message rate of 48 events per 2.048 seconds. For one event 16 x 16 bit words are required.

TELESCOPE PERFORMANCE

The calibration of the fully integrated instrument with radioactive sources and γ-rays produced by nuclear interactions at a van de Graaf-accelerator was performed in 1987 at the Gesellschaft für Strahlen-forschung at Neuherberg, near Munich. The radioactive sources/beam targets were placed at distances of 8.0 to 8.5 m from the COMPTEL D1-detector. After each source run a background run was performed in which the radioactive source was blocked by a thick lead shield. The analysis of the calibration data is now near its completion. Here the preliminary results from a few source runs are presented /2/.

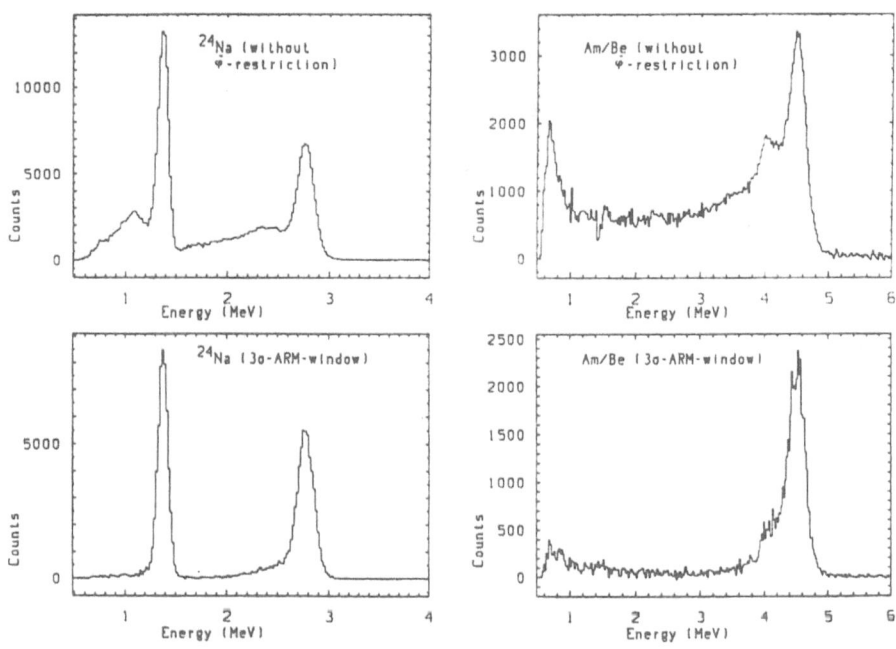

Fig. 3a.   Energy spectra measured with COMPTEL at 2.75 MeV (Na^{24}) and 4.43 MeV (Am/Be). The term ARM-window is defined in the text.

Energy Resolution.

The energy resolution of COMPTEL is determined by the energy resolution of both detector as-semblies. Figure 3a displays the energy spectra mesured with COMPTEL for γ-rays at 2.75 MeV (Na^{24}) and 4.43 MeV (Am ^{241}/Be^9) after background subtraction. The top spectra contains all events above the energy thresholds (50 keV in D1, and 500 keV in D2). The Compton tails in the energy spectra (resulting from incompletely absorbed events in D2) can be suppressed by applying event selection crite-ria, e.g. by limiting the Compton scatter angle φ to values smaller than 30°, or by accepting only events whose event circles pass (within $\pm$ 1.5 σ) through the position of the radioactive source. By this latter event selection a nearly complete suppresion of the Compton tail can be achieved as illustrated in the bot-tom spectra of Figure 3a. This technique can be applied, if the position of a celestial gamma-ray source is known. The photo-peak energy resolutions of COMPTEL are summarized in Table 1.

At photon energies above about 10 MeV , the photopeak fraction becomes two small to make the photopeak visible. But still, multiple Compton collision cause a peak near the photopeak energy, and the width of this peak is about 10 % FWHM between 10 and 20 MeV (see Fig. 3b).

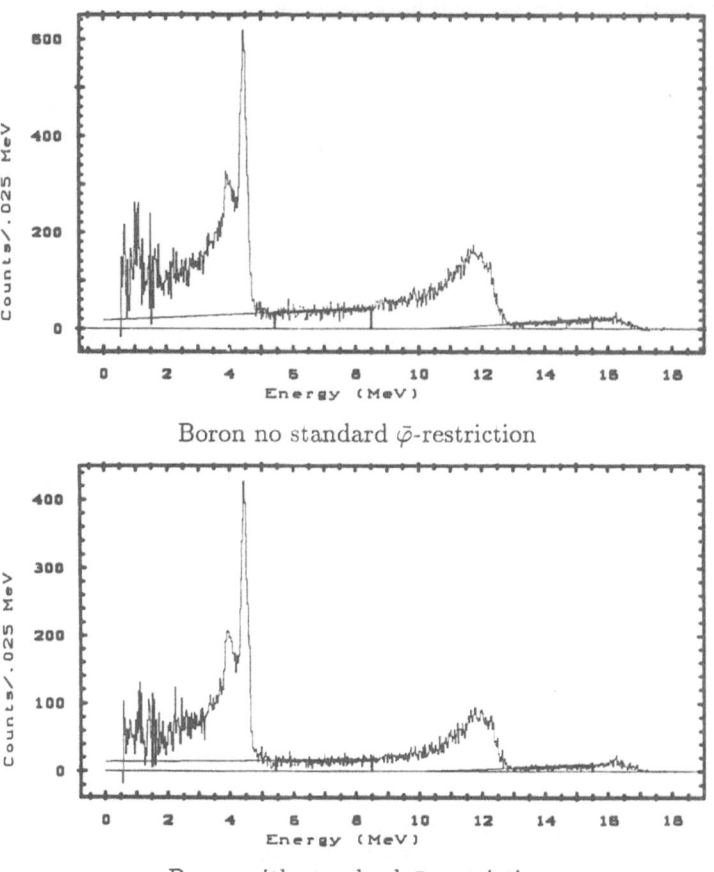

Boron no standard $\bar{\varphi}$-restriction

Boron with standard $\bar{\varphi}$-restriction

Fig. 3b.   Measured energy spectrum from Boron target (12.14 Mev and 4.43 MeV), Background is subtracted. The peak resolution at 11.5 MeV is 9 % FWHM.

TABLE 1. Energy and Angular Resolution of COMPTEL

Energy [MeV]	photopeak-energy resolution [FWHM]	ARM-width [FWHM]
1.27	8.8 %	4.74°
2.75	6.5 %	3.10°
4.43	6.3 %	2.71°

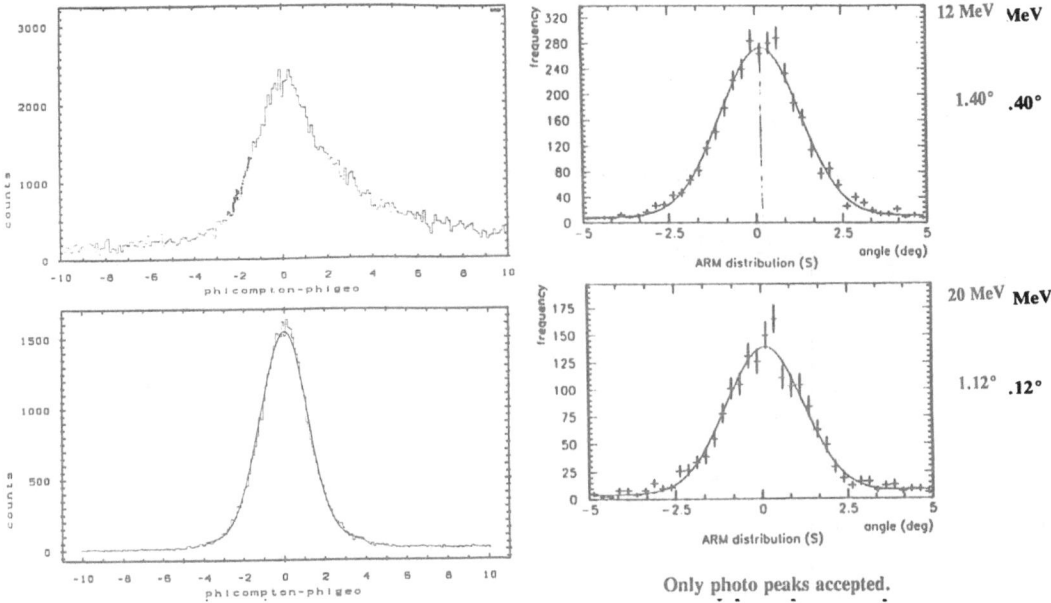

Only photo peaks accepted.

Fig. 4.   Left side: ARM-distribution of 4.43 MeV gamma rays. Top: all events above threshold accepted. Bottom: only photopeak events (FWHM) accepted. Right side: ARM
distributions at 12.14 MeV and 20.52 MeV. Only events within 3 σ energy peaks are
accepted.

Angular Resolution

An appropriate measure of the angular resolution of COMPTEL is provided by the ARM-distribution (Angular Resolution Measure), which is displayed in Figure 4 for 4.43 MeV, 12.14 MeV, and
20.52 MeV gamma rays. Here on the abszissa the difference $\bar{\varphi}$ - φ_{geo} is plotted, where $\bar{\varphi}$ is derived from
the measured energy losses via equation (1), and where φ_{geo} is the geometric scatter angle, which is
derived from the knowledge of the gamma-ray source position and from the location of the interactions in
D1 and D2.

If all events above threshold are accepted, then the incompletely absorbed events in D2 cause an
asymmetric distribution with an excess at positive $\bar{\varphi}$ - φ_{geo} - values (see top of 4.43 MeV distribution).
If, instead, only photopeak events are accepted, then incompletely absorbed events are rejected and the resulting ARM distribution is symmetric (bottom of 4.43 MeV distribution). Table 1 lists the widths of the
ARM distributions of photopeak events for three gamma-ray energies. Figure 5 illustrates how much the
spatial and the energy uncertainty contribute to the total angular resolution.

Absolute Detection Area

The absolute effective detection area for vertical incidence is shown in Figure 6a for various event
selections, for the top curve no $\bar{\varphi}$-restriction was applied to the data, in the bottom curve only events
within the 3 σ ARM window around the source position are accepted. The two curves in between were
derived for a special $\bar{\varphi}$-restriction: φ < 30° below E1 + E2 = 2.6 MeV, $\bar{\varphi}$ < 15° above E1 + E2 =
10 MeV, and E1 < 0.4 (E1 + E2) in between. The dashed curve was predicted in 1982, the solid curve
represents the calibration results.

Field-of-View

The field-of-view of COMPTEL around its axis can be influenced by introducing restrictions on

191

Angular Resolution for Photopeak Selected Data:

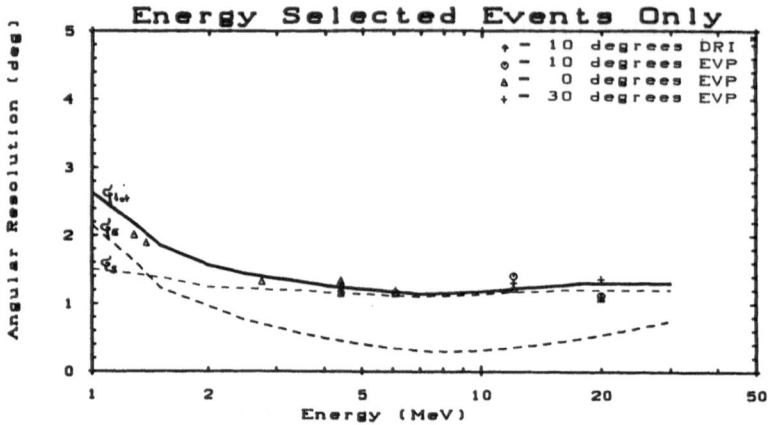

Fig. 5.   1 σ - angular resolution of COMPTEL (ARM - width divided by 2.36) as a function of gamma-ray energy. The contributions to the total resolution from the spatial and energy uncertainty are shown separately.

COMPTEL EFFECTIVE DETECTION AREA FOR VERTICAL INCIDENCE, FOR VARIOUS EVENT SELCTIONS (Veto subsystem in operation)

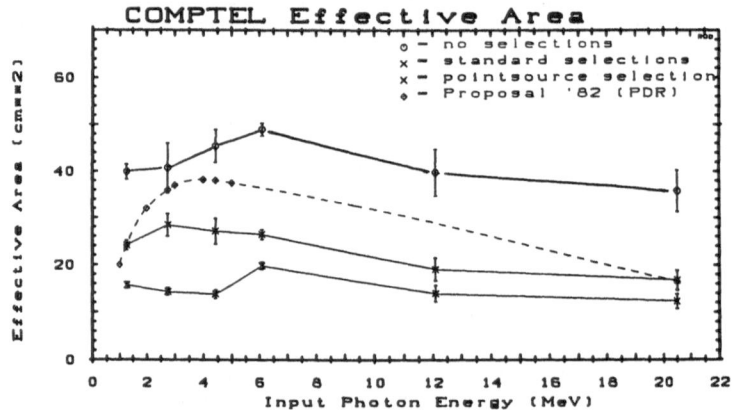

Notice: The "proposal"-curve was for "standard selections".
The values shown at 20 MeV are from Monte Carlo simulation. The effective area at 20 MeV from the COMPTEL calibration is consistent with these values considering the large uncertainties of the actual beam strength.

Fig. 6a.   COMPTEL effective area for vertical incidence.

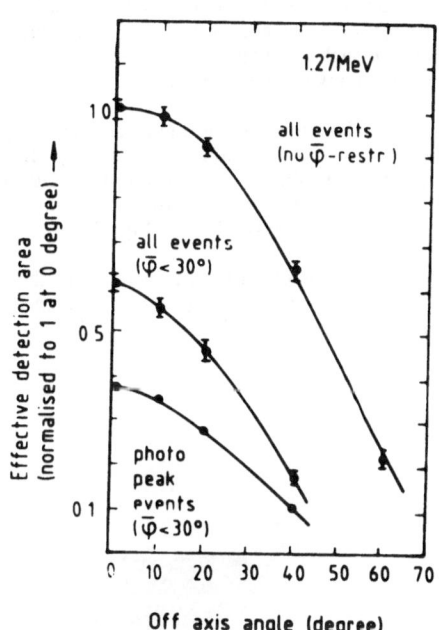

Fig. 6b.   Relative dependence of effective detection area as a function of off-axis angle at
1.27 MeV (Na^{22}). The lines are for eye guidance, only.

$\overline{\varphi}$; also the energy thresholds in D1 and D2 affect the field-of-view. Figure 6b displays the dependence of the relative effective detection area of COMPTEL with off-axis angle at 1.27 MeV (Na^{22}). The widest field-of-view is obtained, if no $\overline{\varphi}$-restriction is applied to the data. A narrower field-of-view can be achieved by applying event selection criteria (e.g. $\overline{\varphi} < 30°$) at the cost of a lower detection area for vertical incidence. A limitation of the field-of-view is necessary in order to reduce the contribution of atmospheric gamma rays from the Earth horizon.

COMPTEL Sensitivity

With the values of Figure 6a of the absolute detection area (reduced by the appropriate event selection factors) and the background rates measured with the MPI balloon borne Compton-telescope (scaled to the COMPTEL size and the GRO orbit) the COMPTEL sensitivities shown in the later Figures 9, 10, and 11 were derived.

Image Deconvolution

In Compton telescopes no one-to-one relation exists between individual sky and data bins. This is due to the fact that for a gamma ray of fixed energy and angle of incidence a wide range of Compton scattering angles in D1 is possible. In such a situation the maximum entropy method is a powerful choice for imaging. This method requires the knowledge of the telescope response to any given sky image.

The telescope response can be described in the simplest way in a three-dimensional data space defined by the scatter direction (χ, ψ) and the Compton scatter angle $\overline{\varphi}$. In the idealized case, in which the scattered gamma ray is totally absorbed in the lower detector, the pattern of all data points originating from a gamma ray source with the coordinates (χ_0, ψ_0) lies on a cone in (χ, ψ, $\overline{\varphi}$) - space, where the cone apex is at (χ_0, ψ_0) and the cone semi-angle is 45° (see Fig. 7).

The response density along the cone is given by the variation of the Klein-Nishina cross-section for Compton scattering. This idealized "cone mantle" response is blurred by measurement inadequacies in the scintillation detectors, especially by incomplete absorption in D2.

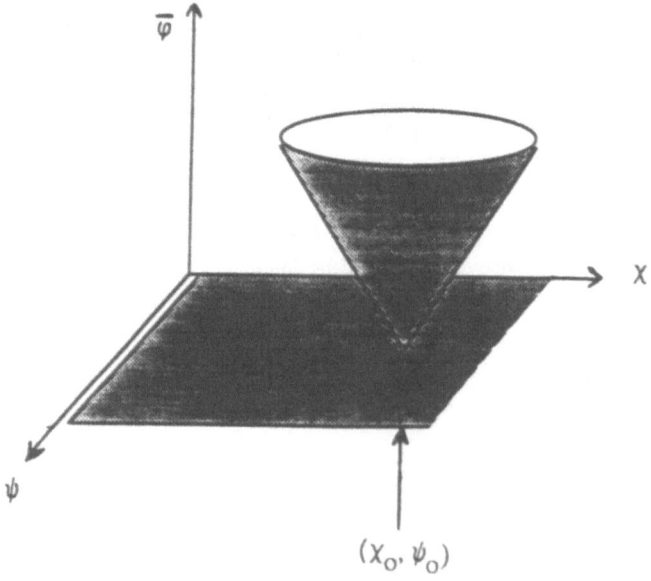

Fig. 7.   Illustration of the COMPTEL response of a celestial point source in the 3-dimensional data space (χ, ψ, $\overline{\varphi}$). The data lie on a cone, the apex of which is at the position of the celestial source (χ_0, ψ_0). The cone semi-angle is 45°. Actually, the cone mantle is blurred due to measurement inaccuracies.

Therefore, the event density in data space can be represented by (see /3/):

(3) $n (\chi, \psi, \varphi) = g (\chi, \psi) \cdot \int \int I (\chi_0, \psi_0) \; A (\chi_0, \psi_0) \; f (\chi - \chi_0, \psi - \psi_0, \overline{\varphi}) \; d\chi_0 \; d\psi_0$

where $I (\chi_0, \psi_0)$ is the infalling sky intensity distribution, $A (\chi_0, \psi_0)$ is the exposure factor (product of effective detection area times observation time), $f (\chi - \chi_0, \psi - \psi_0, \overline{\varphi})$ is the "cone" - function and $g (\chi, \psi)$ is a geometrical function accounting for the incomplete coverage of the upper and lower planes by the detectors.

The maximum entropy image approach is to convolve a (2-dimensional) sky image with the full response of the telescope in the 3-dimensional dataspace $(\chi, \psi, \overline{\varphi})$ and to try to match the "mock data" from the trial image to the measured data in the 3-dimensional dataspace. The trial image yielding a statistically acceptable match to the measured data, and at the same time fulfilling the entropy criterion, is defined to be the maximum entropy image.

An exloratory application to this method has been made to the COMPTEL calibration data. Two calibration runs with 6.1 MeV gamma-ray beams, separated by 10°, were superimposed to simulate a dataset of two sources in the field. The maximum entropy method was then applied. Figure 8 shows the resulting map. The thwo sources are successfully resolved.

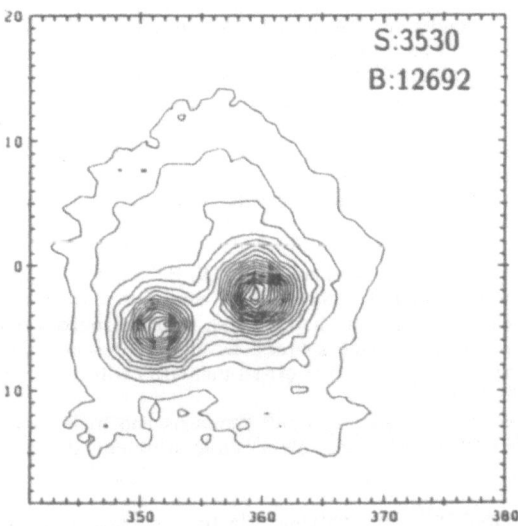

Fig. 8.   Image from COMPTEL calibration data at 6.1 MeV from 2 sources, separated by 10°.

ASTROPHYSICS WITH COMPTEL

Due to the more than 10-times higher sensitivity of COMPTEL in comparison to previously flown instruments, a significant progress in the exploration and understanding of the MeV gamma-ray sky can be expected.

The main targets of interest are listed below. These are:

- discrete objects in the Galaxy that show steady gamma-ray emission;
- the diffuse gamma-ray continuum emission from interstellar space;
- gamma-ray line emission from discrete or extended sources;

- external galaxies, especially Seyferts and Quasars;
- the diffuse cosmic gamma-ray background;
- cosmic gamma-ray bursts (their localization and the study of their energy spectra and time histories);
- the Sun (solar flare gamma-ray and neutron emission).

Each of these topics is briefly addressed below:

Galactic Gamma-Ray Sources

The key question in this field is: "What kind of objects do we see as gamma-ray sources in the Milky Way?"

At present radio pulsars and X-ray binaries are the most promising candidates. Both kinds of objects contain stars in the end stage of their evolution (neutron stars, and in case of binaries perhaps black holes). We know that the Crab and Vela pulsars, both, radiate 5 orders of mangitude more power at gamma-ray energies than in the radioband. If this is true in general for pulsars, then the key for an understanding of the pulsar radiation mechanism may be found at gamma-ray energies. COMPTEL should be able to see more than these two pulsars at gamma-ray energies. The estimates range from a few to about a dozen depending on the pulsar model, which one uses.

X-ray binaries are known to be very common X-ray sources which are powered by accretion. A few of them have also been observed at very hard X-ray energies around 100 keV. Among these are Cyg X-1 - the best black hole candidate, Cyg X-3 - the source which has attracted so much attention by its possible TeV and PeV gamma-ray emission, and other sources like GX 5-1, GX 1+4, GX 339+4. Except for Cyg X-1 and possibly Cyg X-3 none of these sources so far could be detected at gamma-ray energies. From COMPTEL we shall learn how far the emission of these sources extends into the MeV-range, and whether there is a non-thermal component in addition to the thermal one.

Another object of high interest is the Galactic Center - one of the other most prominent black hole candidates. First positive gamma-ray observations from the galactic center have already been made during the last two decades. Further observations will be needed to derive firm conclusions about the nature and the physical processes of this source..

Then, of course, the puzzle of the unidentified COS-B sources has to be solved. Though about half of the originally 22 objects contained in the COS-B source catalogue are now known to be simply regions of enhanced interstellar matter, about half a dozen of these sources remain unidentified till now. Nobody at present knows the nature of these sources! Do they represent a new class of celestial objects which mainly radiate at gamma-ray energies? We expect that GRO will be able to answer this question.

Finally, also extended objects, like Supernova remnants and molecular clouds will be studied by COMPTEL. From previous COS-B gamma-ray observations we know already that both these objects are very promising targets.

The prospects of COMPTEL for studying galactic gamma-ray sources can be judged from the sensitivity diagram in Figure 9. Here the 3σ sensitivities of the 3 instruments OSSE, COMPTEL, and EGRET over their energy ranges are compared with the fluxes of known gamma-ray sources, e.g. the Crab, Cyg X-1 in its low and high intensity state, and the COS-B source Geminga. Between 1 and 30 MeV COMPTEL will be able to detect sources that are roughly 20-times weaker than the Crab. Intensity variations of the high intensity state of the Galactic Center and Cyg X-1 can be recorded with high precision on a very short time scale.

Diffuse Galactic Gamma-Ray Emission

The diffuse galactic gamma-ray emission so far has been studied at high gamma-ray energies (around 100 MeV), only. From the interpretation of the SAS-2 and COS-B sky maps we know that the cosmic ray density is not constant throughout the Galaxy, but higher in the inner part and lower in the outer parts. COMPTEL will extend the survey to lower energies down to 1 MeV and therefore allow to study the electron-induced gamma-ray component. The high sensitivity of COMPTEL combined with the good angular resolution will (hopefully) allow to separate the so far unresolved source component from the really diffuse interstellar component. Measurements at different galactic latitudes will help to separate the two electron-induced components: the bremsstrahlung and the inverse Compton component. We can expect that the COMPTEL measurements will lead to a better understanding of the distribution of low energy cosmic ray electrons throughout interstellar space.

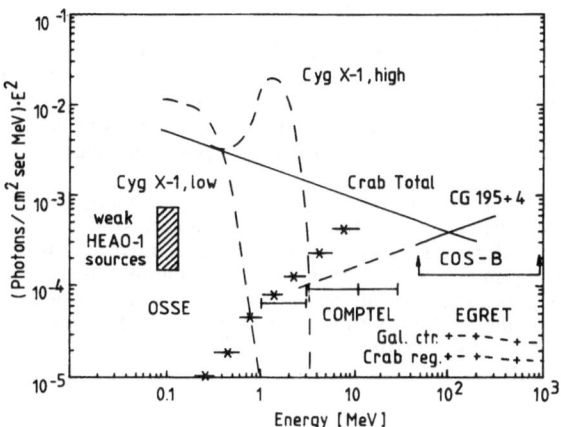

Fig. 9.   3 σ - sensitivity of COMPTEL for the detection of galactic gamma-ray point sources within a 2-weeks observation period.

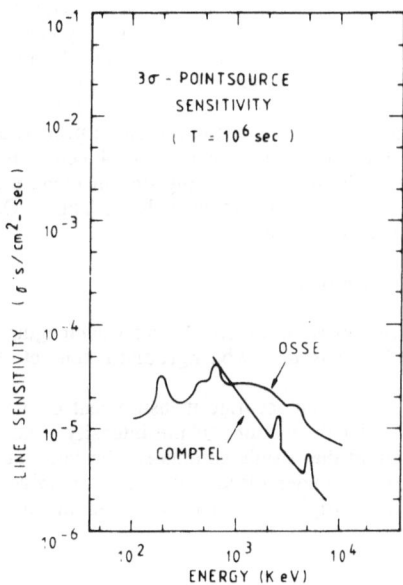

Fig. 10.   3 σ point source sensitivity of COMPTEL to gamma-ray lines.

197

Gamma-Ray Line Spectroscopy

The field of gamma-ray line spectroscopy is closely related to the question: "How were the chemical elements in the Universe synthesized?"

Gamma-ray astronomy provides a powerful tool to answer this question. During the formation of chemical elements not only stable but also radioactive ones were produced. Some of them are gamma-ray emitters and these can be detected by means of gamma-ray telescopes. The first two cosmic gamma-ray lines that were detected are the 511 keV annihilation line and the 1.8 MeV line from radioactive ^{26}Al. The origin of both these lines is not yet really understood. Whereas the ^{26}Al was probably synthesized in galactic objects that are concentrated close to the Galactic Center (like novae or special massive stars), the 511 keV line seems to consist of two components - one which is more widely spread (probably produced in supernovae), and another one which seems to come from a point source near the Galactic Center.

COMPTEL will be able to map the entire galactic plane in the light of the 1.8 MeV gamma-ray line. These measurements can then be used as tracers of those objects in the Galaxy which produced these lines - very much like the 21 cm line at radio-wave lengths is used as a tracer of interstellar neutral hydrogen. Probably other gamma-ray lines from nucleosynthesis processes will be observed by COMPTEL in addition, e.g. from the Ti^{44}-decay or from nucleosynthesis products in supernovae in the Virgo cluster. Unfortunately, the Co^{56}-lines from the SN 1987a in LMC will be too weak at the time of the GRO launch to be still seen by COMPTEL. The COMPTEL point source sensitivity for gamma-ray lines is shown in Figure 10.

External Galaxies

In the extragalactic sky there is a certain chance that continuum emission from LMC will be seen by COMPTEL in a deep exposure.

The most interesting objects in the extragalactic gamma-ray sky are, however, the nuclei of active galaxies. At least some of the known active galaxies and quasars do have their maxima of luminosity at gamma-ray energies. The situation is illustrated in Figure 11, where the sensitivities of OSSE, COMPTEL, and EGRET for detecting AGN's are compared with the observed spectra of the radio galaxy Cen A, the quasar 3C273 - which both peak at gamma-ray energies - and the X-ray spectra of 12 AGN's - mostly Seyferts - that were observed by HEAO-A1. From the diagram we can estimate that COMPTEL will be able to study - say a dozen or even more AGN's - if all AGN's have spectra similar to Cen A and 3C273. We may get the first measurement of the luminosity function of AGN's at gamma-ray energies. This function - together with the measured properties of individual galaxies - may lead to a better understanding of the engine that powers the objects. The question of the nature of the central source in AGN's is one of the most fascinating one in modern astronomy. Many theoreticians believe that the energy source is a mass accreting black hole. The COMPTEL observations may be crucial for an understanding of the central source.

The Diffuse Cosmic Gamma-Ray Background

A cosmic background radiation exists at practically all wavelengths. Best studied is the microwave background at 2.7 K. The origin of the cosmic ray background is not yet really understood.

COMPTEL will not only provide an accurate measurement of the background spectrum at MeV-energies, it will also search for angular fluctuations of the intensity. The COMPTEL data should allow to address the question of the origin of the cosmic gamma-ray background and to decide between the two classes of models which presently are under discussion: an unresolved source origin (e.g. from un-resolved AGN's) or a really diffuse origin (e.g. from matter-antimatter annihilation in a baryon symmetric universe).

Gamma-Ray Bursts

Though the gamma-ray bursters are the strongest gamma-ray sources in the sky during their short outburst, their nature is not yet known. Most people believe that - at least for a large number of bursts - a neutron star is somehow involved. This thinking is mainly based on lines found in some of the burst spectra - though the physical trigger for the outburst is not yet known. Under discussion are at present starquakes, the impact of an asteroid or comet onto the neutron star, magnetic instabilities or accretion of matter onto the neutron star. COMPTEL can locate bursts that happen to be within the COMPTEL field-of-view ($\approx$ 1 ster) with an accuracy which is better than 1°. We expect to see about 1 burst per

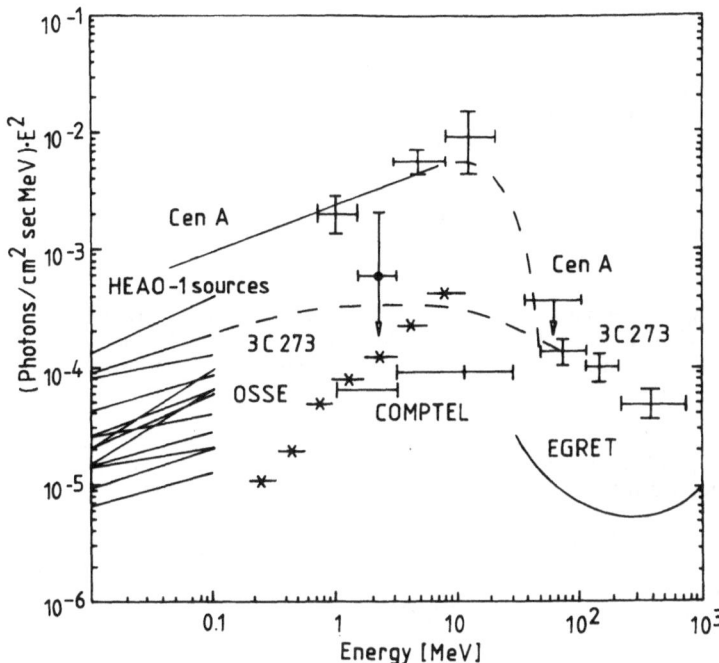

Fig. 11.   3 σ - sensitivity of COMPTEL to detect active galactic nuclei within a 2-weeks observation time.

month within the field-of-view with a fluence S (E > 1 MeV) $\geq$ 2 · 10^{-6} erg/cm^2. In addition, the single detector mode of COMPTEL with its 2.5 ster unobstructed field-of-view allows to measure the energy spectra and time histograms of bursts seen by BATSE in the energy range 0.1 to 12 MeV. This mode is especially important for the search for gamma-ray lines in burst spectra /4/.

The Sun

The Sun is not a prime target of GRO. COMPTEL - like the other 3 GRO instruments does have, however, certain capabilities to measure gamma rays and also neutrons from the Sun /5/. First, COMPTEL can study solar flare gamma-ray emission in the telescope mode, if the Sun happens to be within 30° off-axis. Second, in case of a solar event COMPTEL obtains a solar burst indication trigger from BATSE. As a consequence COMPTEL switches into a solar-neutron mode, which allows measurements of solar neutrons in addition to gamma rays. Third, the single detector mode of COMPTEL allows to measure energy spectra and time histories of solar flare gamma-ray emission. This mode again is especially important for the search of gamma-ray lines in solar flare spectra.

The sensitivities of COMPTEL for observing gamma-ray emission from solar flares are comparable to the sensitivity of the SMM-spectrometer. During the GRO mission we therefore can expect to study a large number of additional flares. The gamma-ray and neutron measurements in conjunction with those of other instruments (interplanetary particle detectors, ground based neutron monitors) provide the best possible observation of production, acceleration, and propagation aspects during solar flares.

CONCLUSION

Significant progress in the exploration and understanding of the MeV gamma-ray sky is to be expected from COMPTEL. The results of COMPTEL and the other GRO instruments, especially if combined with observations in other spectral ranges from both space and ground based observatories, will certainly make the field of gamma-ray astronomy attractive to the whole community of Astronomers and Astrophysicists as can already now been judged from the wide response to the NASA Research Announcement for the GRO Guest Investigator Program.

REFERENCES

1. V. Schönfelder, R. Diehl, G.G. Lichti, H. Steinle, B. Swanenburg, A. Deerenberg, H. Aarts, J. Lockwood, W. Webber, J. Ryan, G. Simpson, B. Taylor, K. Bennett, and M. Snelling, IEEE Trans. Nucl. Sci., NS-31 (1): 766-770 (1984)

2. R. Much, V. Schönfelder, H. Aarts, K. Bennett, A. Deerenberg, R. Diehl, G. Eymann, J.W. den Herder, W. Hermsen, G.G. Lichti, J. Lockwood, J. Macri, M. McConnell, D. Morris, J. Ryan, G. Simpson, H. Steinle, A.W. Strong, B.N. Swanenburg, C. de Vries, C. Winkler, Proc. of 21th ICRC, Adelaide, 4, 158 (1990)

3. A.W. Strong, K. Bennett, P. Cabeza-Orcel, A. Deerenberg, R. Diehl, J.W. den Herder, W. Hermsen, G.G. Lichti, J. Lockwood, M. McConnell, J. Macri, D. Morris, R. Much, J. Ryan, V. Schönfelder, G. Simpson, H. Steinle, B.N. Swanenburg, C. Winkler, Proc. of 21th ICRC, Adelaide, 4, 154 (1990)

4. C. Winkler, V. Schönfelder, R. Diehl, G. G. Lichti, H. Steinle, B.N. Swanenburg, H. Aarts, A. Deerenberg, W. Hermsen, J. Lockwood, J. Ryan, G. Simpson, W.R. Webber, K. Bennett, A.V. Dordrecht, and B.G. Taylor, Adv. Space Res., 6 (4), 113-117 (1986)

5. J.M. Ryan, v. Schönfelder, R. Diehl, G.G. Lichti, H. Steinle, B.N. Swanenburg, H. Aarts, A. Deerenberg, W. Hermsen, G. Kiers, J. Lockwood, J. Macri, D. Morris, G. Simpson, K. Bennett, G. Eymann, M. Snelling, C. Winkler, R. Byrd, C. Foster, and T. Taddeucci, Proc. 20th Int. Cosmic Ray Conf., Moscow, 4, 425 (1987)

RESPONSE DETERMINATIONS OF COMPTEL FROM CALIBRATION MEASUREMENTS, MODELS, AND SIMULATIONS

R. Diehl [1], H. Aarts [2], K. Bennett [4], W. Collmar [1], H. deBoer [2],
A.J.M. Deerenberg [2], J.W. den Herder [2], C. deVries [2], W. Hermsen [2],
M. Kippen [3], J. Knödlseder [1], L. Kuiper [2], G.G. Lichti [1], J.A. Lockwood [3],
J. Macri [3], M. McConnell [3], R. Much [1], D. Morris [3], J. Ryan [3],
V. Schönfelder [1], G. Simpson [3], H. Steinle [1], A.W. Strong [1],
B.N. Swanenburg [2], T. van Sant [1], W.R. Webber [3], C. Winkler [4]

[1] Max Planck Institut für Extraterrestrische Physik D-8046 Garching, FRG
[2] Space Research Leiden, Huygens Laboratorium, NL-2333 Leiden, NL
[3] Space Science Center of University of New Hampshire Durham, N.H, USA
[4] ESA/ESTEC Space Science Department NL-2200 Noordwijk, N.L.

INTRODUCTION

The COMPTEL telescope onboard the NASA Gamma Ray Observatory is an instrument with imaging properties quite different from telescopes at longer wavelengths: Focusing of the incoming radiation is impossible due to the penetrating power of the gamma ray photons, the photons rather interact with the electrons and nuclei of the detector material in a particle-like manner. This affects the way the response of the instrument is described, determined, and used. The instrument consists of 2 planes of detector modules, which measure a photon via two consecutive interactions of probabilistic nature: a Compton scatter in the upper detector plane ('D_1'), and an absorption in the lower detector plane ('D_2') (see figure 1). The active scintillation detectors are surrounded by plastic anticoincidence detector domes to reject events originating from charged particles. Background events are suppressed by measurement of the pulse shape of the scintillation light flash in the upper detectors, and by measurement of the time of flight of the scattered photon from upper to lower detector plane.

The instrument electronics records a set of parameters per each event triggering the active telescope detectors in the desired coincidence criteria; these data are transmitted to ground to be binned into measured data arrays and to be deconvolved using appropriate instrument response matrices.

A detailed description of the instrument can be found in (1). In this paper we describe the methods that have been applied by the COMPTEL Collaboration to determine the instrument response matrices; also the resulting instrument response characteristics are summarized.

INSTRUMENT RESPONSE DESCRIPTION

Ideal Instrument Response

The response of the instrument is defined as the distribution of measured event parameter values for incident photons of specific energy and direction:

$$R = R\left(p_{(measured)} \mid p_{(input)}\right)$$

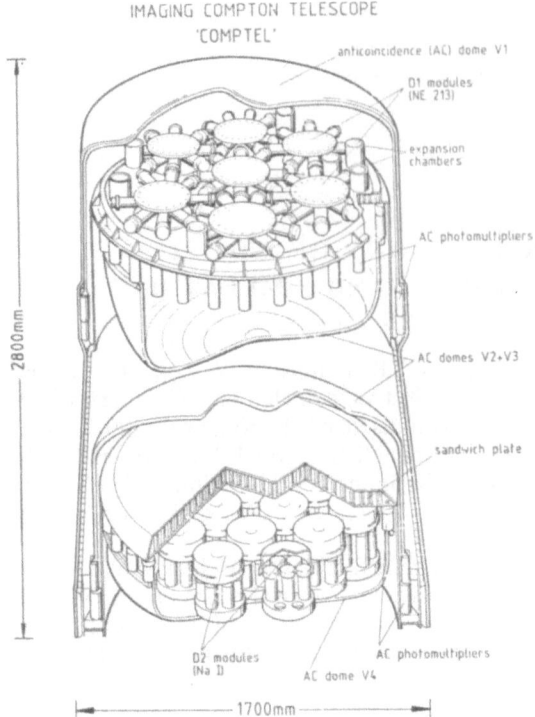

Fig.1. Schematic diagram of the COMPTEL instrument.

The instrument measures directly 21 values per detected photon:
- pulseheights of the set of photomultipliers per D_1 and D_2 module
- detector module identifiers in D_1 and D_2
- time difference between interactions in D_1 and D_2 ('time of flight')
- pulse shape in upper detector module (for neutron event rejection)
- event time and event type flags.

These measurements are converted to the raw event parameters:
- the location of the interactions in both detector planes (measured via the relative signal amplitudes in the set of photomultiplier tubes arranged around the active scintillator volume of a detector cell) $(x_1,y_1,z_1)(x_2,y_2,z_2)$
- the energy deposits of the interactions in both detector subsystems (E_1,E_2)
- secondary event parameters for background signal suppression and sorting (time tag, time-of-flight, pulse shape, event type flags).

As the last set of parameters serves merely for event selections to suppress background events, these are ignored for the discussion of the instrument response due to photon interactions (in this section of the paper).

The interaction location of the 2 successive interactions is determined as exact location within the detector planes. However the symmetry of the instrument components suggests that the response of the individual detector modules in each detector plane is identical: thus the number of event parameters can be reduced by integration over the 2 detector surfaces from 8 to 4, namely:
- two direction angles of the photons on its path from D_1 to D_2 (χ, ψ)
- energy deposit in upper and lower detector module (plane) (E_1,E_2)

The latter two parameters can be converted into the equivalent set of the two parameters "total energy deposit" and "Compton scatter angle", which can be attributed to the physical process of Compton scattering in the instrument more easily:

$$E_{tot} = E_1 + E_2$$

$$\overline{\varphi} = \arccos\left[1 - m_0 c^2\left(\frac{1}{E_2} - \frac{1}{E_1 + E_2}\right)\right]$$

This yields the response expression:

$$R = R\left(E_{tot}, \overline{\varphi}, \chi, \psi \mid E_\gamma, \alpha, \delta\right)$$

In this representation, the response of an ideal Compton telescope to a beam of monochromatic photons from a specific incidence direction is determined from the Klein Nishina formula for the Compton scatter process as a distribution of scatter angles and scatter directions along a cone-shaped feature in the dataspace spanned by these 3 angles (see figure 2). The cone opening half angle is 45 degrees, as the angular distance of the scattered photon's direction from the source direction is identical to the Compton scatter angle in the ideal case; the cone apex is the direction of the incident photon beam. 'Ideal Compton telescope' here means that all photon interactions are Compton scatters in upper detector plane with totally absorbed scattered photons in the lower detector plane.

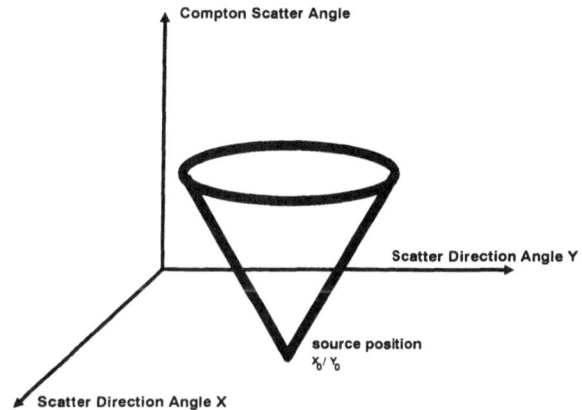

Fig.2. Schematic diagram of the COMPTEL spatial response in the dataspace spanned by scatter direction angles (χ, ψ) and scatter angle $\overline{\varphi}$. The cone apex is the source direction. The blurring due to measurement imperfections widens the cone as a whole, the energy thresholds cut the tip of the cone at the bottom.

Blurring of Instrument Response

In the real situation, imperfections in the location and energy measurements of the interactions, as well as contributions from other (secondary, non-Compton) types of interaction processes result in a broadened distribution of measured event parameters around this ideal cone - we call this the 'point spread function' of the real Compton telescope, as the sharp conelike feature of an ideal point source is spread out by the imperfect measurements of event location and energies.

The blurring components of the response of the Compton telescope are illustrated in figures 3-5:

1) the scintillation detectors of upper and lower plane reveal different measured energy responses to incident photons of specific energy E_γ, due to their different thicknesses for gamma rays:
   - the D_1 NE213A detectors normally show a Compton type of spectrum of measured energy deposits for a monochromatic photon input, as the primary interaction of gamma rays with the detector material is a single Compton scatter interaction with escape of the scattered photon from the detector material. As the scatter angles in the

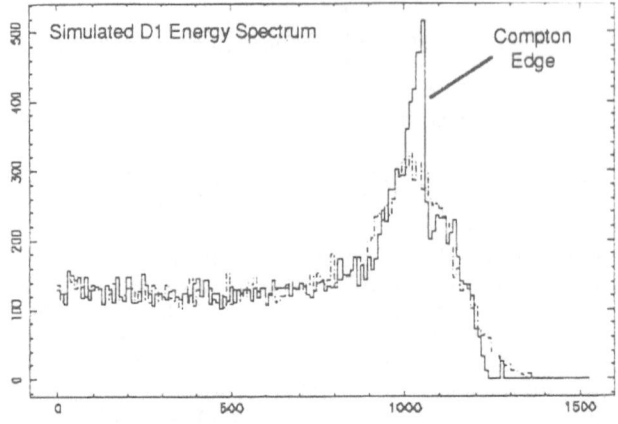

Fig.3a. Energy deposit spectrum as resulting from monochromatic photons at 1.3 MeV (without any additional coincidence requirements) (simulated data)

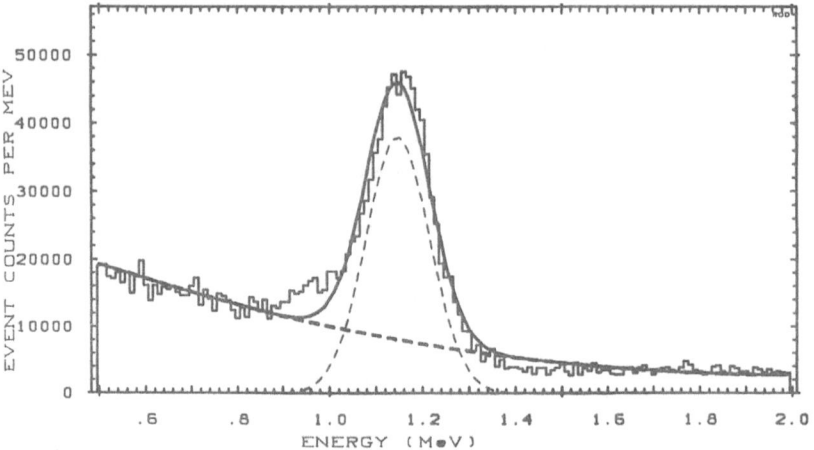

Fig.3b. Response measurement of D_1 detector to a specific energy deposit from a scattering process at a specific scatter angle ('backscatter arrangement', see figure 6b)

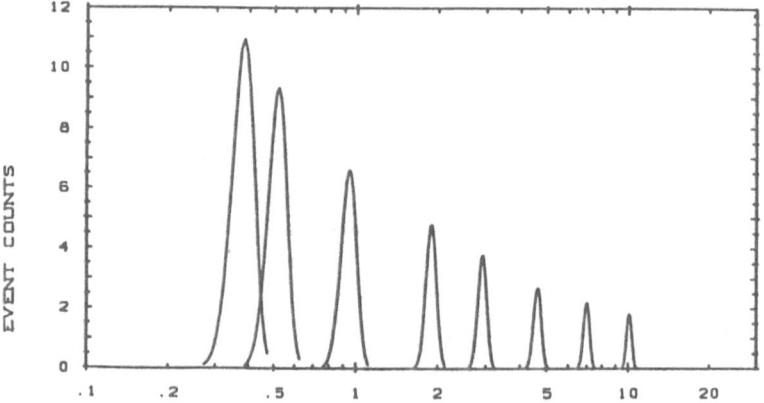

Fig.3c. Response of D_1 detector as a function of input energy (interpolated response model as determined from calibration source energy measurements)

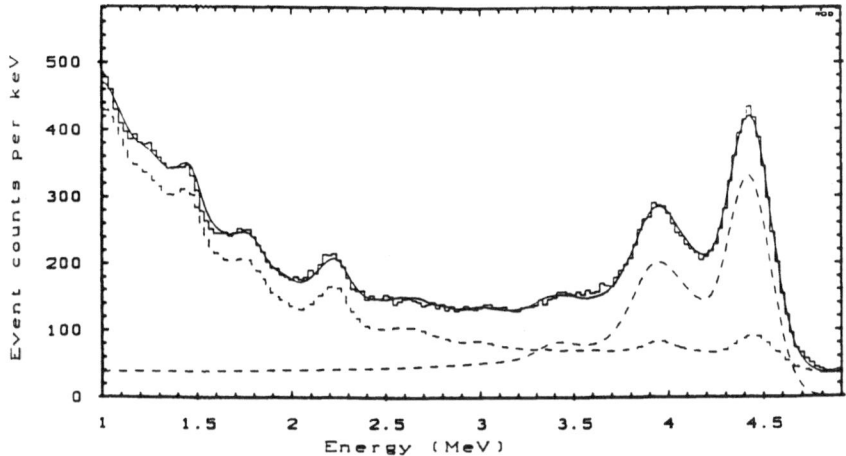

Fig.4a. Energy deposit spectrum of D_2 detector as resulting from monochromatic photons at 4.4 MeV superimposed onto room background: the components from photopeak, escape peaks, and Compton tail are indicated

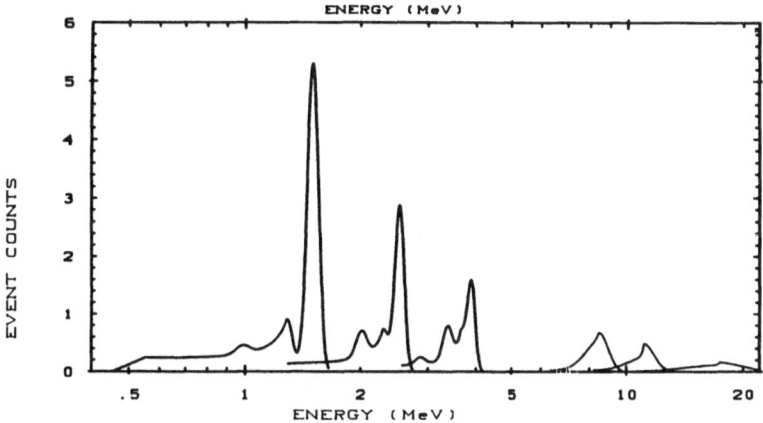

Fig.4b. Response of D_2 detector as a function of input energy (interpolated response model as determined from calibration source energy measurements)

Compton process cover the full range from small angle forward scattering to backward scattering, the measured energy deposits vary from very small energies of a few KeV to the maximum energy deposit for backscattering, the 'Compton edge' energy (example of D_1 energy deposit spectrum see figure 3);

the D_2 NaI detectors have a thickness to gamma rays which results in a series of interactions such as Compton scatterings, pair creation processes in the Coulomb field of detector material nuclei, and photoelectron ejection processes in the shell of these nuclei: as a result of this, at least at low energies the probability of depositing the total energy of the photon within the detector is quite large, resulting in a dominating 'photopeak' in the spectrum of measured energies (see figure 3c). The tail in the spectrum of measured energies is composed from interaction events where the Compton scattered photon escapes the detector ('Compton tail'), or where one ore more of the annihilation photons created as a result of electron-positron pai production interaction escapes the detector ('escape peaks' of first or higher orders, at energies of n *0.511 MeV below the photopeak energy). (example of D_2 energy deposit spectrum see figure 4a; the charateristic components of the spectrum are indicated).

The electronics measuring the scintillation light requires a certain minimum pulse height - this results in an effective energy measurement threshold (about 40 KeV in D_1 and 500 KeV in D_2).. The thresholds are not sharp energy values due to the electronic

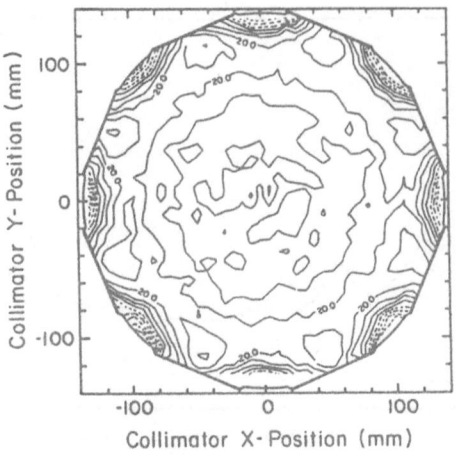

Fig.5a. Interaction location accuracy for D_1 detector modules; photomultiplier tubes are indicated with thick lines; (the 1.7 cm contour line is marked)

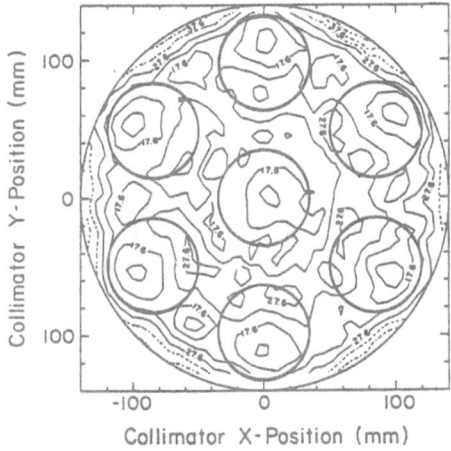

Fig.5b. Interaction location accuracy for D_2 detector modules; photomultiplier tubes are indicated with thick lines circles; (the 1.7 cm contour line is marked)

noise superimposed on the pulseheight signal; this results in rather broad threshold transition regions, on the order of 10 KeV for D_1 and 100 KeV for D_2.

2) the scintillation detector modules of COMPTEL employ the principle of the Anger camera to locate the interaction position within the module. This method is based on a measurement of the scintillation light within a detector by a set of photomultiplier tubes arranged at the detector outer housing; the different distances of the scintillation flash to

the different photomultipliers results in relative signal amplitudes that can be converted into information about the location of the scintillation flash within the detector module. The response of the photyomultiplier sets to photons incident at specific locations of the detector modules was calibrated with a radioactive source located in a massive lead collimator which formed a 'pencil beam' of 1.3. MeV photons with 5 mm beam diameter. The location algorithm to determine event locations from relative signal amplitudes is based on a neural net method. For details of the Anger camera response and determination of the interaction locations see A. Connors et al., this volume. Figure 5 displays the location characteristics that are achieved with this method. The blurring attributed to this interaction location measurement of COMPTEL has been determined to be 2.0-2.6 cm (σ-value for an assumed Gaussian spread of derived locations around the true location) for D_1 modules at 1 MeV, and 0.5-2.5 cm for D_2 modules at 1.27 MeV.

The spatial response (imaging response) of the telescope is derived for intervals of measured energy deposits E_{tot} as a distribution as a distribution in the 3 'spatial dataspace' coordinates (χ, ψ, $\bar{\phi}$) for the specified spatial photons incidence ection direction coordinates (χ_0, ψ_0) of photons with specific energy Egamma.

The spectral response (energy response) of the telescope is derived for regions of spatial dataspace parameters (χ, ψ, $\bar{\phi}$) as a distribution in measured energies E_{tot} for specified energy E_γ of the input photon.

The response of the instrument is used in the analysis of instrument data via a deconvolution of measured event distributions to derive a distribution of parameters for incoming photons (see e.g. discussion of the 'Maximum Entropy' deconvolution by Strong et al., this volume). This response must be determined independently prior to data analysis via calibration measurements and simulations.

INSTRUMENT RESPONSE DETERMINATION AND RESULTS

There are 3 independent (but related) ways to determine the instrument response:
1) calibration measurements of instrument data for know monochromatic photon sources at specified positions within the instrument field of view
2) calibration measurements of instrument detector module data for know monochromatic photon sources, and composition of these detector module responses to a telescope response via (assumed) physical characteristics of the telescope
3) simulation of the instrument detector module physical processes that are experienced by interacting photons.

All 3 response determination methods have their distinct advantages and disadvantages, and are compared in the following.

The COMPTEL response determination was based on a combination of the above approaches:
- The responses of the upper and lower detector modules were calibrated at photon energies from 0.3 to20 MeV
- The telescope was irradiated with photon sources at a range of incidence angles, also testing (and utilizing) instrumental symmetries
- Photon interaction computer software from high energy nuclear physics laboratories was adapted to theCOMPTEL instrument to generate simulated event messages.

Response Calibration of Telescope

This appears to be the straight forward measurement of the real instrument response to photons of specified incoming energy and direction. In reality, calibration measurements suffer from imperfections, however, which one wishes to not directly degrade the accuracy of the response knowledge. Such imperfections are e.g.:
- non-monochromatic photon sources (more than one gamma ray line emitted by various calibration sources)
- background photon sources, the contribution of which cannot accurately be assessed; background subtractions are affected by the background measurement statistics and sometimes yield 'negative counts' in some bins.
- calibration photon source position notcharacteristic for the photon source position during the real measurement (e.g. not at virtually infinite distance)
- calibration photon sources only available at a few energies.

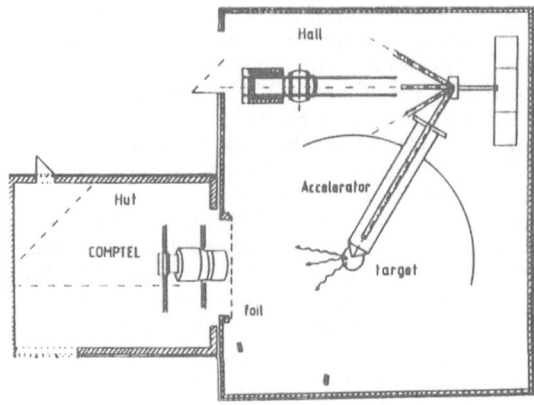

Fig.6a. Experiment setup at GSF Neuherberg accelerator facility; the calibration photons are generated in a target at the end of the beam pipe in the center of the experiment hall; COMPTEL is mounted on the 2 axis-rotation capable dolly in the attached experiment hall

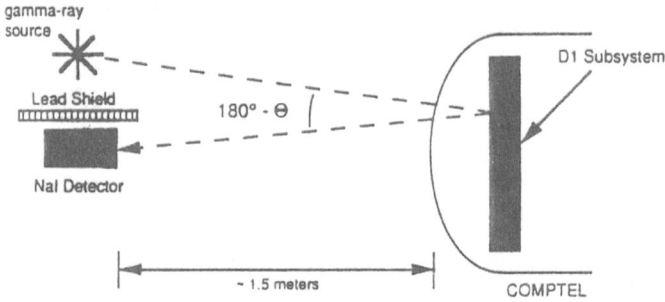

Fig.6b. Backscatter setup for calibration of the D_1 response with radioactive sources

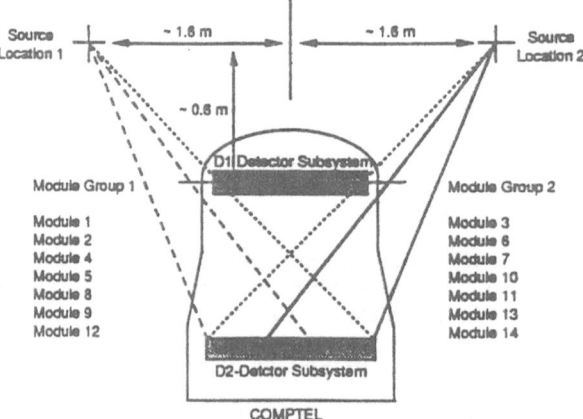

Fig.6c. Setup for calibration of the D_2 response with radioactive sources

In summer 1987 the telescope manufacturing was complete, and the flight instrument was mounted in a special experiment hall at the accelerator site onto a calibration dolly which allowed manipulation of the instrument such as to position the photon generating target of the accelerator beam pipe in any position within the telescope field of view (see figure 6). Radioactive sources were also used for calibration, they were mounted at the backside of the reaction target in order to have a compatible setup and local background environment. The calibrated energies and positions were:

Source	Energy	Zenith Angles
Mn^{54}	0.84	0 / 20 / 40* / 60
Na^{22}	1.27	0 / 10* / 20 / 40 / 60* / 80 / 102
Na^{24}	2.75	0 / 10* / 20* / 30 / 40 / 60*
Am^{241}/Be^9	4.43	0 / 10* / 20 / 40 / 60
O^{16}	6.13	0 / 5 / 10* / 15 / 20* / 30 / 40* / 50 / 60
C^{12}	12.14	0 / 5 / 10* / 15 / 20 / 30 / 40* / 102
Tritium	20.25	0 / 15* / 40

* At these zenith angles, more than one azimuthal orientation of the instrument was calibrated

Alltogether about the equivalent of 1.5 years of mission data were recorded in the 2 month calibration activity, before the telescope was shipped to the GRO satellite manufacturer for integration onto the observatory together with the 3 other GRO instruments.

The results of the instrument response as a whole are presented in figure 7 (spectral response) and figure 9-10 (imaging response) for different energies. It is noted that the response measurements at energies above 10 MeV suffer from the difficulties to provide clean monochromatic photon sources from nuclear reactions at an intensity which guarantees a satisfactory signal-to-noise ratio for the calibration data.

Response Calibration for Telescope Components

This approach makes use of calibration measurements and known instrument characteristics in a way to avoid the direct impact of calibration inadequacies onto response inaccuracies, however at the price of involving some prior (maybe imperfect) knowledge about the instrument's imaging response. In the COMPTEL Compton scatter telescope, this approach assumes the telescope response to detect an event via the Compton scatter process, where the incoming photon is Compton scattered in the upper detector plane and (approximately) absorbed in the lower detector plane; the instrument response is assumed to be determined mainly by the characteristics of the individual scintillation detector modules, rather than by the characteristics of any process resulting in quasi simultaneous scintillation events in the two detector planes.

The response of the D_1 and D_2 detector components had been calibrated with the experimental setup as shown in figure 6b,c. The setup for D_1 calibration had to introduce an additional coincidence with an external detector to constrain the energy deposit in the module ('backscatter arrangement'). For D_2 response calibration, 2 different source positions were required due to the shadowing effect of the D_1 detector platform.

The assumption of a Compton scatter event in the upper detector plane results in specification of known energy deposit and scatter angle in the upper detector plane from the incoming photon's energy, which is equivalent to known energy deposits in both detector planes as specified by the incoming photon's energy and the interaction positions in both detector planes. From this, one may calculate the telescope response as a result of the individual scintillation detector module responses to specified energy deposits via:

$$\hat{f}\left(\varphi_{geo}, \overline{\varphi}; E_\gamma\right) = \left(1 - e^{-\tau\mu\left(\hat{E}_2\left(\varphi_{geo}, E_\gamma\right)\right)}\right) \cdot P\left(\varphi_{geo}; E_\gamma\right) \cdot a\left(\overline{\varphi} \mid \varphi_{geo}; E_\gamma\right)$$

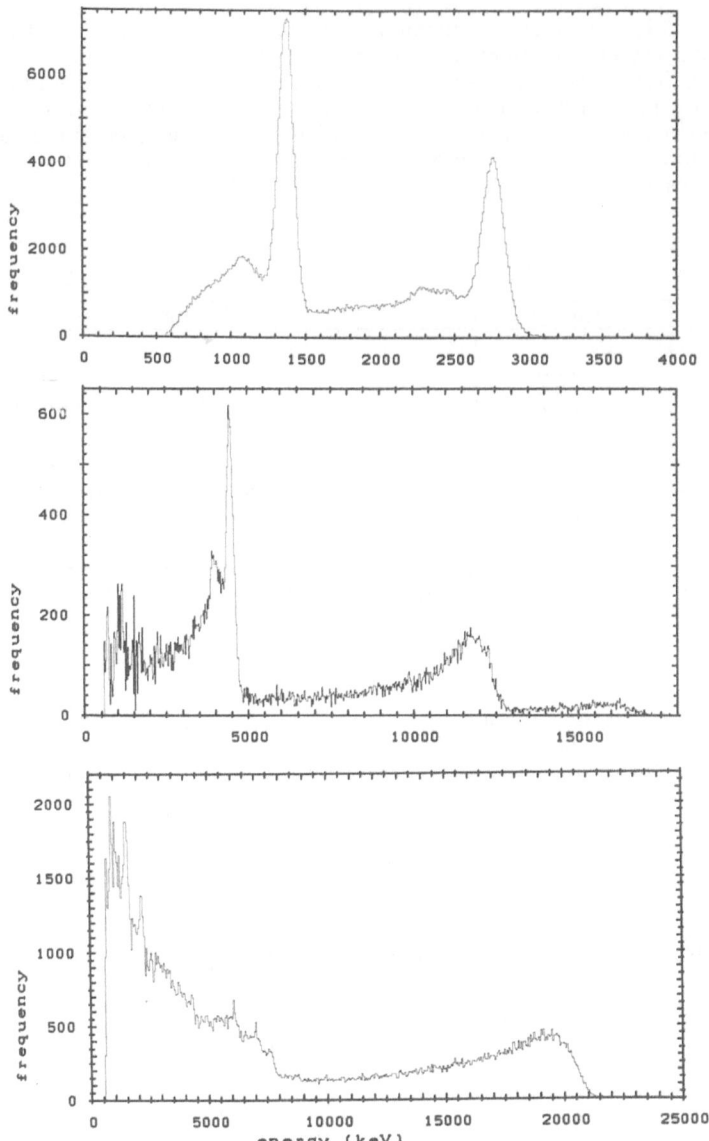

Fig.7. Energy response measurements for the telescope with calibration photons from radioactive ^{24}Na, (1.37 and 2.75 MeV), from the Boron target (12.14, 4.43, and 16.57 MeV) and Tritium target (20.25 MeV) nuclear reactions

as a probability per angular interval in φ_{geo} and $\bar{\varphi}$. Here 'P' is the probability density of φ_{geo}, which is given by the Klein-Nishina formula, and 'a' is the conditional probability density of $\bar{\varphi}$ for a given φ_{geo}. $(1 - e^{-\bar{\tau}\mu(\hat{E}_2)})$ is the probability for an interaction in D_2, determined by the average D_2 thickness $\bar{\tau}$ and the interaction coefficient $\mu(\hat{E}_2)$. The essential term is the convolution of the 2 energy responses of the 2 detectors in the regime enclosed by the detector thresholds E_l, E_u:

$$a\left(\overline{\varphi} \mid \varphi_{geo} ; E_\gamma\right) = \int_{E^{\cdot(\overline{\varphi}, E_{1,1}, E_{1,})}}^{E^{\cdot(\overline{\varphi}, E_{u,1}, E_{u,})}} g_1\left(E_1 \mid \widehat{E}_1\left(E_\gamma, \varphi_{geo}\right)\right)$$

$$\cdot g_2\left(E_2\left(E_1, \overline{\varphi}\right) \mid \widehat{E}_2\left(E_\gamma, \varphi_{geo}\right)\right) \cdot J\left(E_1, \overline{\varphi}\right) \cdot dE_1$$

E_2 can be expressed as a function of E_1 and $\overline{\varphi}$:

$$E_2\left(E_1, \overline{\varphi}\right) = -\frac{E_1}{2} + \sqrt{\frac{m_0 c^2}{\sin^2\left(\frac{\overline{\varphi}}{2}\right)} \cdot \frac{E_1}{2} + \left(\frac{E_1}{2}\right)^2}$$

In this approach the calibration measurements of the individual scintillation detector responses are much less critical with respect to the photon source position: background photon radiation can more easily be subtracted, as the characteristics of such scintillation detectors can be fitted by much more reliable models from text books or other laboratory measurements (compared to the complex Compton telescope coincidence instrument's response). The disadvantage of this approach, however, is that non-Compton scatter event types are not represented by the resulting telescope response function. We know that such event types exist and contribute significantly to the instrument response at energies above 10 MeV. Another disadvantage is the empirical inclusion of the interaction location uncertainty as a parameter, with which the calculated PSF is convolved via an assumed Gaussian:

$$f_s\left(\chi, \psi, \overline{\varphi} ; E_\gamma\right) = \int_{-\infty}^{\infty} \int_{-\infty}^{\infty} s_x\left(\chi, \chi'\right) s_y\left(\psi, \psi'\right) \cdot \tilde{f}\left(\chi', \psi', \overline{\varphi} ; E_\gamma\right) d\psi' d\chi'$$

where

$$s_x\left(\chi, \chi'\right) = \frac{1}{\sqrt{2\pi\lambda\left(E_\gamma, \varphi_{geo}\right)}} e^{\frac{-(\chi - \chi')^2}{2\lambda\left(E_\gamma, \varphi_{geo}\right)}}$$

$$s_y\left(\psi, \psi'\right) = \frac{1}{\sqrt{2\pi\lambda\left(E_\gamma, \varphi_{geo}\right)}} e^{\frac{-(\psi - \psi')^2}{2\lambda\left(E_\gamma, \varphi_{geo}\right)}}$$

is the function of the scatter direction uncertainty due to the imperfect positional resolution of the interaction location in D_1 and D_2, which is convolved with the $\overline{\varphi}$ response (characterized by the energy resolution in D_1 and D_2).

The final conversion of this model to yield the 3-dimensional model in spatial dataspace is:

$$\tilde{f}\left(\chi, \psi, \overline{\varphi} ; E_\gamma\right) = \hat{f}\left(\varphi_{geo}(\chi, \psi), \overline{\varphi} ; E_\gamma\right) \cdot \frac{\widehat{\Delta\Omega}}{\widetilde{\Delta\Omega}}$$

where the discrepancy between spherical and radial grid is taken into account via the solid angle weighting.

The response of scintillation detectors (both organic scintillation detectors with NE213 as active material, as well as NaI scintillators) is quite well known from previous measurements and literature. Therefore the measurements of the COMPTEL detector modules can be fitted with significant prior knowledge/good models, such that the impact of the particular detector module geometries are calibrated to precision. Nevertheless, also in this case the high energy regime above 10 MeV presents problems, with significant uncertainty of the response details at these energies (see figures 3 for D_1 and 4 for D_2)

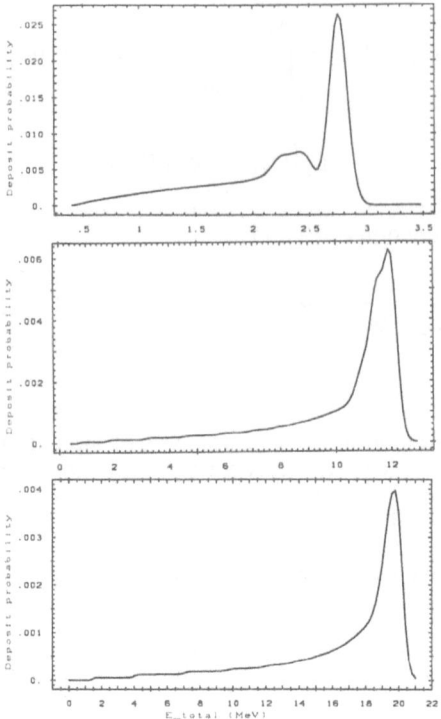

Fig.8a. Energy response as modelled from single detector characteristics, evaluated at energies similar to the telescope calibration energy measurements shown in figure 7

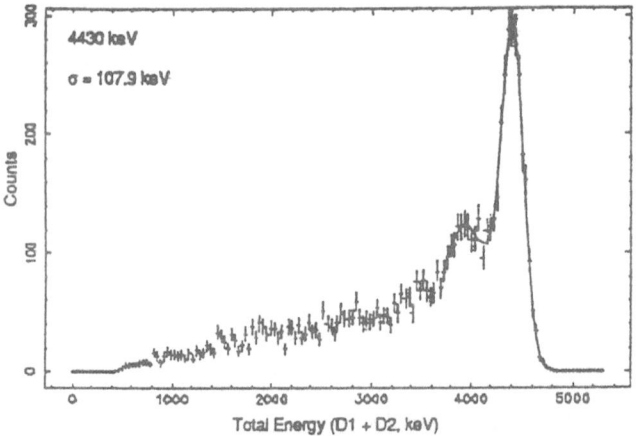

Fig.8b. Energy response as simulated from assumed incidence of 4.4 MeV photons at 10 degrees incidence angle

The telescope response derived from these detector characteristics can be seen from the modelled energy response (see figure 8a) and the 'point spread functions' dispayed in figures 9b and 10. The measured responses in energy domain (figure 7) and point spread function (figure 9a) show similar features, although in detail the distributions are more erratic and broader in the telescope run measurements. The 2-dimensional PSF's cannot directly be compared, as the modelled PSF does not contain the contribution from location uncertainty

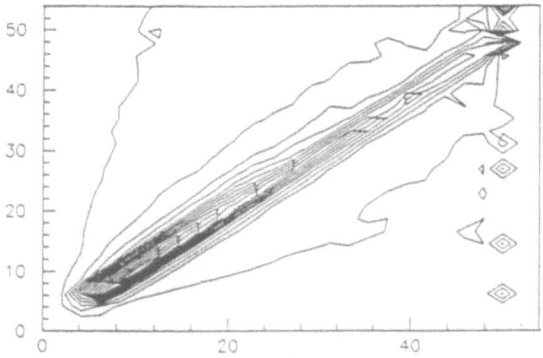

Fig.9a. Imaging response from telescope calibration measurement at 6.13 MeV

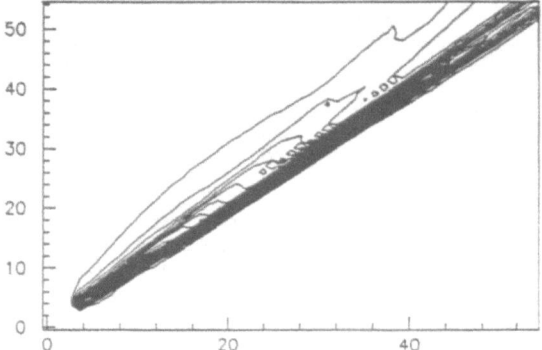

Fig.9b. Imaging response from modelling based on single detector characteristics, at 6.13 MeV. This response does not contain yet the effect of interaction location smearing, therefore it appears sharper than the measured response

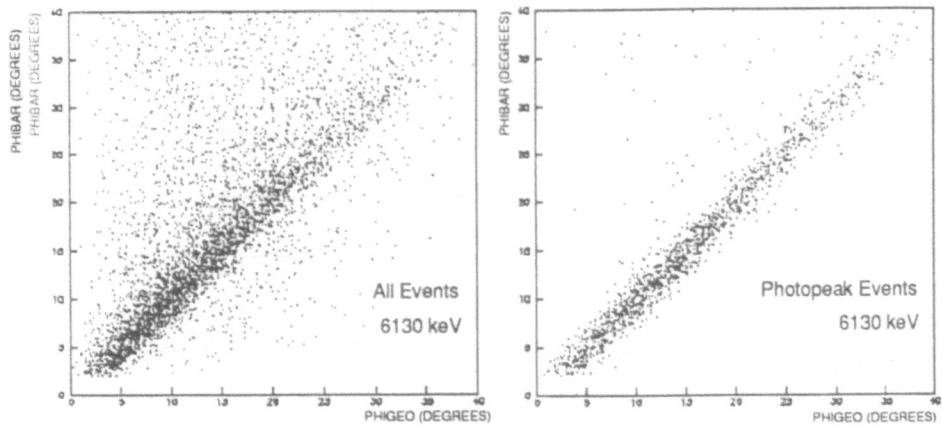

Fig.9c. Imaging response from simulations of the telescope with different selections (6.13 MeV incident photons)

213

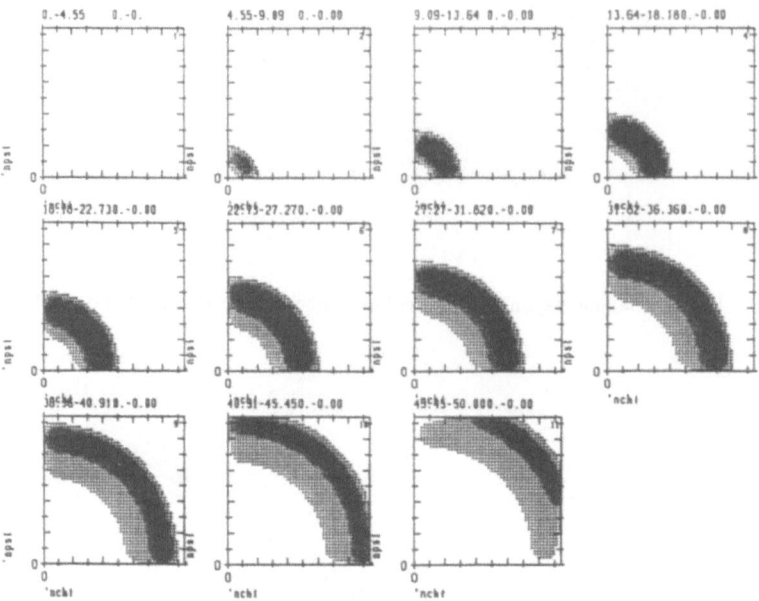

Fig.10. Imaging response in 3-dimensional dataspace; this matrix represents the point spread function assembling all blurring components of an image in COMPTEL data. It is based on the model response shown in figure 9 center, with an assumed (worst case) location broadening of 2 degrees

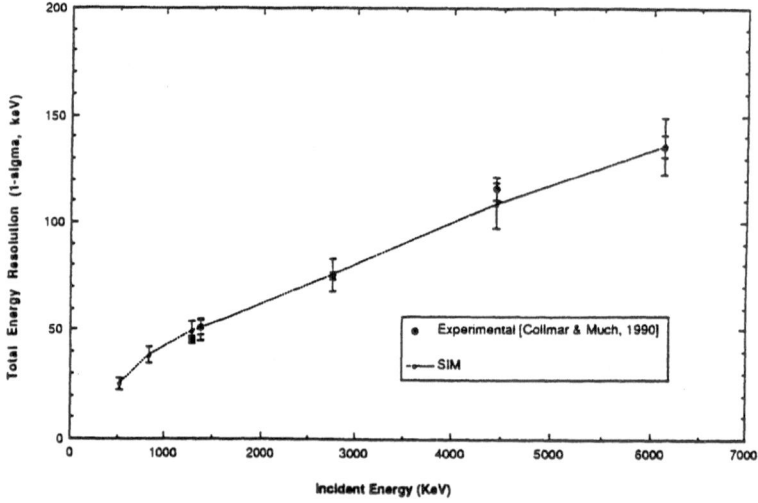

Fig.11. Telescope energy resolution summary (as determined from calibrations and simulations)

yet: rather, the differences in this case point to a contribution to blurring from the location uncertainty of about 1 degree. The 3-dimensional response displayed in figure 10 shows the cone-like feature of the spatial response, and the sidelobes due to incomplete photon absorption in the lower detector plane as secondary features within the main cone response at larger scatter angles.

Response Simulation

The telescope response can be determined also from a computer simulation of the photon interaction physics in the entire telescope, to as much detail as required by the response accurary. If this can be achieved, the telescope response can be regarded as entirely under-

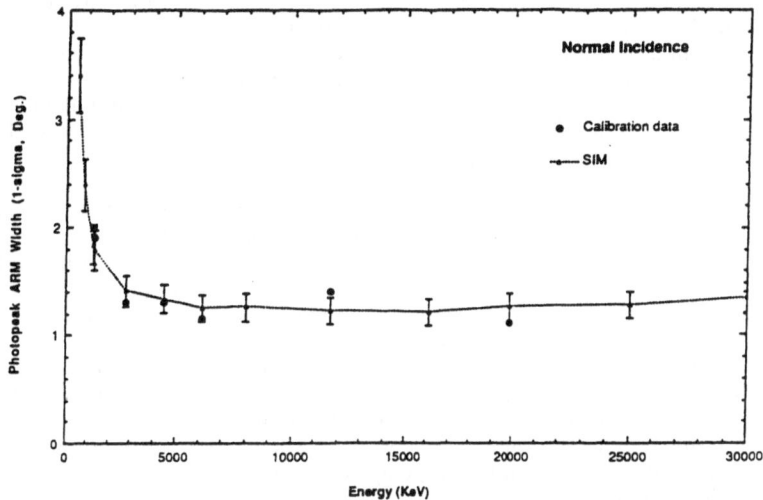

Fig.12. Telescope angular resolution summary (as determined from calibrations and simulations)

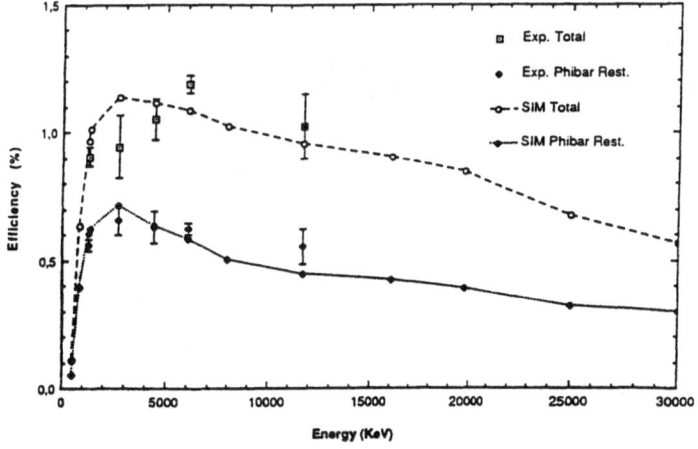

Fig.13. Telescope detection efficiency summary (as determined from calibrations and simulations)

stood. In reality, the simulated data must be compared to calibration measurements of high precision at a large set of photon energies and positions; agreement of simulations with calibrations at many calibration points with the above mentioned caviots provides growing levels of confidence for the simulation software.

The simulation software for COMPTEL is based on the GEANT software package (originating from CERN, Geneva). The instrument geometry had been incorporated, as well as the observed resolution broadening characteristics of the detector modules. We have simulated single detector responses and all calibrated telescope calibration runs at the different angles and energies (see figures 8b and 9c). Comparison of the energy spectra and angular resolution measures confirm the measured values to the expected accuracies; more detailed comparison of point spread functions is done when more statistics is accumulated from simulated event message.

SUMMARY

In conclusion, the spatial response of COMPTEL is available from different approaches, as presented in the figures of this paper. The summary spectral resolution and angular resolution is presented in figures 11 and 12. Figure 13 summarizes the effective area of the telescope as derived from simulations and calibration measurements.

Note that these results are not final yet, as the response determination methods and software are still being optimized, and the simulation database being extended with additional data. Therefore the characteristics presented in this paper document the result status as of the time of the launch of GRO. The final assessment of the optimum approach of response determination is pending. However, the goal of the COMPTEL collaboration is to improve the simulation tools until very accurate agreement with the measurement of the response (both approaches) is assured, as the parameter regime for incoming radiation could not nearly be scanned adequately in the 3 month calibration measurement campaign at theNeuherberg nuclear accelerator facility - COMPTEL's response details will rely on simulation software for a large region of incidence angles and energies.

Meanwhile the observatory had been launched into orbit successfully, and in the initial phase we will determine if the detector characteristics are still the same as at the time of calibration. If the detector response characteristics should have changed, the analysis software allows to fold in updated resolution figures into the response determination tools, so that the in-flight response at the desired angles and energies can be calculated based on calibration, simulation, and in-fligh measurements.

LITERATURE

Schönfelder et al., IEEE Trans Nucl.Sci., NS-31(1), 766 (1984)
Connors et al., this volume (1991)
Strong etal., this volume (1991)
Strong et al., int.resport COM-TN-MPE-K70-53, (1990)
Diehl et al.,NASA Calibration Review (int.report), (1990)
Kippen, Masters Thesis, Univ. of New Hampshire, USA (1991)

COMPTEL PROCESSING AND ANALYSIS SOFTWARE SYSTEM: COMPASS (REQUIREMENTS AND OVERVIEW)

J.W. den Herder [1], H. Aarts [1], K. Bennett [2], R. Diehl [3], W. Hermsen [1],
M. Johnson [3], J. Lockwood [4], M. McConnell [4], J. Ryan [4], V. Schönfelder [3],
G. Simpson [4], H. Steinle [3], A.W. Strong [3], B.N. Swanenburg [1],
C. de Vries [1], C. Winkler [2], I. Wood [2]

[1] Laboratory for Space Research Leiden, P.O. Box 9504, 2300 RA Leiden,
The Netherlands
[2] Space Science Department of ESA/ESTEC, Noordwijk, The Netherlands
[3] Max-Planck-Institut für Extraterrestrische Physik, Garching, FRG
[4] University of New Hampshire, Durham, USA

ABSTRACT

In this paper the COMPTEL Processing and Data Analysis System COMPASS is described. A general overview of the system software is given with a focus on the software engineering aspects.

INTRODUCTION

COMPASS is the software system which will support the analysis of the data from COMPTEL, one of the four astronomical experiments aboard NASA's Gamma Ray Observatory. COMPTEL, a Compton telescope, is the joint product of the Max Planck Institut für Extraterrestrische Physik in Garching-bei-München (BRD), the University of New Hampshire (USA), the Space Science Department of the European Space Agency and the Laboratory for Space Research Leiden (The Netherlands) (Schönfelder et al., 1984). During its first one and a half years in orbit COMPTEL map the sky in the energy range from 1 to 30 MeV. Thereafter in-depth studies of selected objects or regions of the sky are planned. In addition, COMPTEL is able to collect spectra of gamma-ray bursts (after receiving a trigger from BATSE, one of the other GRO instruments) and to collect neutrons from solar flares. In figure 1 the main scientific output is shown schematically. The expected lifetime of GRO is at least 5 years.

A special software environment to process and analyse the COMPTEL data was required because of the following reasons:

* Analysis of the data from a Compton telescope is rather complex and CPU intensive. Duplication of work between the participating institutes should be avoided and it should be possible to exchange the data between the sites;

* The size of the programs needed to perform the scientific analysis was estimated to be 100.000 lines of executable Fortran code (it turns out to be closer to 200.000). As all institutes are involved in the science, it was agreed that the development of the software should be shared between them;

Data Analysis in Astronomy IV, Edited by V. Di Gesù et al.
Plenum Press, New York, 1992

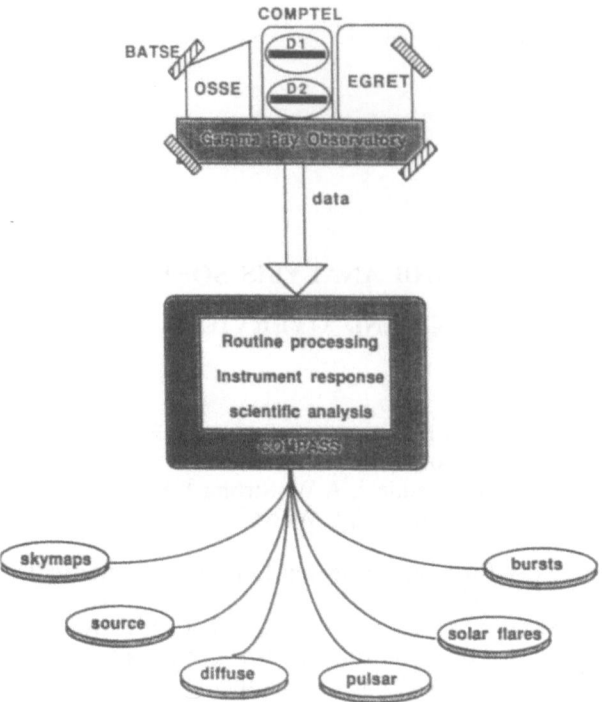

Fig.1. Schematic overview of the COMPTEL processing and analysis software system COMPASS

* All the institutes have different computers and operating systems. In order to facilitate exchange of software and to protect it against changes in the operating environment the scientific software should be independent of the actual host computer.

* The volume of data to be processed and analysed is large (25 Gbyte of raw data per year). This requires special software to handle and to control such volume;

* It is expected that the methods will evolve during the mission and that re-processing of the data can be important. At least it should be known under which conditions (software versions and control parameters) certain data were produced.

In this paper some aspects of COMPASS are described with a focus on software engineering and data handling. After describing the development approach, the system requirements and the system design, a global overview of the scientific components is given including the routine processing, the software to determine the instrument characteristics and the scientific analysis.

DEVELOPMENT APPROACH

In a project expected to last over more than 15 years (the development time and the expected life time of GRO) and with an involvement of over 30 people situated at different locations, it is crucial to have a common development approach. A top-down development approach was selected because this would allow the collaboration to define in a early stage the different components (subsystems) to be developed by each of the sites.

This top-down approach started with the definition of the development approach itself and the definition of the software requirements for the system. Subsequently the requirements for all its components (subsystems) were specified. Next an architectural design was made (for the full system and for each subsystem) and finally the detailed design together with the code. Each part of COMPASS was completed by an appropriate test plan and report as well as a user manual. Configuration control was strictly exercised for any of the generated documents as well as for the program units. Also the interfaces between the various subsystems were controlled.

The top-down approach made it possible to assign special parts of the software to each institute. However, modifications to some of the application subsystems required at a later stage in the project, could have been avoided if work on the central supporting software had been completed at an earlier stage. Also the fact that not all scientific methods were fully established in the early days (e.g. a lack of detailed scientific prototyping), resulted in some cases in software which did not meet its requirements (e.g. the throughput of some software was not sufficient) and which made significant modifications in a late stage necessary.

In addition to the development approach some other aspects of the project had to be fixed. Fortran 77 was selected as coding standard and none of the computer specific extensions was allowed for. Nevertheless, certain computer specific limitations were still encountered during the porting of the software from one computer type to another. Also a number of external software packages were selected, mainly on the basis of the availability on a large number of platforms. This included the CERN graphical display packages HBOOK and HPLOT, the relational database ORACLE, a maximum entropy package MEMSYS and the NAG mathematical libraries. Furthermore, where applicable, the data are stored in FITS format in order to allow other packages such as AIPS and IRAF to display COMPTEL data.

SYSTEM REQUIREMENTS

As part of the top-down development approach the requirements for COMPASS and for each of its subsystems were specified. The toplevel requirements include the following items:

* The system should be functionally equivalent at all sites;

* There should be a uniform interface to the host computer allowing for an easy exchange of the software between the sites;

* There should be a uniform user interface to all COMPASS software which also takes care of the differences between the Job Control Language at each of the host computers;

* The should be proper software configuration management. The tools should allow for evolutionary changes to the software and the status of the software should be traceable to any point in time;

* The data generated by COMPASS should be properly ordered and catalogued using the same identification code at all sites;

* The heritage of generated data within COMPASS should be traceable including the original data, the applied programs and the relevant control parameters;

* The exchange of data between the sites should be feasible allowing for different local file systems and data representations at any of the sites;

* There should be test environments which provide the same functionality as the production environment, and which permit the development and testing of new scientific applications, before they are released to the production environment.

As will be shown belolw, these requirements are implemented in a number of supporting system components. In addition, the primary purpose of COMPASS is, of course, to process and assist in the analysis of the scientific data. For this the following top level requirements were specified:

* Provide tools to process the raw data including corrections to remove time variable effects on the instrument and to express the data in meaningful physical units;

* Provide tools to determine the instrument response (module response, telescope response, response to instrumental and atmospheric background);

* Provide tools to perform specific scientific analysis in order to extract meaningful scientific results (e.g. timing analysis, sky mapping, source recognition, burst analysis, diffuse gamma-ray emission etc.);

* Provide tools to display scientific results achieved in COMPASS.

These requirements will be met by various scientific subsystems which will be discussed in some detail once the design of the supporting system components has been presented.

SYSTEM DESIGN

The design of the COMPASS system will be presented using three different views: the conceptual design gives a global view of the system, the data model provides an overview of the entities managed and controlled by the software and the activity model gives an overview of the various scientific functions. Together these provide an overview of the system.

Conceptual Design

In figure 2 the conceptual design of COMPASS is given. COMPASS is represented as a number of shells around the host computer. All independent functions are handled by separate layers. This makes each part of the software relatively immune to changes in another part of the software which is a big asset during maintenance of the system.

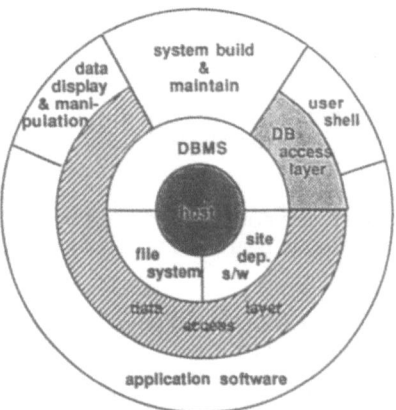

Fig.2. Conceptual design of COMPASS

The inner shell isolates COMPASS from the specific host computer and the data storage system. This shell contains three parts: program units which take care of site dependent system calls (time, bit shifting, terminal and printer access, etc.), program units which take care of the site dependent part of the local file system (such as filename convention, differences between the Fortran OPEN calls) and the database management system (DBMS). The DBMS contains, from the software perspective, a full image of the COMPASS system. It contains descriptions of all the available software at any of the four sites, of all data used and/or generated within COMPASS and all control parameters for the software.

The second shell, which is independent of the specific host computer, provides a number of services to the application developer and to the users. The data access layer separates the application programs from the data storage and allows modifications in the physical data structure without the need to modify the application software. This is a major benefit for a project of this size and duration. It also allows for the combination of data measured during different periods without modifying the application software (different physical datafiles are combined in the data access layer into a single logical datafile). The database access layer defines a number of operations to retrieve and store control parameters in the database.

The <u>third shell</u> is that which is seen by the users. The application layer contains the programs which perform the various steps in the data processing and analysis. There are 18 subsystems in this layer and their functions are described in more detail below. The "user shell" is a menu driven system which allows the user to select a job to run, to provide its control parameters, to select data from the database and to submit the job to the batch queue or to execute the job interactively. The user shell will check on the availability of the data and, when needed, retrieve the data first from archive. In addition, the user shell provides some additional functionality to the user: one may select control parameters from a previous run of the same task (program) or one may combine a number of tasks into one job in which the output of one program is used as input for the next program.

The data display and manipulation subsystem consists of three parts: a number of special display tasks for a given data-type (such as events), a task to download any of the data generated within COMPASS to a PC or workstation and commercial software to manipulate the data. There has been no selection of a single package for this purpose as this depends heavily on the available hardware at the four sites.

There is one zone in the diagram which is referred to as "system buond and maintain". Its function is twofold: it allows a privileged user to add components (application software or external data) to the system by inserting the appropriate records in the DBMS and it allows all users to query the database. In addition it handles all the communication between the different sites for database entities, software and data.

Data Model

A simplified data model is given in figure 3 using Entity Relationship Diagrams.

Each of the entities in the shaded boxes is stored in the DBMS which contains a full image of COMPASS. A USER may start one or more JOBs where each JOB may consist of one or more (connected) TASKS (programs). For each run of a task the CONTROL PARAMETERS are stored as well as the used and/or created data. This data is described by a DATA DESCRIPTOR in the DBMS. Each TASK specified by a single SOFTWARE DESCRIPTOR consists of a number of SOFTWARE LIBRARIES linked in a specified order. An important asset of this image in the database is that at any point in time the heritage of data can be traced and that previous versions of software can be re-installed and previous control parameters can be queried (making use of the functionality of the DBMS).

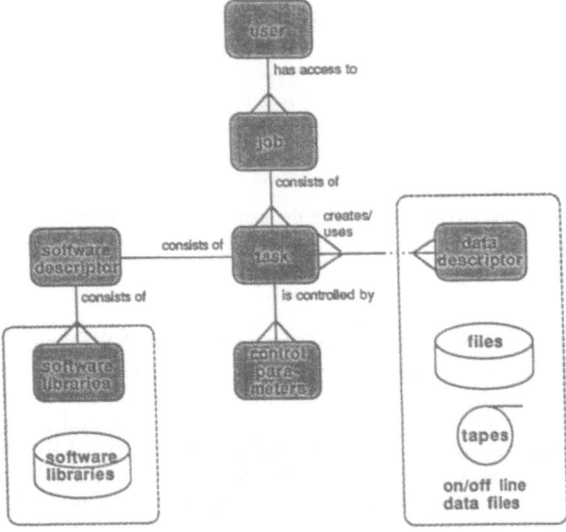

Fig.3. Simplified data model (entities stored in the DBMS are specified in the shaded boxes). Dotted lines indicate optional relations, solid lines mandatory relations between two entities, a single line denotes a one-to-one relation and a fork a one-to-many relations or many-to-one relation.

DATA DESCRIPTORS are the key entity used to facilitate data access and to protect COMPASS from site specific conventions for file names. Each DATA DESCRIPTOR has a number of attibutes which may facilitate its usage:

* a unique identifier consisting of the generating site, its type and a sequential number. This allows unique references to the data throughout COMPASS at all four sites;

* a time validity interval specifies the time for which the data are valid;

* the creating task, time and user help to separate valuable data from data produced by a too old version of the software and to trace back its heritage (creating task and related control parameters);

* a user supplied title may contain any scientific relevant information;

* a quality flag which can be used to specify the quality of the generated data;

* an access flag which can be used to provide certain users with a limited access to the data (e.g. guest investigators will only be interested in higher level data).

For the other entities stored in the database, such as the software descriptors, the task descriptors and the related control parameters, similar attributes are stored.

In order to facilitate the development of new scientific methods and to allow users to test new software without interfering with the production environment test environments can be created. For each test environment a set of new database tables is created. A user working in a test environment has read only access to the production environment (and thus data) and read and write access to his test environment (see figure 4). These test environments were implemented by exploiting the functionality of the DBMS (synonyms and views) (Johnson, 1989).

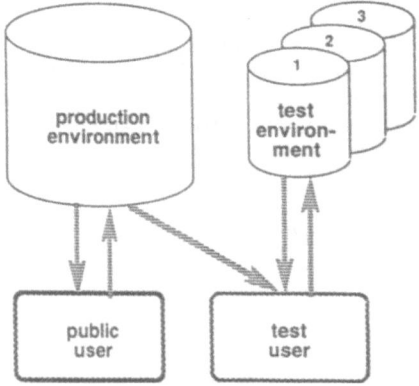

Fig.4. Implementation of test environments with the same functionality as the production environment

Activity Model

In figure 5 a simplified activity model for COMPASS is presented and three main components can be distinguished: the processing of data in order to correct the measured data for time dependent variations in the instrument and convert the raw data in physical units; the determination of the instrument response including the experimental background and finally the scientific analysis. Indicated in this activity model is that some of the activities are exercised mainly at one of the institutes (processing) or distributed over all institutes (scientific analysis). In addition it is shown that the COMPASS system is used for the analysis of the calibration data and the generation of simulated data.

A global overview of the three groups of activities is presented below.

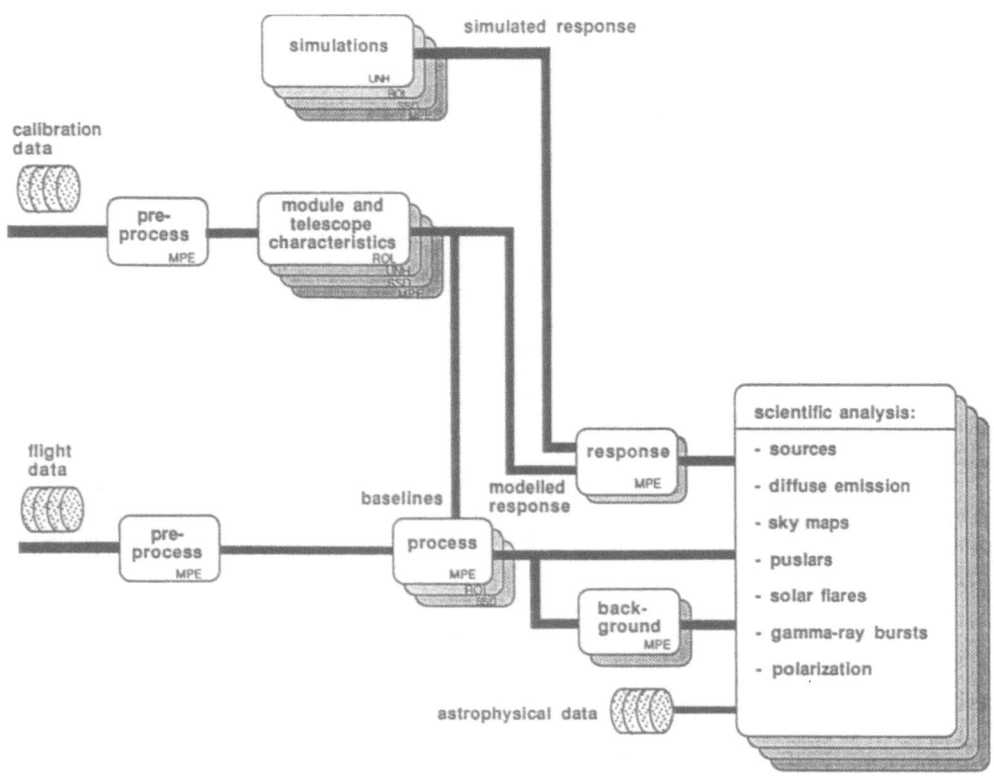

Fig.5. Simplified activity model for COMPASS indicating the main activities (boxes), the datastream (solid lines). The dataflow is from left (top) to right (bottom).

DATA PROCESSING

COMPTEL generates various types of raw data including in-flight calibration data, insttrument state of health data, burst data and event data (see also Schönfelder, these proceedings). The raw events are characterized by various pulse height outputs, a time-of-flight measurement, a pulse shape discriminator measurement and a time tag (with a accuracy of 1/8 msec). In addition COMPTEL may, when triggered by BATSE, generate burst spectra.

As part of the routine processing the state of health of the instrument is checked and the various operational modes of COMPTEL are separated (e.g. during the some parts of the orbit the instrument is switched off). Next the raw data are converted into physically meaningful units. The establishment of correction factors for gain fluctuations in the 154 photo multiplier tubes is a major goal in this step ("in-flight calibration"). Finally the energy deposit and location of each event in the detectors and the angle between the incoming gamma and the scattered gamma in the upper detector is determined ("event & burst processing"). The raw burst spectra are also converted into a proper energy scale. After this step, the resultant data products can be interpreted without particular knowledge about the operational conditions of COMPTEL and data measured during different periods can be combined.This is visualized in figure 6.

As is indicated in figure 6 the data processing which is planned for the flight data, was also applied to the prelaunch telescope calibration data.The main benefit of this is twofold: significant experience with routine data processing could be obtained before the actual launch and there is a consistent handling of telescope calibration data and the flight data. In addition COMPASS is used for the determination of the characteristics of COMPTEL.

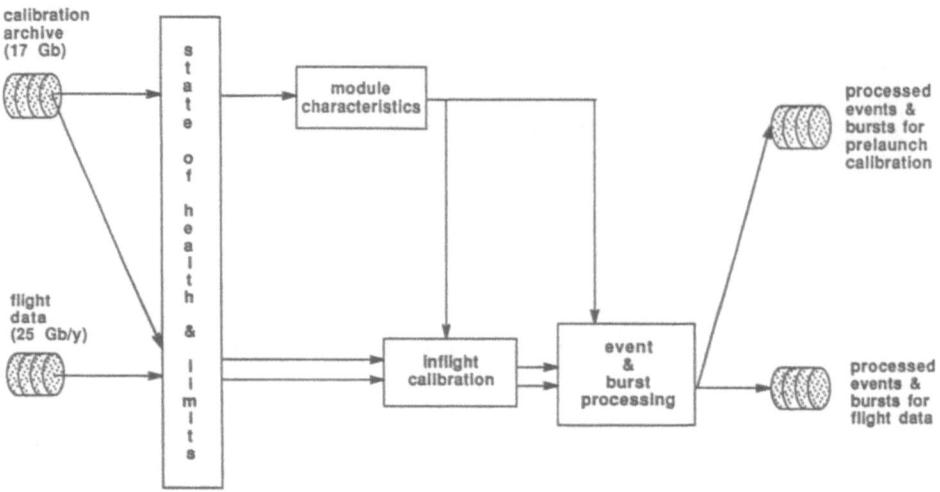

Fig.6. Routine data processing. The same approach is followed for prelaunch calibration data as for flight data.

INSTRUMENT RESPONSE

The second step is to determine the instrument response. This includes the determination of the point spread function, the energy response of the telescope and the energy response of the burst system using prelaunch calibration data (see also Diehl, these proceedings). As is indicated in figure 7 the telescope response can be extracted from different sources, e.g. based on the single detector response the telescope response can be calculated ("modelled telescope response"). This can be compared to the measured response during the prelaunch calibration period for selected energies and positions. Finally this information is supplemented by simulation data to interpolate and extrapolate to energies and angles not covered during the prelaunch calibration ("simulated response"). This Monte Carlo code has been based upon the CERN package GEANT. Special software was required for the energy response of the burst system as there is a clear angular dependence in this response (due to intervening material between the burst location and the detectors).

In figure 7 the point spread function is represented in a 3 dimensional "data space" given by the direction of the scattered photon between the upper and lower detector and the Compton scatter angle in the upper detector (Diehl, 1989). Since only the scatter angle in the upper detector is know, a source generates a cone-like pattern in this dataspace. In this space an important fraction of the scientific analysis will be performed.

Apart from this instrument response the instrumental and atmospheric background requires special attention. Various sources of information will be combined to make the best estimates for the instrumental background including:

* internal symmetries of the instrument assuming that the instrumental background is symmetrical;

* smoothing of measured data and the use of special observations (high latitude with a virtually empty sky);

* application of special data selections to enhance background features;

* study of features in the data as a function of the position in orbit including special modes of operation;

* simulations of radiation induced in the spacecraft.

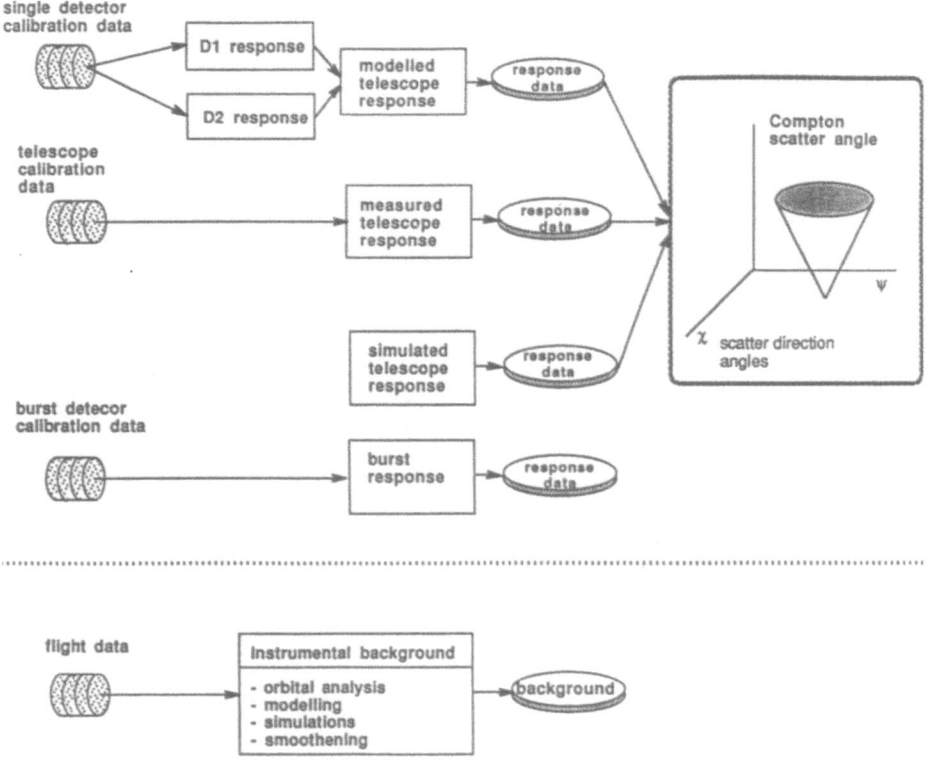

Fig.7. Instrument response determination

SCIENTIFIC ANALYSIS

For the various scientific objectives software has been developed and a global overview of these subsystems is presented in figure 8.

Sky imaging and source detection

Using the imaging capacity of the Compton telescope an image of the sky can be extracted from the event data using the maximum entropy method (Strong et al., 1990 and these proceedings). In addition sources can be searched for using the likelihood method. Because the point-spread function is sensitive to polarization of the gamma-ray radiation the degree of polarization of strong sources might be determined also. Once the strongest sources are recognized the remaining gamma-ray radiation will be studied in terms of the diffuse galactic gamma-ray emission (and of course an instrumental and atmospheric background component) using the maximum likelihood approach. Various model components will be included in the latter analysis including standard CO and HI maps and models forthe inverse Compton radiation.

Pulsar analysis

In addition to this analysis the event data can be searched for pulse emission (1/8 msec resolution). Various standard techniques are employed including the Z^2 test (Buccheri and de Jager, 1989). In order to enable the detection of pulsed gamma-ray emission several hundred pulsars will be regularly monitored by a number of radio observatories to provide timing parameters used in the COMPTEL analysis (Busetta, these proceedings).

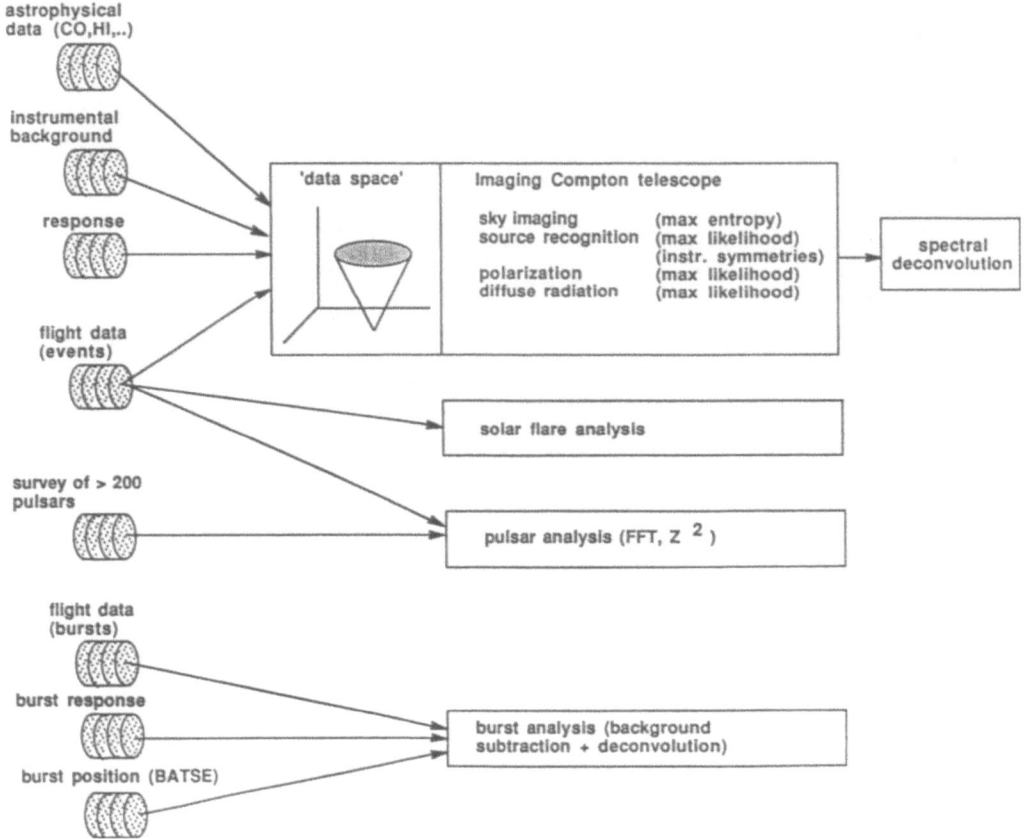

Fig.8. Global overview of the COMPTEL scientific analysis

Transient sources

When BATSE detects a gamma-ray burst a trigger will be sent to COMPTEL and COMPTEL will record (using two of its detectors) the energy spectra in the range from 0.1 to 12 MeV (Winkler et al., 1989). These spectra will be background subtracted and deconvolved using the maximum entropy method. For a proper deconvolution the position of the transient source is required.

Solar flares

When such a burst is coming from the direction of the sun it is assumed to be a solar flare and COMPTEL will, in addition to the recording of the gamma-ray burst, change its mode to the solar neutron mode (Lockwood et al., 1989). In this mode the time-of-flight measurement will be used to determine the neutron energy and results (energy spectra and flux time profiles) will be extracted.

REFERENCES

Buccheri, R. and de Jager, O.C., 1989, Detection and description of periodicities in sparse data, in "Timing Neutron Starts", H.Ogelman and E.P.J.van den Heuvel ed., Kluwer, Dordrecht

Busetta, M., et al., Pulsar Analysis within COMPASS, in these proceedings

Diehl, R., 1989., Data Analysis of the COMPTEL instrument on the NASA Gamma Ray Observatory, in "GRO science workshop", Greenbelt, USA

Diehl, R., et al., Response Determination of COMPTEL using Calibration Measurements, Models and Simulations, in these proceedings

Johnson, M., 1989. Die Verwendung von ORACLE als Systemkomponenten für Astronomie und Raumfahrt, in "Deutsche ORACLE Anwender Gruppe""

Lockwood, J.A., et al., 1989. COMPTEL as a Solar Gamma-Ray and Neutron detector, in "GRO science workshop"", Greenbelt, USA

Schönfelder, V., et al., 1984. IEEE Trans. Nucl. Sci. NS-31(1)

Schöonfelder, V., et al. The GRO-COMPTEL Mission: Instrument Description and Scientific Objectives, in these proceedings

Strong, A.W., et al., 1990. Maximum Entropy Imaging with COMPTEL data, in Proceedings of the XXI Cosmic Ray Conference, Adelaide, Australia

Strong,A.W., et al.., Maximum Entropy Method for Sky Imaging and Spectral Deconvolution, in these proceedings

Winkler, C., et al., 1989, Observing Cosmic Gamma-Ray Bursts with COMPTEL, in "GRO science workshop", Greenbelt, USA

PULSAR ANALYSIS WITHIN C O M P A S S

M.Busetta[1], K.Bennett[1], R.Buccheri[2], J.Clear[1],
R.Diehl[3], J.W. den Herder[4], W.Hermsen[4],
L.Kuiper[4], J.Lockwood[5], M.McConnell[5], J.Ryan[5],
V.Schönfelder[3],A.Strong[3],B.G.Taylor[1], C.Winkler[1]

1) Space Science Department of ESA/ESTEC,
   P.O. 299, 2200 AG Noordwijk ZH,
   The Netherlands
2) I.F.C.A.I. / C.N.R., Palermo, Italy
3) Max-Planck-Institut für Extraterr. Physik,
   Garching, Germany
4) Laboratory for Space Research Leiden,
   The Netherlands
5) University of New Hampshire, Durham, USA

Abstract

The COMPASS system is a large software package, involving
several subsystems, whose development has been 'shared' by
the four institutes of the COMPTEL Collaboration.
A principle goal of the COMPTEL mission is the observation of
point sources whose emission, in gamma or in other energy
bands, is pulsed. In order to search for periodicity in the
gamma-ray emission from these sources and from the new
COMPTEL-detected sources, the COMPASS subsystem PUL (for
"PULsar analysis") has been designed. In this paper a
description of the PUL performance, with the adopted analysis
methods, is given.

Data Analysis in Astronomy IV, Edited by V. Di Gesù *et al.*
Plenum Press, New York, 1992

Introduction

The **COMP**tel **A**nalysis **S**oftware **S**ystem COMPASS[1] has been designed and developed to pre-process, select and analyze the 1-30 MeV events collected by the Compton Telescope COMPTEL[2], on the NASA spacecraft GRO[3], the Gamma Ray Observatory. Elsewhere, in these proceedings, a detailed description of this system is given.

Within COMPASS, PUL is the software subsystem devoted to the "Pulsar Analysis" (periodicity searches, light-curve analysis and modelling, etc...). More specifically, PUL v.1 (March 1991) includes the following major tasks:

1)   analysis of signal from already "known" pulsars
2)   analysis of signal from newly discovered COMPTEL-sources
3)   multiple observation analysis
4)   global analysis

Then, the architectural design of PUL involves four "analysis functions":

1) **PULIND** which performs the analysis of single observations of the known pulsars, e.g.:

▸   search for periodicity in a "small" period range around the value extrapolated at the COMPTEL observation epoch and

▸   analysis of the pulsar light curve in the case of positive detection or

▸   evaluation of the upper-limit for the source flux in the case of no detection.

The need for pulsar parameter extrapolation is minimized because of simultaneous radio observations[4],[5]

2) **PULCOM** which performs a "large band" search for periodicity in case of new gamma-ray sources detected by COMPTEL (e.g. whose parameters are not known). Once again, in case of positive detection, a specific analysis of the "new-pulsar" light-curve will be performed.

3) **PULMLA** which performs analysis of several observations of the same source (so improving the statistics in the case of

"weak" pulsars, whose pulsation features could be undetectable in a single observation).

4) **PULGLA** for the "global analysis" of a sample of pulsars, which single observations have given a null detection. In this case, the combination of the significances of the single observations can supply information about the behaviour of the entire source sample.

PUL data-flow

PUL subsystem requires "external" inputs; these external interfaces and some intermediate steps (before the actual analysis procedure) are shown in Fig.1.
The Event Message dataset (EVP) contains the information (arrival time at the spacecraft, direction, energy released) for each of the events in one observation day. The first step performed by PUL is the "rough" pre-selection of events: this step reduces the data volume by:
- selection of the "good" time intervals (by removing those parts of the observations which are affected by South Atlantic Anomaly (SAA), Earth in the field of view, ...;
- selection of the global energy range of interest;
- selection of the events by photon arrival direction.
The output EPP dataset reproduces the same structure of the EVP dataset (all information are kept for all events), but only for the potentially "good" events.
Subsequently, the actual "scientific" selection discriminates between events in terms of more scientific criteria (e.g. a restricted energy range, by taking into account the Point-Spread Function). Eventually, only the list of arrival times for the selected events will be used for the analysis (PSE dataset).
PSE contains the time values, as measured at the spacecraft, which are to be corrected, during the analysis, by referring them to the Solar System Barycentre. This correction (dependent from the source coordinates which are supplied, as

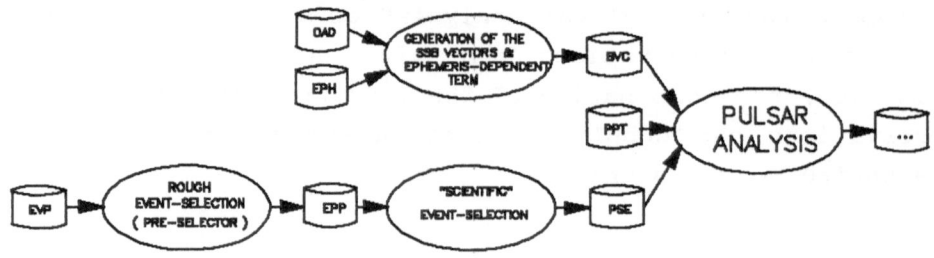

EVP : Event Messages
EPP : Pre—selected Event Messages
PSE : Selected Event Arrival Times

OAD : Orbit/Aspect Data
EPH : Ephemeris JPL DE200
BVC : Barycentric Vector Components

PPT : Pulsar Parameters

Fig. 1

Tab. I

INPUTS:	1.	UTC arrival time
	2.	GRO_{geo} vector components
	3.	source direction
	4.	JPL DE200 Ephemeris[6]

DATA PROCESSING:
- Time units transformation from UTC to ephemeris time
- Evaluation of the components of the GRO_{SSB} vector using the information on the Solar System Barycentre position supplied by the Ephemeris
- Evaluation of light travel time to the Solar System Barycentre
- Computation of relativistic delay due to gravitational field of the Sun

ANALYSIS METHOD:
The term to be added to UTC time to get the "corrected" SSB time is computed by adding the above defined terms[7]

well as the other pulsar parameters, by the Pulsar Parameter Table) uses the JPL Ephemeris at epoch 2000.0 and the geocentric spacecraft position which is supplied elsewhere by COMPASS.

The "source" and "off-source" event time lists, PSE, and the list of parameters required for the barycentric correction, BVC, together with PPT information, are the inputs to the PUL functions for the analysis of the signal from the source.

Steps of Pulsar Analysis by PUL

The main purpose of PUL is to search for periodicity in the gamma-ray emission of point-sources: the two main analysis methods available in PUL are described in detail.
The other analysis procedures directly follow and depend upon the results previously obtained. The barycentric correction is always required at the beginning of the analysis, before starting the search for periodicities in the COMPTEL detected gamma-ray signal.

Barycentric correction of the photon arrival times is required in order to eliminate the effect of the "clock motion" (due to the orbiting observatory) affecting the time measurements at GRO. Each time value is corrected by referring it to the Solar System Barycentre. This reference-frame transformation requires the spacecraft coordinates, the source coordinates, and the Solar System Ephemeris: PUL, as well as the other GRO systems for timing analysis, adopts JPL DE200 Ephemeris. The routine to properly read and handle the ephemeris information was kindly supplied by L.Rawley. In Tab.I a schematic description of the procedure is given.

Search for periodicity methods are employed to detect the presence of pulsed signal in the gamma emission of the sources observed by COMPTEL. Different approaches are to be followed in the two cases: "already known pulsars" and "new sources".

For the <u>already known pulsars</u> (see Tab.II), PUL is required to verify the presence of a pulsation at the expected frequency in the signal from a "pulsar" source. In this case, the first step is the extrapolation of the pulsar parameters to the COMPTEL observation epoch and then the evaluation of the scanning steps required to span the period range. The pulsar parameter reference time should be close to the COMPTEL observation, or the accuracy in the parameter values very high, to avoid scanning in period and period derivative which would reduce the sensitivity of the method.

If a positive detection is obtained, a check against system frequencies has to be performed: in order to state "positive detection", analysis of arrival times from an "off-source" event list, at the same period value where the positive detection from the source signal has been obtained, must give a negative result.

Large-band scanning is required to detect the presence of a pulsed component in the gamma-ray emission of a <u>new COMPTEL source</u> (see Tab.III): PUL includes the facility to perform this large-band search by FFT analysis.

For each COMPTEL observation (14 days) the events are grouped in time blocks (whose length is defined by user) and the FFT procedure over each data block generates the corresponding power spectrum. The resulting overall power spectrum (obtained by adding the single spectra), has to be properly "normalized" to estimate the significance of any power peak. This procedure has to be iterated for each of the scanning steps in Right Ascension (RA) and Declination (D) that are required to span the uncertainty in the source location. The "significance" of each power peak is verified in terms of comparison with a chosen threshold value. Also, a check to discriminate against "system frequencies" has been foreseen: if a 'significant' peak has been detected, a search for periodicity at this frequency value is also performed in an "off-source" list of events to verify the presence of the same signal.

Light-curve analysis and **modelling** procedures can be performed in case of 'positive detection'.
The representation of the pulsar light-curve produced by

Tab. II

<table>
<tr><td>INPUTS:</td><td>1.</td><td>UTC arrival time list from the source direction</td></tr>
<tr><td></td><td>2.</td><td>UTC arrival time list from the off-source direction</td></tr>
<tr><td></td><td>3.</td><td>List of GRO_{SSB} components</td></tr>
<tr><td></td><td>4.</td><td>Pulsar parameters</td></tr>
<tr><td>DATA PROCESSING:
<u>list 1.</u></td><td>–</td><td>Determination of scanning steps in period P, period derivative Pdot, Right Ascension RA and Declination D, required to span the pulsar parameter uncertainty regions</td></tr>
<tr><td></td><td>–</td><td>For each scanning step:
▸ barycentric correction of the arrival times
▸ time correction by referring it to binary focus (if the source is a binary one)
▸ computation of residual phases
▸ Evaluation of the phase distribution statistical significance</td></tr>
<tr><td></td><td>–</td><td>Determination of the set of pulsar parameters (<u>P</u>, <u>Pdot</u>, <u>RA</u> and <u>D</u>) which maximize the statistical variable adopted to estimate the significance of the phase distribution</td></tr>
<tr><td><u>list 2.</u></td><td>–</td><td>Barycentric correction of the arrival times</td></tr>
<tr><td></td><td>–</td><td>Computation of the residual phases at P=<u>P</u> and Pdot=<u>Pdot</u></td></tr>
<tr><td></td><td>–</td><td>Evaluation of the statistical significance for the phase distribution</td></tr>
<tr><td>ANALYSIS METHOD:</td><td></td><td>Significance evaluation of residual phase distribution by Z_n^2 [8] variable</td></tr>
<tr><td>ESTIMATION CRITERIA:</td><td></td><td>Comparison between results from the analysis of signals (at <u>P</u> and <u>Pdot</u>) detected in the two lists 1. and 2.: a <u>positive detection</u> means that the signal detected in list 1 does not appear in list 2.</td></tr>
</table>

Tab.III

INPUTS:	1.	UTC arrival time list from the pulsar direction
	2.	UTC arrival time list from an off-source direction
	3.	List of GRO_{SSB} components
DATA PROCESSING: list 1.	–	Determination of scanning steps in RA and D to span the source parameter uncertainty regions
	–	For each of N_S scanning steps
	▸	group the UTC times in blocks which time length has been chosen
	▸	barycentric correction of the arrival times
	▸	For each time block: – binning of the SSB times – evaluation of power spectrum by FFT
	▸	Sum of power spectra
	▸	Transform the spectral densities into chi-sq. distributed variables
	▸	Determination of maximum power value and relative signal frequency
	–	Maximum power determination among the maxima found at the different scanning steps and relative values of $\underline{P}, \underline{RA}$ and $\underline{D}$
	–	Evaluation of the probability relative to the detection of this signal after N_S trials
list 2.	–	Barycentric correction of the arrival times
	–	Evaluation of residual phases at $P=\underline{P}$
	–	Evaluation of the statistical significance for the residual phases distribution
ANALYSIS METHOD:		For list 1 the transformation from spectral densities to chi-sq. variable in order to evaluate the probability to get the found power peak from a poissonian distribution. For list 2 the evaluation of the significance of residual phase distribution is made by computing of Z_n^2 [8]
ESTIMATION CRITERIA:		Comparison between results for the two list at $P=\underline{P}$

classical histogramming is affected by the a-priori choice of the number of bins. A "free-binning" representation of the light-curve is achieved by the Kernel Density Estimator analysis[9]: this method supplies the pulse shape after smoothing statistical fluctuations affecting its structure (at least, at the defined confidence level).

Further analysis of the pulsar light-curve may be performed by modelling the single structures (peaks): for each phase window (defined by user as "peak region") a fitting procedure may be performed to describe the peak in terms of a gaussian or sinusoidal function and a more detailed analysis of the pulse structure is possible in terms of an exponential model to fit limited regions of the peak.

The **multiple analysis** task has been included in PUL in order to fully exploit the available COMPTEL data; the analysis of several observations of the same target could make a "weak" pulsed signal detectable. This analysis can only be performed for sources whose parameters are well known (no scanning required); after a "negative" detection from single observation analysis, a binning of the residual phases is obtained and the resulting 2000 bin histogram is stored. The procedure can handle up to 10 such histograms: they are "added" and the statistical significance of the resulting "light-curve" is evaluated.

The **global analysis** task involves a sample of pulsars which have given no positive detection: sum of single significances may bring to visibility an overall signal as obtained from many small, statistically undetectable, signals. The algorithms implemented in PUL are elsewhere[10] described.

Perspectives

The present version of the PUL subsystem has no pretence to be exhaustive: it will be subject to improvement and increments in terms of possibility of analysis sophistication offered to users.

With PUL v.2 we are planning to enlarge the fields of interest covered by COMPASS: PUL should include also the possibility to perform timing analysis of signal from binary sources. This means introducing the possibility of handling signals at higher period values ($P \approx 10 \div 100$ s.) than the typical values foreseen by the present version ($P \leq 1$ s.), or with a "quasi-periodic" behaviour; these tasks will require different analysis instruments and methodologies.

A good knowledge of the actual instrumental performances of COMPTEL, following the final validation and calibration of the software with the in-flight data, will probably suggest further improvements and corrections.

Acknowledgments

The authors acknowledge the other colleagues of the COMPTEL Collaboration. M.Busetta acknowledges the receipt of an ESA Research Fellowship.

References

[1]   J.W. den Herder, H. Aarts, K.Bennett, R.Diehl, W. Hermsen, M. Johnson, M. McConnell, J. Ryan, V.Schonfelder, G. Simpson, H. Steinle, A.W. Strong, B.N. Swanenburg, C. de Vries, C. Winkler, I. Wood; "COMPTEL Processing and Analysis Software System: COMPASS", (this volume)

[2]   R. Diehl; Sp. Sc. Rev. 49, 85, 1988

[3]   V. Schönfelder; Adv. Space Res., vol.10, no.2, p.243, 1990

[4]   R.Buccheri, V.Schonfelder, R.Diehl, G.Lichti, H.Steinle, B.N.Swanenburg, H.Aarts, A.Deerenberg, W.Hermsen, J.Lockwood, J.Ryan, G.Simpson, J.Macri, K.Bennett, J.Clear, C.Winkler, B.J.Taylor, A.G.Lyne, R.N.Manchester: "Gamma-Radio coordinated pulsar observations with GRO-COMPTEL and ground based radio observatories"; Proceedings of an International Conference, Strasbourg, 1987

[5]   J. H. Taylor: "Radio pulsar timimg observations for GRO"; EGRET Science Symposium Proceedings, Greenbelt, 1989

[6]   E.M.Standisch, X.X.Newhall: "The JPL export planetary ephemeris"; Pasadena, 1988 (Tab.I)

[7]   L.Fairhead, P.Bretagnon; Astron.Astrophys. 229, 249, 1990

[8]   R.Buccheri, B.Sacco; "Data Analysis in Astronomy", p.11, 1985

[9]   C.O. de Jager, J.W.H. Swanepoel, B.C. Raubenheimer; Astr.Astroph. 170, 187, 1986

[10]   R.Buccheri, K.Bennett, G.F.Bignami, J.B.M.Bloemen,
         V.Boriakoff, P.A.Caraveo, W.Hermsen, G.Kanbach,
         R.N.Manchester, J.L.Masnou, H.A.Mayer-Hasselwander,
         M.E.Ozel, J.A.Paul, B.Sacco, L.Scarsi, A.W.Strong;
         Astr.Astroph. 128, 245, 1983

MAXIMUM LIKELIHOOD METHOD APPLIED TO COMPTEL

SOURCE RECOGNITION AND ANALYSIS

H. de Boer[2], K. Bennett[4], H. Bloemen[2], J.W. den Herder[2],
W. Hermsen[2], A. Klumper[2], G. Lichti[1], M. McConnell[3],
J.Ryan[3], V. Schönfelder[1], A.W. Strong[1], C. de Vries[2]

[1] *Max Plank Institut für Extraterrestische Physik, D-8046 Garching, FRG*
[2] *Space Research Leiden, P.O. Box 9504, NL-2300 Leiden, NL*
[3] *Institute for the Study of Earth, Oceans and Space,*
    University of New Hampshire, Durham NH, USA
[4] *ESA/ESTEC Space Science Department, NL-2200 Noordwijk, NL*

1. INTRODUCTION

As the instrumental resolution of high-energy astronomical experiments increases, the observer is confronted with a 'space of outcomes' (hereafter *dataspace*) with at most a few events per bin (particularly if time resolution comes into play). This implies that the sought signal is not only contaminated by additive noise components (*e.g.* instrumental, earth's atmosphere), but is also masked by relatively large intrinsic statistical fluctuations. In the case of the 1-30 MeV imaging telescope COMPTEL, the number of counts per bin is of the order 1, if the full resolution is to be explored.

The canonical approach to data analysis in such a situation is to model the probability distribution (pdf) of the measured quantities, based on knowledge of the instrumental response and photon intensities. Such models may contain free parameters (say θ) which one wants to constrain by the experimental result (in our case *e.g.* source position or flux). For any such set of parameters, one may calculate the probability of the measurement at hand. If one considers this probability as a function θ for a given experimental result, the appropriate name is likehood function and is denoted by $L(\theta)$. Given two hypotheses H_0 and H_1 (which may differ only in parameter values), the likelihood ratio $L(H_0)/L(H_1)$ is to be interpreted as the degree to which the data support H_0 against H_1 (Edwards, 1972). It is evident that the likelihood ratio is a convenient statistic for composite hypothesis testing and parameter estimation. In this paper we shall overview some of its properties and its application to γ-ray astronomy.

Data Analysis in Astronomy IV, Edited by V. Di Gesù *et al.*
Plenum Press, New York, 1992

In section 2 we briefly outline the interpretation of test-statistics in general, and discuss how the likelihood ratio method (LRM) can be applied in practice . We briefly describe the COMPTEL dataspace in section 3, and present some preliminary results of the method for COMPTEL (based on Monte Carlo simulations of the dataspace) in section 4.

2. BASIC THEORY

In general, the observer postulates a model of reality, which summarizes his *a priori* knowledge. This is translated into a probability distribution function (pdf) for the anticipated events encountered in the experimental set-up at hand. In astronomy the straightforward approach is to make an input intensity model map (composed of a finite number of well defined components) and fold it with the point-spread function of the telescope. A particular set of parameters defining the model map constitute a simple hypothesis, H.

It is customary to divide dataspace into bins for computational convenience and to allow for certain well-known *distribution-free* tests. Although this implies a small loss of information, it will not significantly affect parameter determination as long as the coordinates defining the dataspace (*e.g.* time, angular position, energy) are binned according to the corresponding instrumental resolution. For example, if the number of events per bin is larger than ~ 10, we may apply the minimum χ^2 test (see *e.g.* Lampton *et al.*, 1976) for parameter estimation, tacitly assuming a gaussian distribution of the number of events per bin.

Generally, the model will predict a certain continuum intensity for the number of photons per time unit per bin, say $\lambda_i(\theta)$ where i denotes the bin number. If the events have no "memory" then the probability of finding n_i counts after integration time T is distributed like $P(n_i) = e_i^{n_i} \exp(-e_i)/n_i!$, where $e_i = \lambda_i T$. If the probability for bin j is independent of the probability for bin i ($\forall i, j$), we can straightforwardly assign a likelihood to a dataspace of n bins under a given hypothesis H:

$$L(\{n_i\}|H) = \prod_{i=1}^{n} e_i^{n_i} \exp(-e_i)/n_i! \tag{1}$$

We tend to express most believe in those hypotheses which maximize L. Such a maximum will be denoted by $\hat{L}$, and the corresponding (maximum likelihood-) estimates of the involved parameters θ by $\hat{\theta}$. It is here that parameter estimation and model testing (following the terminology of Cash (1979)) seemingly go separate ways. In principle, model testing may incorporate the determination of L under various functional *shapes*, *e.g.* power law versus exponential energy density spectrum. Parameter estimation concentrates on the variation of L (or another statistic) with θ under an accepted model "shape", thereby introducing a 'Bayesian preference' towards the anticipated models (Eadie *et al.*, 1971). However, in the practical cases encountered the Bayesian approach leads to a confusion between model and parameter estimation. To illustrate this, consider a pdf of the shape $\sum_k a_k S_i$, where the a_k are the amplitudes of various shapes S_k. The S_k may be rooted in physically different processes (*e.g.* pulsar radiation versus diffuse emission) and themselves contain other free parameters. One tends to consider that estimate of a_k which maximizes L as the "best" estimate, say $\hat{a}_k$. From a non-Bayesian point-of-view we cannot truly derive the probability distribution of L for we do not know the true a_k and so we cannot assign a statistical significance to the result. However, one may derive the pdf of the $\hat{a}_k$ for supposedly true a_k^s, $\hat{P}(\hat{a}_k|a_k^s)$ and accept only those a_k^s for which $\hat{P}$ is larger then a certain value (defining the confidence interval for a_k^s). In the above example this literally means generating confidence intervals on independent models, thus blurring the distinction between model and parameter estimation. The conservative approach to the decomposition problem is to use Occam's razor: starting out from the most simple hypothesis that one still has believe in (H_0) towards more and more complex hypotheses. The LRM, although computationally slow for large dataspaces, is ideal to do the job in the case of a small number of events per bin (non-gaussian distribution) and simultaneously allows for model testing and parameter estimation. Let the general hypothesis H_g involve p undetermined parameters, $\theta = (\theta_1, \ldots, \theta_p)$. Suppose that the true values of q parameters are θ_q^t (implicitly assuming that the functional shape of H_g is appropriate). Although these values are unknown, we formally introduce a "sub-hypothesis" H_s which is H_g with the former q-parameter values set to their true values. The likelihood ratio $R(q)$ is defined by

$$R(q) = \hat{L}(\{n_i\}|H_g)/\hat{L}(\{n_i\}|H_s) \tag{2}$$

Obviously, $R \geq 1$ because H_g includes the most likely H_s. The theorem of Wilks (1938,1963) establishes that $\lambda = 2\log R(q)$ will adopt a χ^2 probability distribution with q degrees of freedom as $\sum_i n_i \to \infty$. For simple hypothesis testing, the λ statistic provides the most powerful test. So the procedure will be:

a. State the most simple hypothesis (H_0) which one has confidence in (in the example above for instance, put $a_k = \delta_{kl}$ so that all but model component l are excluded). This fixes q parameters of the general hypothesis H_g.

b. Next state a more complex hypothesis (H_1) which incorporates H_0 and which specifies r parameters of H_g, $f = q - r > 0$.

c. Calculate $R(f) = \hat{L}(\{n_i\}|H_1)/\hat{L}(\{n_i\}|H_0)$. If H_0 is true, λ will have a χ_f^2-pdf so that it is highly unlikely to find a value much in excess of $\sim 2f$. If we reject H_0 when $\lambda > \lambda_c$ the confidence level is $P(\chi_f^2 < \lambda_c)$.

d. If we reject H_0, we generate confidence intervals on the maximum-likelihood estimates of the f parameters by putting $\tilde{H}_0 = H_1(\hat{\theta}_f + \delta\theta_f)$ and $\tilde{H}_1 = H_1(\hat{\theta})$. If r of the $\delta\theta_f$ are non-zero, $\tilde{H}_0$=true implies $\tilde{\lambda} = 2\log L(\{n_i\}|\tilde{H}_1)/\log L(\{n_i\}|\tilde{H}_0)$ has a χ_r^2-pdf and we will accept those $\delta\theta_f$ for which it is not too unlikely to find $\hat{\theta}$ as maximum likelihood estimates.

The above procedure again illustrates the full equivalence of model and parameter estimation from the likelihood perspective. Obviously, having accepted H_1 the procedure may be repeated (with more complex hypotheses) if there is a physical reason to do so and if simultaneously a statistically significant improvement is achieved.

The LRM was succesfully applied in the analysis of the observations made by the γ-ray telescope COS-B ($E \approx 50\,\text{MeV} - 5\,\text{GeV}$). At first it was used to confirm the detection of extragalactic γ-ray sources (Pollock et al., 1981). It later proved to be a convenient method for studying the properties of the galactic diffuse γ-ray emission

(Lebrun *et al.*, 1983; Bloemen *et al.*, 1986; Strong *et al.*, 1988; Bloemen 1989) and the superimposed point-like sources (Pollock *et al.*, 1985).

However, the COS-B dataspace could be described in 2 dimensions so that the input sky image was of the same dimension as the dataspace. In the case of COMPTEL, 3 dimensions are in principle required due to the nature of the measured quantities (see below). The complexity of the resulting dataspace makes a study of the likelihood method results based on simulations desirable.

3. COMPTEL DATASPACE

The Compton telescope is described in detail elsewhere in these proceedings (Schönfelder *et al.*; Diehl *et al.*). The instrument utilizes the most efficient γ-matter interaction process at few MeV energies for light nuclei, namely Compton scattering. This scattering takes place in the first layer of scintillator detectors D1 (the 'lense') after which the scattered photon is absorbed in an underlying layer of detectors D2 (the 'film'). The Compton scattering is that part of the telescope response which allows for imaging, as it is direction sensitive. However, because for unpolarized photons the corresponding cross-section depends only on the scatter angle and not on its azimuth, the reconstruction of the underlying image is not straightforward.

If the energy deposit in D1 is E_1 and in D2 is E_2, Comptel's formula for the scatter angle $\bar{\phi}$ reads

$$\cos \bar{\phi} = 1 - mc^2 \left(\frac{1}{E_2} - \frac{1}{E_\gamma} \right) \quad \text{where} \quad E_\gamma = E_1 + E_2 \qquad (3)$$

(m is the electron rest mass). The photon interaction positions inside the detectors are combined into a vector $\vec{r}$: the direction into which the photon has been scattered. The cone centered on $\vec{r}$ with opening angle $2\bar{\phi}$ is the collection of possible photon arrival directions. If we map the sky in an arbitrary spherical coordinate system, denoted by (χ, ψ) (which could be for example equivalent to (l, b)), the cone projects as a circle centered on the direction $(\chi(\vec{r}), \psi(\vec{r}))$ with angular radius $\bar{\phi}$. However, for every source a range of $\bar{\phi}$ will occur (distributed according to the Klein-Nishina cross-section) and to exploit this degree of freedom $\bar{\phi}$ is added as the third dimension of the dataspace in which each photon may now be described by $(\chi, \psi, \bar{\phi})$. If we take the origin of (χ, ψ) somewhere near the COMPTEL pointing axis, then for the photons of interest we can use the 'locally flat approximation' (LFA): if the source is at (χ_0, ψ_0) and not much more than $\sim 10°$ from the pointing axis, $\bar{\phi} \approx \sqrt{(\chi - \chi_0)^2 + (\psi - \psi_0)^2}$. In the LFA, the dataspace response to a source is an event-cone with apex at (χ_0, ψ_0), running at an angle of 45° with the (χ, ψ)-plane. Because of the finite resolution in $\vec{r}$ (event location within the modules) and in $\bar{\phi}(E_1, E_2)$, the cone transforms into a mantle. For each energy E_γ this mantle defines the PSF, and is denoted by $f(\chi, \psi, \bar{\phi}; \chi_0, \psi_0, E_\gamma)$. In the LFA the PSF dependency on the image coordinates is only through $(\chi - \chi_0, \psi - \psi_0)$. The effective area for COMPTON scattering may be calculated for a given pointing direction and photon arrival direction (χ', ψ'), say $A(\chi', \psi'; E_\gamma)$. Because of the finite dimensions of the instrument only a fraction of the scattered photons is absorbed in D2 with geometrical absorption probability $g(\chi, \psi)$. The expected number of events for integration time T invoked by a *mono-energetic* sky intensity distribution $I(\chi', \psi')$

can be written as

$$\tilde{e}(\chi, \psi, \bar{\phi}) = g(\chi, \psi) \int d\chi' \, d\psi' I(\chi', \psi') A(\chi', \psi') T f(\chi, \psi, \bar{\phi}; \chi', \psi') \qquad (4)$$

where explicit reference to energy has been dropped. If we denote a dataspace bin by d, a sky pixel by s and the exposure AT by X, the discretized form of (4) becomes

$$\tilde{e}(d) = g(d) \sum_s f(d, s) I(s) X(s) \qquad (5)$$

Additional dataspace structures, such as due to background lines (activated within the instrument) and random coincidence photons, can in general not be cast in the form of equation (5). If we assume that we can use a time averaged shape for this contribution (B) in reducing data of a given exposure, the final expectation value becomes $e(d) = \tilde{e}(d) + a_B B(d)$, where a_B can be adopted as a free parameter. Complicated dataspace selections, which must exclude most of the events arriving from the earth's atmosphere, can in principle be incorporated in the matrix g, so that the dataspace description is not altered. If all detectors are active and no explicit dataspace selections are required, g does not depend on $\bar{\phi}$ and is given by the function displayed in figure 1. The dataspace response to an on-axis point source at 6.13 MeV is given in figure 2 (based on the emperically derived PSF (Strong, 1990)). Note that the probability density is still significant at large $\bar{\phi}$, so that a typical dataspace for a single pointing contains contributions of a sky image of the size $\sim 150° \times 150°$.

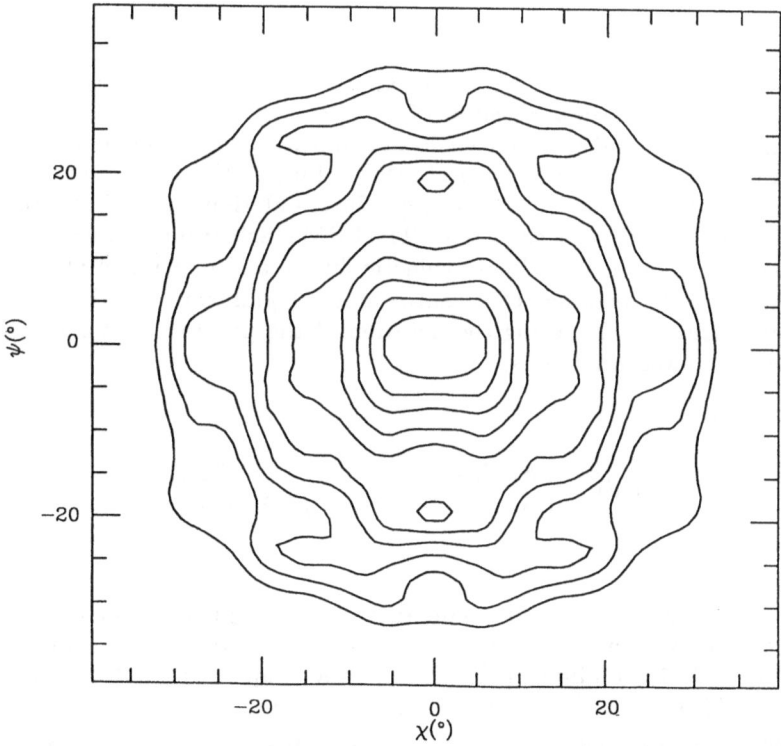

Fig. 1. The variation of the geometry function g with (χ, ψ) when COMPTEL points towards $(\chi, \psi) = (0, 0)$. The fluctuations reflect the positions of the 7 D1 and 14 D2 detectors (see Diehl et al., this volume). Contour levels at $n \times 0.48$, $n = 1, \ldots, 10$.

245

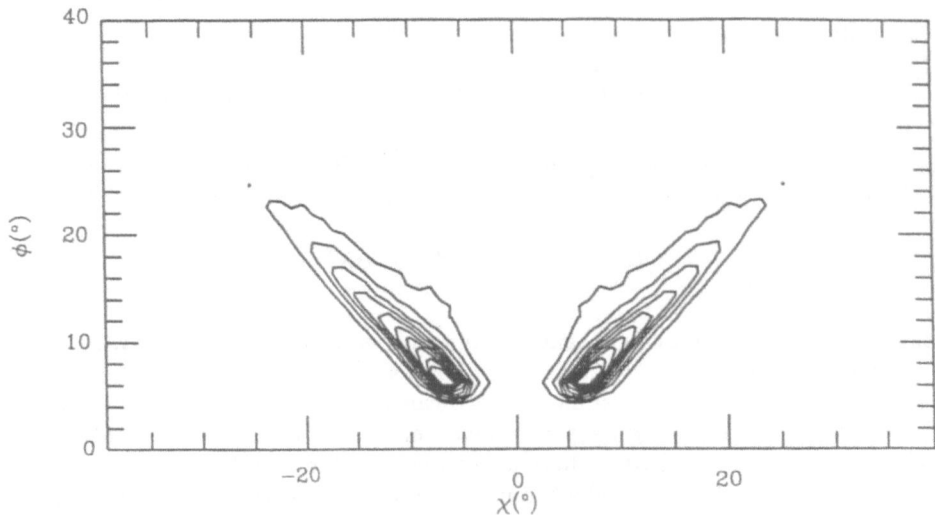

Fig. 2. A cut along $\psi = 0$ of the dataspace response ($e(\chi, \psi, \bar{\phi})$ in the text) for an on-axis source at 6.13 MeV. Contour levels at $n \times 0.16$, $n = 1, \ldots, 10$.

4. RESULTS

We simulated dataspaces for arbitrary model intensities, because calibration data have limited use for the verification of the applied LRM to flight data: the signal-to-noise ratio is large and the sources do not appear as ideal point sources because of their finite distance (*e.g.* Strong *et al.*, this volume). Furthermore we want to be able to control our input image completely. The simulation is based on equation (5), however without the LFA (see section 3). The LRM starts from an *a priori* prediction of the dataspace event density due to galactic diffuse and instrumental emission, say M_1. We test for the presence of a source and if significant, add it to the model. We can summarize this by, using the notation of the previous sections, writing the expectation per dataspace bin as

$$e(d) = \begin{cases} S_1 M_1(d) & \text{if } H_0 \\ S_1 M(d) + S_2 f(d, \tilde{s}) g(d) & \text{if } H_1 \end{cases} \qquad (6a).$$

S_1 is the scaling of the background distribution, S_2 is the source strength (proportional to the exposure) and $\tilde{s}$ is the assumed source position (χ_0, ψ_0). If we calculate λ for each $\tilde{s}$, we obtain a likelihood ratio map $\lambda(\tilde{s})$. If there is no source, then the $\max_{\tilde{s}} \lambda(\tilde{s}) \equiv \lambda_{\max}$ is distributed in a classical interpretation as χ_3^2 so that we have a 99% confidence detection if $\lambda_{\max} > 11.3$. If we accept H_1, confidence levels can be generated on both the source position and the source flux. A source which is significant 'beyond reasonable doubt' will be added to the background model at the most likely position (say $\hat{s}$), with its flux as a free scaling parameter.

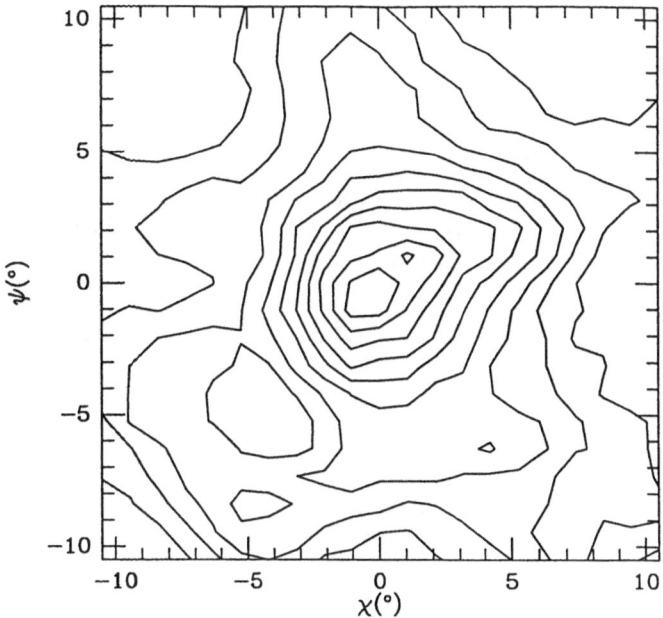

Fig. 3. A sample λ map for a single simulation; contour levels start at 8.8 with steps of 7.8. For details, see text.

The updated hypotheses become:

$$e(d) = \begin{cases} S_1 M_1(d) + S_2 M_2(d) & \text{if } H_0 \\ S_1 M_1(d) + S_2 M_2(d) + S_3 f(d, \tilde{s}) g(d) & \text{if } H_1 \end{cases} \quad (6b),$$

where $M_2(d) = g(d) f(d, \hat{s})$.

The simulations were done with the emperically determined PSF at 6.13 MeV, and counts are typical for the integrated energy range of 3 to 12 MeV. The background is estimated mainly from balloon-flights (see *e.g.* Schönfelder *et al.*, 1980) and probably comprises about 10^5 photons for the quoted energy range and a full observation period ($\sim 4 \times 10^5$ sec effective integration time). The dataspaces used for the likelihood results are $79° \times 79° \times 40°$ in $(\chi, \psi, \bar{\phi})$ with $1°$ bins along each dimension. The likelihood ratio map displayed in figure 3 is for a simulated on-axis source (corresponding to $(\chi, \psi)=(0,0)$), with about 1100 source counts whereas the $\sim 10^5$ background counts are distributed as they would be for an isotropic sky ($I_B(s) =$ constant). About 6000 source counts would approximate the number expected for the Crab total emission from 3 to 12 MeV for a single pointing. Ofcourse, since the data are not consistent with H_0, the distribution of λ in these examples is dictated by the source position and strength instead of by χ^2 statistics. The formal resolution obtained, for 1% source counts on about 10^5 counts in total, is about $2°$ if we adopt a 99% confidence level. The resolution quickly improves with the number of source counts. For instance, for 6% source counts it is $\sim 0.3°$. For such a well-resolved source, the relative error in flux is characteristically less than 0.05.

A likelihood ratio map for a simulation of 2 sources, each containing about 3000 counts on an isotropic background of the same strength as above, is given at left in figure 4. Note that the simulated events exhibit statistical fluctuations, so that the λ-map is not symmetrical with respect to the $\chi = 0$-axis. The sources are separated by 6 degrees, both are 3 degrees off-axis.

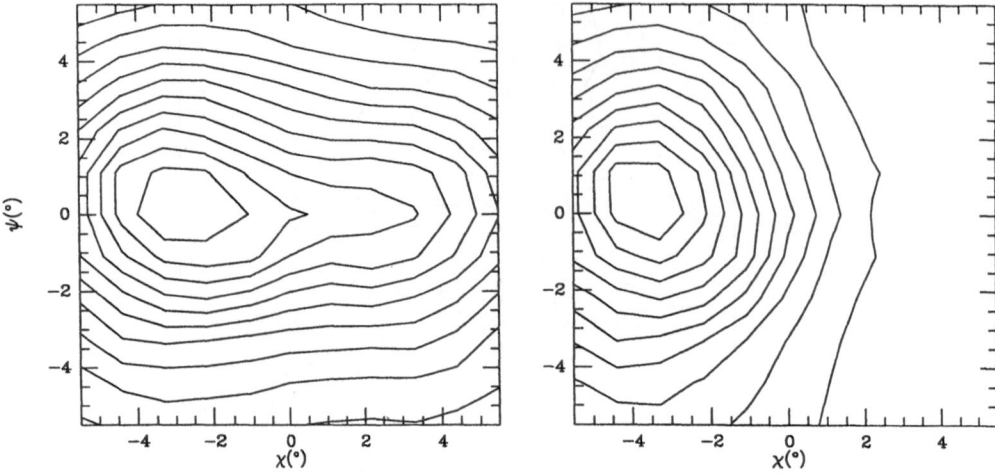

Fig. 4. *Left*: λ-map for 2 sources, at (χ, ψ)=(-3,0) and (3,0) respectively; levels start at 127, increment=35. *Right*: λ-map for the same 2 sources, but with one source included in the background model; levels start at 44, increment=26. Details are in the text.

The applied model hypotheses are given by equation (6a), which we know is wrong for this case, but still there is a clear indication of extended emission which is not consistent with the expectation for a single localized source.

Suppose we knew about a potential γ-ray source near the $\chi = +3°$ position. This would lead us to update the model to that given by equation (6b), with $\tilde{s} = (3, 0)$. We then find the remaining source with the proper number of counts (2800 ± 125) and at the right position, as we can see from the map at right shown in figure 4.

If we take the simulated data of figure 3, and apply the 4.43 MeV PSF as f in (6a), we find that the likelihood ratios have dropped characteristically by about 30%, but the formal angular resolution and source flux are consistent with the results described above. We therefore conclude that at least in this energy range, the derived source parameters are relatively insensitive to energy.

5. CONCLUSIONS

The success of previous applications of the LRM in γ-ray astronomy invites a similar approach to COMPTEL data. To verify if the method works and converges for the latter data, we presented results based on simulations of the COMPTEL dataspace. These results also allow us to anticipate the kind of likelihood ratios we may expect for realistic source detections. This is useful because in general we only have a very global *a priori* model M_B and we do not accurately know the behaviour of the likelihood ratio statistic in the presence of additional unknown sources. The simulated results in principle can tell us, for a given total number of counts, how a λ value corresponds to a signal-to-noise ratio, so that we have a reference model for interpretating flight-data.

REFERENCES

Bloemen, J.B.G.M., Strong, A.W., Blitz, L., Cohen, R.S., Dame, T.M., *et al.*, 1986, *Astron. Astrophys.* **154**, 25

Bloemen, J.B.G.M., 1989, in : *Annual Review of Astronomy and Astrophysics*, 469-516

Cash, W., 1979, *Astrophys. J.* **228**, 939

Eadie W.T., Drijard, D., James, F.E., Roos M., and Sadoulet, B., 1971: "Statistical Methods in Experimental Physics", North-Holland Publ. Comp., Amsterdam.

Edwards, A.W.F., 1972: "Likelihood", C.U.P.

Lampton, M., Margon, B., and Bowyer, S., 1976, *Astrophys. J.* **208**, 177

Lebrun, F., Bennett, K., Bignami, G. F., Bloemen, J.B.G.M., Buccheri, R., *et al.*, 1983, *Astrophys. J.* **281**, 634

Pollock, A.M.T., Bignami, G.F., Hermsen, W., Kanbach, G., Lichti, G.G., *et al.*, 1981, *Astron. Astrophys.* **94**, 116

Pollock, A.M.T., Bennett, K., Bignami, G.F., Bloemen, J.B.G.M., Buccheri, R., *et al.*, 1985, *Astron. Astrophys.* **146**, 352

Schönfelder, V., Graml, F., and Penningsfeld, F.P., 1980, *Astrophys. J.* **240**, 350

Strong, A.W., 1990, COMPASS internal report COM-AL-MPE-RES-016 (algorithm description empirical PSF generation)

Strong, A.W., Bloemen, J.B.G.M., Dame, T.M., Grenier, I., Hermsen, W., *et al.*, 1988, *Astron. Astrophy.* **207**, 1

Wilks, S.S., 1938, *Ann. Math. Stat.* **9**, 60

Wilks, S.S., 1963, *Mathematical Statistics*, Princeton University Press, Princeton.

MAXIMUM ENTROPY IMAGING AND

SPECTRAL DECONVOLUTION FOR COMPTEL

A.W. Strong[1], P. Cabeza-Orcel[4]*, K. Bennett[4], W. Collmar [1], R. Diehl [1],
J.W. den Herder [2], W. Hermsen [2], G. Lichti [1], M. McConnell [3], J. Ryan [3],
H. Steinle [1], V. Schönfelder[1], C. Winkler[4]

1. Max-Planck-Institut für Extraterrestrische Physik, Garching, FRG
2. Laboratory for Space Research, Leiden, The Netherlands
3. University of New Hampshire, Durham N.H., USA
4. Space Science Dept. of ESA, ESTEC, Noordwijk, The Netherlands

ABSTRACT

The method of maximum entropy will be used to generate sky maps from COMPTEL flight data. The application to COMPTEL allows the full instrument response to be included, and extensive tests with calibration data give encouraging results. The sensitivity of the method to background has been investigated. Maximum entropy will also be used for deconvolution of burst and solar energy spectra. Successful tests have been made with calibration and also SMM solar flare data.

1. IMAGING OF COMPTON TELESCOPE DATA

The generation of images from Compton telescope data presents an interesting challenge and one which is timely in view of the COMPTEL instrument on GRO. Application of the Maximum Entropy method (MEM) to γ-ray astronomy was first described by Skilling et al. (1979) for COS-B spark chamber data; Diehl and Strong (1987) discussed both COS-B and COMPTEL applications, and Strong et al.(1990) made simulations of multiple source imaging with the COMPTEL response. Varendorff (1991) has applied MEM to balloon observations of the Galactic centre region with the MPE Compton telescope with excellent results.

The Compton telescope is *a priori* an excellent candidate for MEM, not only because no other method has proved able to provide sky maps in the strict sense of intensity distributions consistent with the data (other methods give various types of probability distributions which cannot be interpreted as intensities), but also because of the nature of the instrumental data (see Schönfelder et al., these proceedings, for a description of the instruments and

---

* present address: Les Edition Belins, Pour La Science, Paris, France

Data Analysis in Astronomy IV, Edited by V. Di Gesù et al.
Plenum Press, New York, 1992

its measurement principles). The data is characterized by a 3D point-spread-function (PSF) with the additional dimension (after the two spatial ones) being provided by the Compton scatter angle. The 'data space' is similarly 3D (or more if energy resolution is included). Since we want to generate 2D skymaps it is clear that our 'data space' is very different from our 'image space'. This is just the kind of problem where MEM is effective; in fact the more different the data and image spaces are the more useful are the so-called 'indirect imaging' methods. In this sense we are nearer to radio interferometry or Fourier spectroscopy where indirect imaging is the natural technique. One interesting aspect of Compton telescope data is that the Compton scatter 'dimension' gives an 'overdetermination' of the image; it is as if we observed the sky with many separate 2D PSFs and attempt to use them all simultaneously for the reconstruction. One advantage of having very diffent image and data spaces and of overdetermination is that unwanted correlations between image and data (e.g. the image following point-to-point statistical variations in the data) are practically eliminated.

2. APPLICATION OF MAXIMUM ENTROPY TO COMPTEL IMAGING

Maximum entropy is a general technique for deconvolution which produces the 'flattest' solution consistent with the given data and response function. Only features for which there is 'evidence' in the data are thereby obtained; at least this is the main property claimed by the advocates of MEM. The principles of the method are described in the literature (e.g. Gull and Skilling 1984), and will not be repeated here. The software package used is MEMSYS2 from Maximum Entropy Data Consultants Ltd of Cambridge, England. The user provides a function (see below) which computes the data space response to any given input image; the package then allows an iterative solution for the maximum entropy image starting from a uniform (or reference) image and iterating towards improving the fit to the data. A particular version of the software allows the correct treatment of data containing small or zero data values; the log-likelihood function for Poisson statistics is then used instead of the χ^2 adopted in the standard version. This is particularly important since we want to to use binsizes small enough to exploit the instrumental angular resolution and this may well imply small statistics per bin.

The COMPTEL image space is defined in a coordinate system (χ_o, ψ_o) analogous to longitude and latitude but quite generally oriented. The COMPTEL data space is defined as $(\chi, \psi, \bar{\varphi})$ where χ, ψ are angles defining the direction of the photon after scattering in the upper (D1) detector layer (using the same coordinate system as (χ_o, ψ_o)) and $\bar{\varphi}$ is the Compton scatter angle computed from the energy deposits in the upper and lower detectors. The full response of the instrument to a given input image is complicated but a representation which is reasonably fast to compute and which preserves all the essential features is:

$$n(\chi, \psi, \bar{\varphi}) = g(\chi, \psi)d\Omega t \int_{\chi_o} \int_{\psi_o} I(\chi_o, \psi_o) A(\chi_o, \psi_o) f(\phi_g(\chi, \psi, \chi_o, \psi_o), \bar{\varphi}) d\chi_o d\psi_o \quad (1)$$

where $n(\chi, \psi, \bar{\varphi})$ is the expected count in a dataspace cell, $g(\chi, \psi)$ is the probability that a photon scattered in D1 into a direction (χ, ψ) encounters a D2 module (a purely geometrical effect which reflects the fact that the D1 and D2 layers consist of discrete modules), $I(\chi_o, \psi_o)$ is the image space intensity distribution, $A(\chi_o, \psi_o)$ is the area of the exposed D1 surface (about 4.2 m^2), $f(\phi_g(\chi, \psi, \chi_o, \psi_o), \bar{\varphi})$ is the probability that a photon encountering D1 makes an interaction in both D1 and D2, scatters through an angle ϕ_g and has a measured Compton scatter angle of $\bar{\varphi}$. It is defined for a 'infinite' extent of the D2 layer, since the discrete nature of D2 is included in 'g'. It includes among other things the energy-dependent interaction probablity in D1 (not therefore included in A) and the effects incomplete absorption in D2. dΩ is the solid angle covered by the bin at (χ, ψ), t is the exposure time. $f(\phi_g, \bar{\varphi})$ is referred to here as the 'PSF' since it has the property of being (almost) independent of the input direction of the photon and here is similar to the normal type of PSF encountered in optics. This is achieved by the definition in terms of an infinite D2 layer and the removal of the geometrical part of the response from the convolution. If this were not done then every point in image space would have a different 'PSF' and the computational requirement would be much too large.

Note that the PSF would be a perfect cone defined by $\phi_g = \bar{\varphi}$ if the energy and position determinations in D1 and D2 were exact. The finite energy and position resolutions lead to a widening of the cone; in particular there is a filling in of regions with $\bar{\varphi} < \phi_g$ due to incomplete absorption in D2 (see Diehl et al., these proceedings).

The response representation given by equation(1) enables a fast computation of the response to any given input image provided the convolution with the PSF can be computed fast; in practice we have used an FFT method (in particular: Winograd FFT which is not restricted to powers of two) which treats the spherical coordinate system as a locally flat Cartesian system. Unfortunately this approximation is increasingly poor as we go to larger scatter angles (characteristic of lower photon energies); but explicit 'brute force' computation of the convolution is at present ruled out as it involves very large computer resources. A fast method for convolving a function on the sphere, if it exists, appears not to be generally known; in the long term, it may be necessary to use 'brute force' with vector processors. However the FFT method is good enough when the scatter angle is not too large, as the practical results below bear out.

For clarity I have omitted the energy dependence of terms in equation (1); however it should be clear that in a real application an additional convolution over the input intensity spectrum is required, and that the PSF is an energy-dependent function. Methods for obtaining the PSF from calibration data are discussed by Diehl et al. (these proceedings).

3. RESULTS USING CALIBRATION SOURCES

The calibration of COMPTEL included sources with energies from 0.83 to 20 MeV. These data can be used as a test of the MEM deconvolution, although the finite distance of the source (about 8m) means some spreading of the image. This can be minimized by using data from only one D1 detector; the spread is then about $2°$. This problem does not of course arise in flight. Since the sources used were quite strong the signal-to-background ratio is not typical of flight situations, so it is necessary to add in data from background runs to obtain more realistic tests. Data for different source positions can be combined to simulate more than one source in the field-of-view and test the spatial resolution for source separation as a function of signal-to-background.

We concentrate on one source energy, 6.1 MeV and a configuration of 3 sources separated by $10°$, and 2 sources separated by $5°$. The source counts are typical of what might be expected for a Crab-like source and a COMPTEL observation of 2 weeks. Table 1 summarizes the cases presented here.

For the ideal case of no added background (Fig 3.1) the sources are completely resolved and in the correct positions, with no artefacts; this establishes the basic correct operation of the method.

If the background is increased to 5 times the counts for one source (Fig 3.2) the sources are still well imaged but the contrast is smaller. Edge effects become noticeable: some parts of the image near the edge of the field are not constrained by the data and are assigned the 'default' value defined by the user.

Further increase of the background to 16 times the counts for one source (Fig 3.3) reduces the source visibility and contrast is further reduced. Some extra features appear; however it should be noted that the 'background' data used consists of runs where sources in

Table 1. data used for maximum entropy images of calibration sources

Figure	Source energy	Source configuration Angles from axis	Counts per source	Total counts
3.1	6.1 MeV	0° ,10° ,20°	1470	4424
3.2		0° ,10° ,20°	1470	12408
3.3		0° ,10° ,20°	1470	28367
3.4		0° , 5°	1470	3480
3.5		0° ,10° ,20°	1470	12408

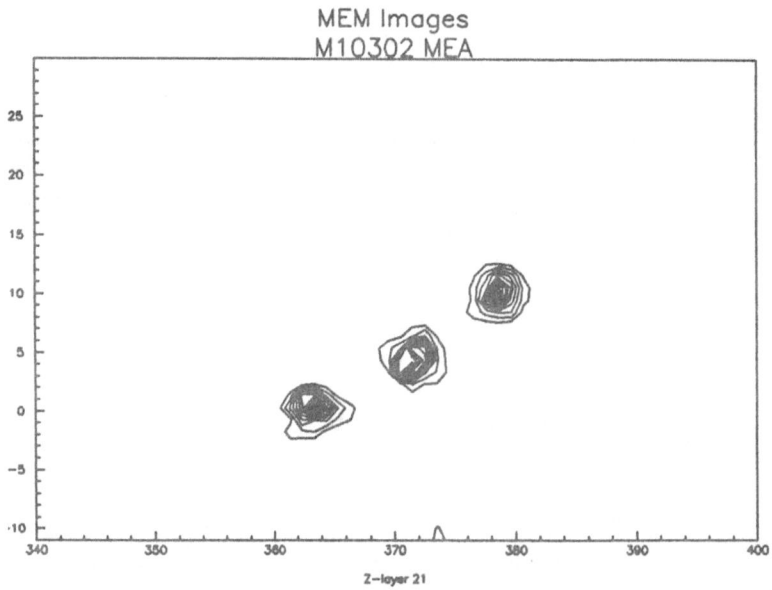

Fig 3.1. MEM images of three 6.1 MeV calibration sources separated by 10°, without added background. Details in Table 1.

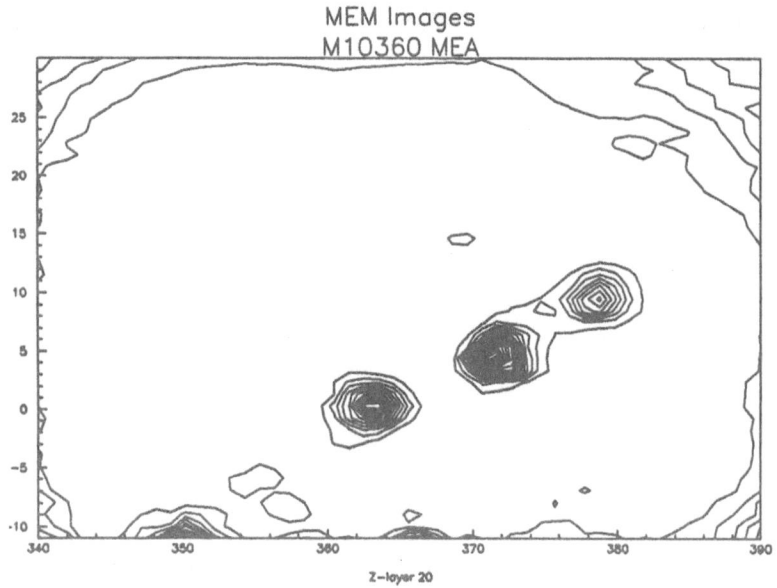

Fig 3.2. Adding a significant amount of background reduces the quality of the image, but the sources are clearly visible.

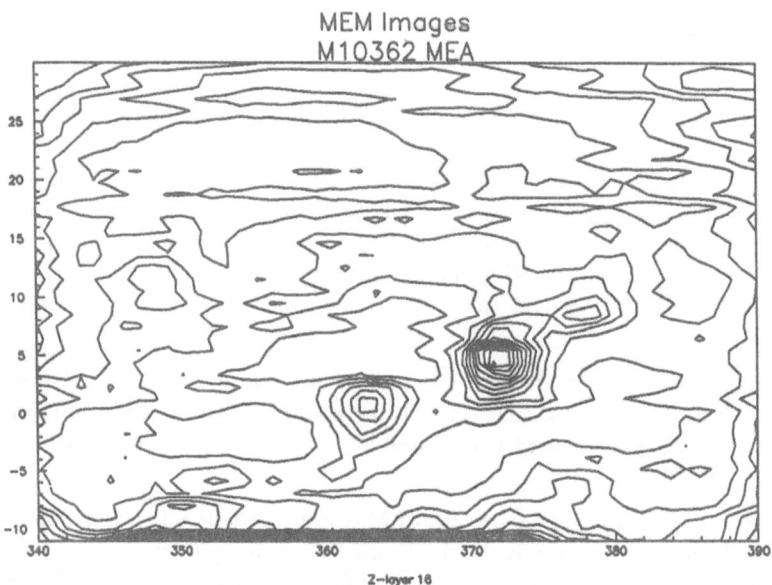

Fig 3.3. Increasing the background to 16 times tne counts from a single source reduces the source visibility significantly.

various positions were occulted by a lead shield which was not completely efficient; therefore there is real structure present in the background, and the extra features are at least in part due to this structure. Pure 'room background' data would have been preferable for this test but not enough of this is available. The effect of undefined parts of the image is even more noticeable here.

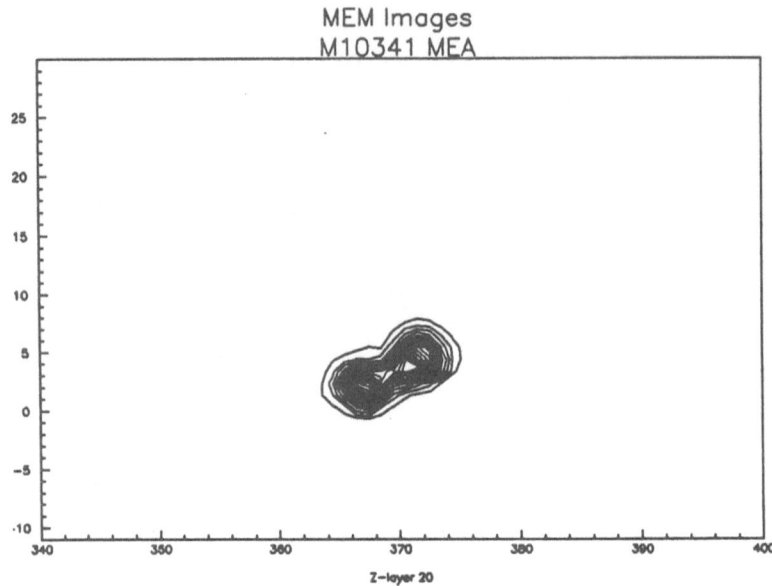

Fig 3.4. Two sources separated by $5°$ are resolved; the intrinsic source diameter of $2°$ contributes to the overlap, so the true resolution is better than apparent here.

A test of the angular resolution is shown in Fig 3.4, where two source separated by $5°$ are used. The sources are clearly resolved, but there is significant fill-in between the sources; the intrinsic angular width ($2°$) due to source proximity (see above) contributes to this, as does the binsize used, $1°$ in (χ, ψ) and $2°$ in $\bar{\varphi}$ which is really too large fully to exploit the PSF. Considering both these effects the source resolution for strong sources is probably $2\text{-}3°$.

All of these tests were made with just one D1 module to avoid source spreading; this also has the effect of reducing the coverage of the geometrical response g in Equation 1. In flight all D1 modules can be used (sources at infinity) and the geometrical response is in this case more uniform.

One problem which always arises in MEM applications is when to stop the iteration process, i.e. which of the images is the 'best' one ? In principle we can use the first image which satisfies some statistical test at some confidence level, but in practice the fit may never be acceptable (e.g. by χ^2) due to imperfect knowledge of the instrumental response and also imperfections in the data. The best way to present the results is to show a series of iterations ranging from 'too flat' to 'too structured'. The 'best' map lies somewhere in between and usually it is clear from the series what features are to be trusted. As an example, Fig 3.5 shows the iterations for the case shown earlier in Fig 3.2. The range of images up to 14 are too flat, while images after 22 are breaking up into 'noise'; images between these limits are all acceptable, and indeed they all look much the same.

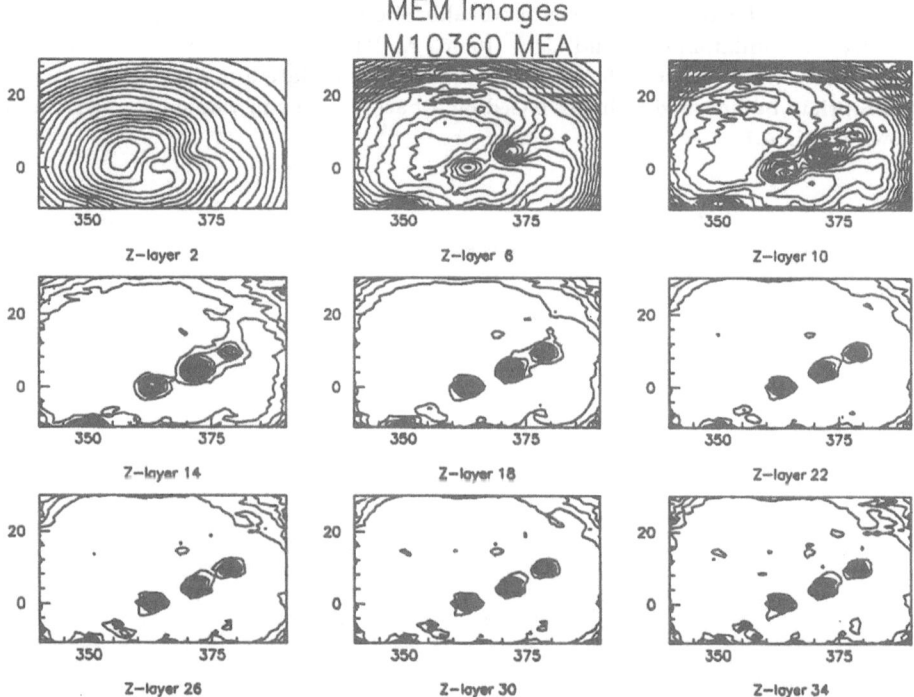

Fig 3.5. Iterations of the maximum entropy algorithm for the same data as Fig 3.2. Starting from a flat map, the fit to the data is improved in each iteration; in the limit the data is overfitted and too much structure appears. The 'best' image lies somewhere between, but presentation of a series of iterations is the best way to indicate the uncertainty in the structure.

4. APPLICATION TO SPECTRAL DECONVOLUTION

One of the scientific objectives of COMPTEL is to study solar flare γ-rays and cosmic γ-ray bursts. The burst detector uses two of the lower (D2) detectors and is described by Winkler et al. (1989). Because the response function contains a significant 'Compton tail' in addition to the photopeak the measured spectrum is significantly modified relative to the input spectrum; deconvolution is therefore very desirable particularly for spectra showing absorption/and or emission lines. The alternative method is to fit an explicit model to the data; but a 'hypothesis free' method is always useful in order to suggest what should be included in the model.

Cabeza-Orcel. et al (1989) have studied the application of the MEM COMPTEL burst data and SMM solar flare data. They used calibration data from Ba^{133} (0.356 MeV), Na^{22} (0.511 MeV annihilation line) and Cs^{137} (0.662 MeV) sources together with a provisional response matrix. In one test the data were added to simulate a multi-peak spectrum and an appropriate background estimate based on a background run was subtracted; the result is shown in Fig 4.1.

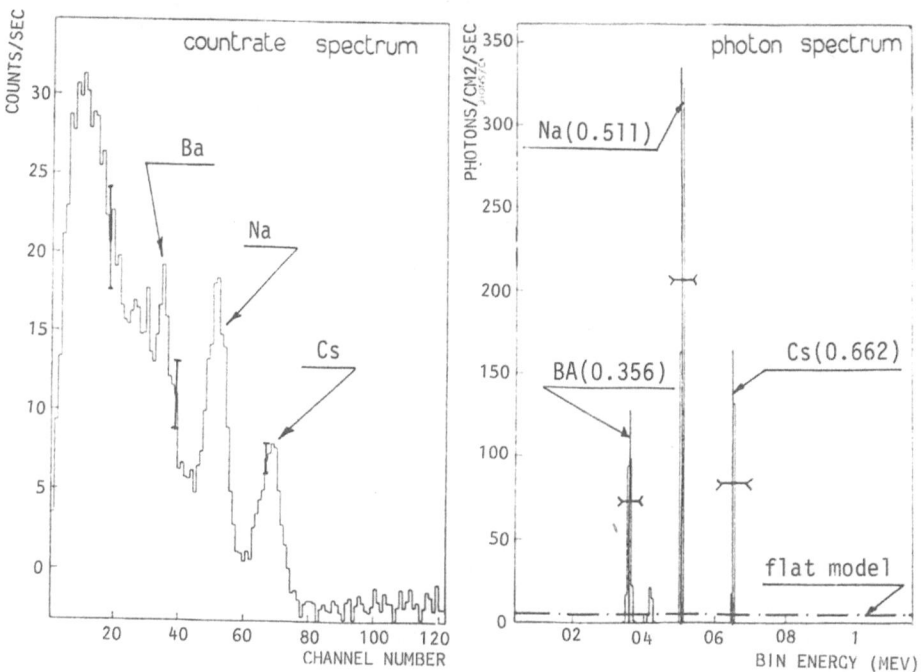

Fig 4.1. MEM deconvolution of a combination of calibration sources

The results of several tests indicated that the source spectrum is correctly recovered as long as the signal/background is at least 10% corresponding to an incident flux of a few 10^{-5} ergs/cm^2/sec, and the noise is no more than 35% of the background-subtracted signal amplitude. The lines were correctly placed to within 1% and the full width at half maximum never exceeds the detector resolution. Below this flux however the method was found much less successful.

Cabeza-Orcel. et al (1989) also used solar flare data from the Solar Maximum Mission Gamma-Ray Spectrometer (SMM GRS); the SMM detector is quite similar to COMPTEL burst modules so this provides a realistic test on data of astrophysical interest. The response matrix consists of 476*500 elements. The data used were for the famous solar flare event

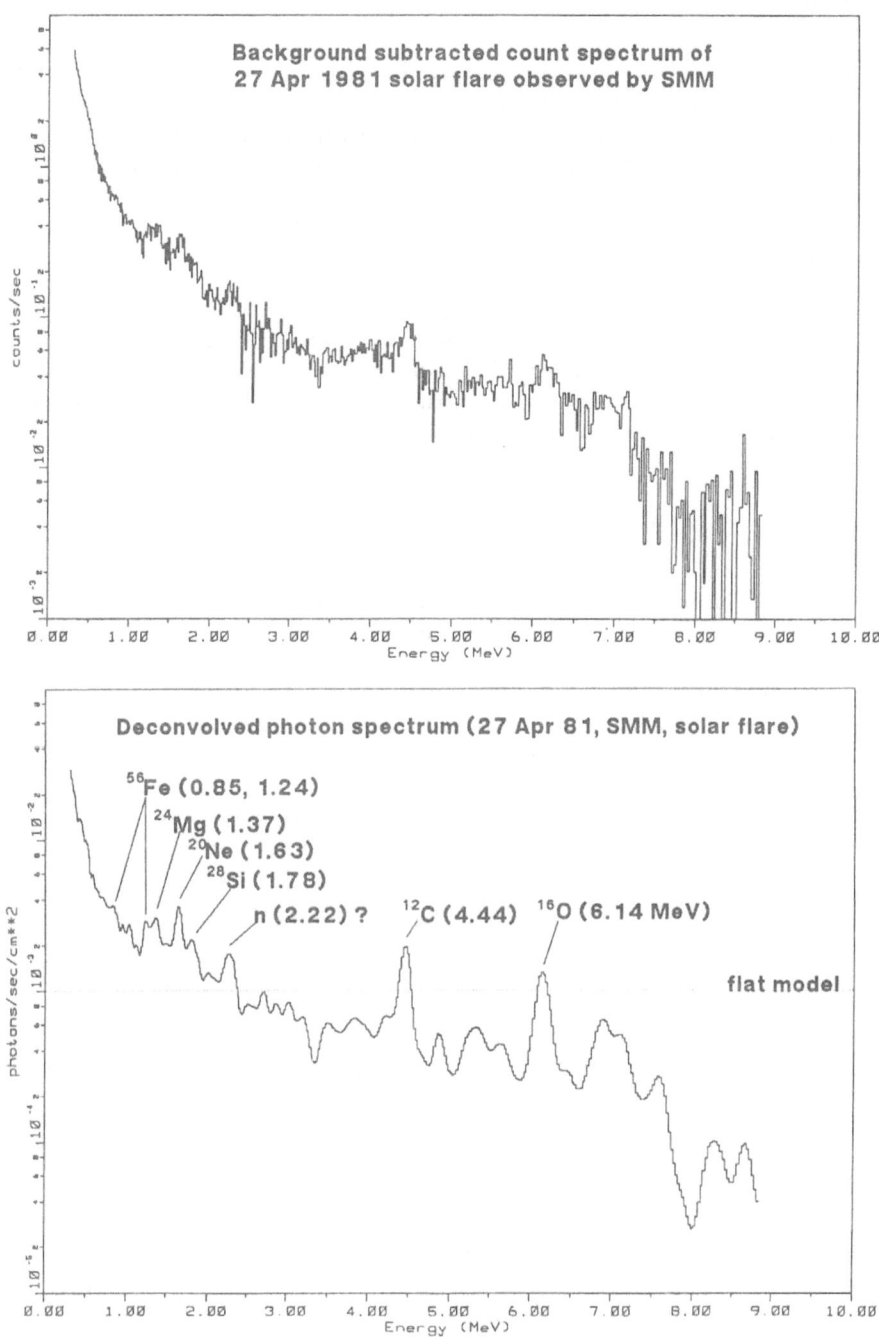

Fig 4.2: SMM solar flare data and MEM deconvolution

of 27th April 1981 (1722 sec duration). The deconvolution (Fig 4.2) shows most of the lines expected in a specific model (Murphy et al. 1985), with the exception of the e^+e^- annihilation line.

The method has been adopted as the standard deconvolution technique for COMPTEL burst data. At present the MEMSYS2 package has been used, but in future it is hoped to take advantage of recent (Bayesian) enhancements of MEMSYS which allow (for example) error estimates on spectral features to be obtained in addition to simply the deconvolution itself.

5. CONCLUSIONS

Maximum entropy promises to be an important component of COMPTEL data analysis, both in the skymapping and spectral deconvolution areas. There is no doubt that much experience will have to be accumulated on in-flight data before the method is optimized, but the calibration data have provided an excellent basis for evaluating the practicability of the method with the real instrument response, and getting the software into a form where it can address real data.

REFERENCES

Cabeza-Orcel P, Bennett K, Winkler C, (1989),
       Proc. GRO Science Workshop , NASA/GSFC, 4-512
Diehl R, Strong A W , (1987), Data Analysis in Astronomy III , Plenum, New York, 55
Gull S F , Skilling J , (1984), IEE Proceedings , 131, 646-659
Murphy R et al., (1985), Proc. 19th Cosmic Ray Conf. , 4 , 249
Skilling J, Strong A W, Bennett K, (1979), Mon. Not. R. astr. Soc., 187, 145-152
Strong A W et al., (1990), Proc. 21st Int. Cosmic Ray Conf. , 4 , 154
Varendorff M, (1991), Thesis, TU München
Winkler C et al., (1989), Proc. GRO Science Workshop , NASA/GSFC , 4-470

COMPTEL AS A SOLAR GAMMA RAY AND NEUTRON DETECTOR

J.M. Ryan[3], H. Aarts[2], K. Bennett[4], R. Byrd[5], C. de Vries[2], J.W. den Herder[2], A. Deerenberg[2*], R. Diehl[1], G. Eymann[4], D.J. Forrest[3], C. Foster[5], W. Hermsen[2], J. Lockwood[3], G. Lichti[1], J. Macri[3], M. McConnell[3], D. Morris[3], V. Schönfelder[1], G. Simpson[3], M. Snelling[4], H. Steinle[1], A. Strong[1], B.N. Swanenburg[2], T. Taddeucci[5], W.R. Webber[3], C. Winkler[4]

[1]Max-Planck Institut für Extraterrestrische Physik, Garching, Germany
[2]Research Organization of The Netherlands, Leiden
[3]Institute for the Study of Earth, Oceans and Space, University of New Hampshire, Durham, NH
[4]Space Science Division, ESTEC, Noordwijk, The Netherlands
[5]Indiana University Cyclotron Facility, Bloomington, Indiana

ABSTRACT

The imaging Compton telescope COMPTEL on the Gamma Ray Observatory has unusual spectroscopic capabilities for measuring solar γ-ray and neutron emissions. Flares can be observed above the 800 keV γ-ray threshold of the telescope. The telescope energy range extends to 30 MeV with high time resolution burst spectra available from 0.1 to 10 MeV. Strong Compton tail suppression facilitates improved spectral analysis of solar flare γ-ray emissions. In addition, the high signal-to-noise ratio for neutron detection and measurement provides new neutron spectroscopic capabilities. For example, a flare similar to that of 1982 June 3 will yield spectroscopic data on > 1500 individual neutrons, enough to construct an unambiguous spectrum in the energy range of 20 to 150 MeV. Details of the instrument response to solar γ-rays and neutrons are presented.

INTRODUCTION

COMPTEL, the Imaging Compton Telescope, has unique capabilities for measuring the flux and energy of both the solar γ-ray and neutron emissions. With its field-of-view (FOV) of about 60° for γ-rays and 90° for neutrons the Sun will often be in the γ-ray or neutron FOV when the Observatory is in the sunlit portion of the orbit. According to a recent viewing plan for Phase I of the mission, COMPTEL will have approximately 20% of the solar exposure of a dedicated solar observatory such as the Solar Maximum Mission. The Sun will be in the larger neutron FOV more frequently. In this paper we briefly describe the instrument and the detection technique and then present a preliminary response of COMPTEL to γ-rays and neutrons. Finally, we estimate the COMPTEL response to the γ-ray and neutron emissions from large solar flares.

---

*Deceased

Data Analysis in Astronomy IV, Edited by V. Di Gesù *et al.*
Plenum Press, New York, 1992

COMPTON TELESCOPE BASICS

A Compton telescope is a γ-ray detector in which the Compton scattering process is used to measure the photon's energy and incident direction. In this type of detector a γ-ray must scatter in two physically independent detecting elements or planes, nominally a forward and rearward detector. A precise time-of-flight (TOF) measurement establishes the general forward or backward direction of the scattered photon. For γ-rays the measured TOF must be consistent with the light travel time, in the process identifying scattered neutrons. Pulse shape discrimination techniques can further isolate the effects of neutrons. This delayed coincidence requirement efficiently suppresses contributions from internal radioactivity and backward scattered particles. These selection criteria can be adjusted to select neutrons, as opposed to photon events, as discussed below. Charged particles are normally rejected via thin active charged particle shields. The Compton scatter kinematics impose an additional geometrical constraint upon the scattering process using the energy deposits in the two detectors to provide the photon scatter angle as described below and illustrated in Fig. 1.

$$\phi = \cos^{-1}(1 - \varepsilon/E_2 + \varepsilon/(E_1 + E_2)) \tag{1}$$

Here, ε is the electron rest mass energy, E_1 is the energy deposit in the forward detector, E_2 is the energy deposit in the second or rearward detector and ϕ is the Compton scatter angle provided $E_1 + E_2$ is the full incident γ-ray energy. Although the constraints of coincidence, timing and inferred scatter angle greatly reduce the background contribution to the instrument count rate, they simultaneously reduce the detecting efficiency. Consequently, Compton telescopes are generally large in order to compensate for the reduced efficiency while retaining a high signal-to-noise ratio. Because of the large size, it is fortunate that Compton telescopes often have a large ratio of active to passive material, minimizing background generation. The necessity of collimation, it can be argued, is obviated by the inherent directionality of the TOF and scatter angle criteria, eliminating the mass associated with active or passive γ-ray shielding.

The basic scattering process for photons is illustrated in Fig. 1. The quantities that one measures are the location of the Compton scatter in the forward detector, the energy E_1 of the scattered electron, the location of the scatter in the rearward detector and the energy deposit E_2 in that detector. One computes from these interaction locations the direction of the scattered photon's velocity vector. From the energy deposits, one computes the scatter angle ϕ and the total γ-ray energy ($E_1 + E_2$). Without measuring the direction of the scattered electron in the forward detector, only the polar angle and not the azimuth angle of the scatter is known. The azimuth information is solely contained in the direction of the scattered electron in the forward detector; this direction is not measured in COMPTEL.

COMPTEL γ-RAY RESPONSE

COMPTEL, designed for the Gamma Ray Observatory, is a Compton telescope as described above (Schönfelder et al. 1991). It is the first Compton telescope to be placed on-orbit and will provide the opportunity to perform observations solar flares in addition to other cosmic sources. The mechanical design of COMPTEL is illustrated in Figure 1. An incoming γ-ray scatters off an electron in one of 7 D1 detectors and proceeds down to one of 14 D2 detectors scattering again. Such events constitute the ideal type of γ-ray interaction. The material in D1 is a liquid organic scintillator, NE213A, with the properties of low density and low Z (H/C ratio = 1.213). The material scatters both γ-rays and neutrons elastically, off atomic electrons and hydrogen nuclei, respectively. The detector is a fraction of a mean free path thick, meaning that the incident γ-ray (or neutron) can scatter in D1 and usually leave D1 without scattering again. For the case of small angle γ-ray scatters (< 10°), the incident γ-ray can deposit a large part of its energy in the D2 detector which is composed of NaI (high density, high Z).

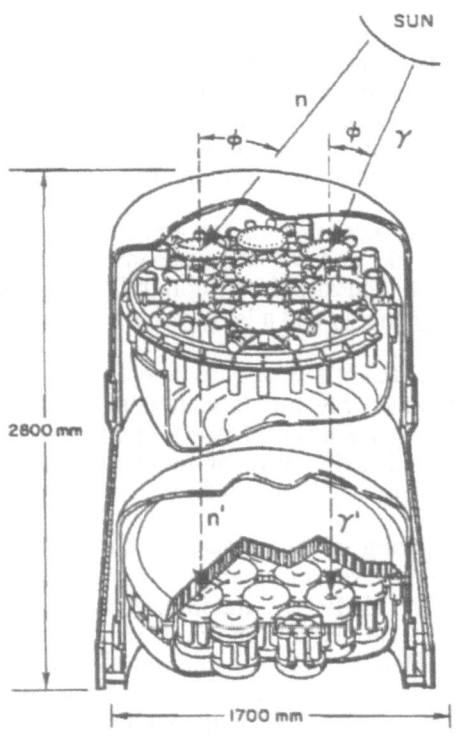

COMPTEL
IMAGING COMPTON TELESCOPE

Figure 1.
Schematic of COMPTEL with typical γ-ray and neutron interactions.

The liquid organic scintillator in D1 (NE213A) possesses pulse shape discrimination properties, in that energetic protons produce light pulses with longer rise times than those of electrons (and other minimum ionizing particles). This capability allows for efficient identification of signals from recoil protons produced by fast neutrons elastically scattering off hydrogen in D1. Any reaction producing a recoil proton can be identified by these means, such as inelastic scattering of fast neutrons off carbon producing a γ-ray and either a knock-on proton or neutron (which can then elastically scatter off hydrogen). Pure γ-ray producing neutron reactions off carbon in D1 are also possible and represent an intrinsic background in both γ-ray and neutron measurements.

The D1 and D2 subsystems of the telescope are each completely surrounded by charged particle detectors (see Fig. 1). These 4 domes of plastic scintillator NE110 are 1.5 cm thick and do not significantly attenuate the incident γ-ray or neutron fluxes, yet are virtually 100% efficient in identifying charged cosmic rays. The charged particle shields and other intervening material will heavily attenuate the solar flare hard X-ray flux, minimizing pulse pile-up effects in the D1 and D2 detectors.

The incident photon direction is constrained to a cone mantle of half angle ϕ, a result of the azimuthal symmetry in the detection process. In order to translate this geometrical feature to the the coordinate system of the telescope we require knowledge of the positions of the γ-ray or neutron scatters in D1 and D2. This is accomplished not only by knowing in which detectors the scatters occurred, but also by locating an event within a detector by comparing the relative pulse heights of the attached photomultiplier tubes. This provides spatial information within the triggered detectors in D1 and D2.

For γ-rays, errors in the measured energy and scatter angle occur via uncertainties in the measured energies in D1 and D2 and uncertainties in the measured interaction positions in D1 and D2. Partial energy absorption in D2 (from an escaping γ-ray) yields a low value for the total γ-ray energy and a large value for the scatter angle ϕ.

The total geometrical area of the forward detector array D1 is approximately 4300 cm^2, while the effective area for γ-ray double scatters in the range of 1 to 10 MeV is < 40 cm^2. The spatial resolution (1 σ) in a D1 module (~ 1 MeV) is ~ 2 cm, while that in a D2 module is ~ 1 cm. Although a function of incident energy and angle, the energy resolution of the system for γ-rays is between 5 and 10% (Fig. 2).

COMPTEL as an imaging solar γ-ray telescope relies on the full energy deposit of the γ-ray to correctly estimate the scattering angle ϕ of the photon in the instrument. For a solar flare γ-ray interacting in COMPTEL, the inferred scatter angle ϕ about the vector of the scattered γ-ray must be such that the photon is assigned a solar origin as indicated schematically in Figure 1. Hence, we know that the photon deposited its full energy in the detector. The response of the telescope to such events is simple. The energy or pulse height distribution is basically Gaussian in shape with a heavily suppressed Compton tail at low energies. Since the solar γ-ray spectra are rich in lines from C, N, O, Ne, Mg etc., a simple instrumental response function will facilitate correct de-convolution of the pulse height spectra. The response of a telescope prototype to monoenergetic 1.375 and 2.75 MeV photons from a ^{24}Na source is shown in Figure 2.

The effective detector area of COMPTEL to 5 MeV solar photons incident at 30° is about 25 cm^2, decreasing to about 10 cm^2 at 1 MeV and 20 MeV (Schönfelder et al. 1991). For the solar γ-ray flare, which occurred on 1982 December 7, the average γ-ray flux > 1 MeV was about 1 photon cm^{-2} s^{-1}, including bremsstrahlung and nuclear emissions, over a period of ~ 1000 s (Vestrand et al. 1987). With COMPTEL's sensitive area this would result in a count rate of 25 s^{-1}, slightly exceeding the telemetry rate of COMPTEL of 20 s^{-1}. For a 1000 s event duration this yields ~ 20 000 counts for good spectroscopy statistics.

COMPTEL BURST MEASUREMENTS

The burst mode of the COMPTEL instrument can also be used to detect the first fast burst of solar γ-rays. The programmable Burst Spectrum Analyzer (BSA) continually integrates γ-ray spectra from two separate D2 detector modules in a background mode at a programmable cadence and integration time (nominally 30 s). One detector covers the energy interval from 0.1 to 1 MeV and the other the interval from 1 to 10 MeV. Each detector has an unobscured field-of-view of about 2.5 sr and an area of ~ 600 cm^2. Outside this field-of-view varying amounts of intervening material exist

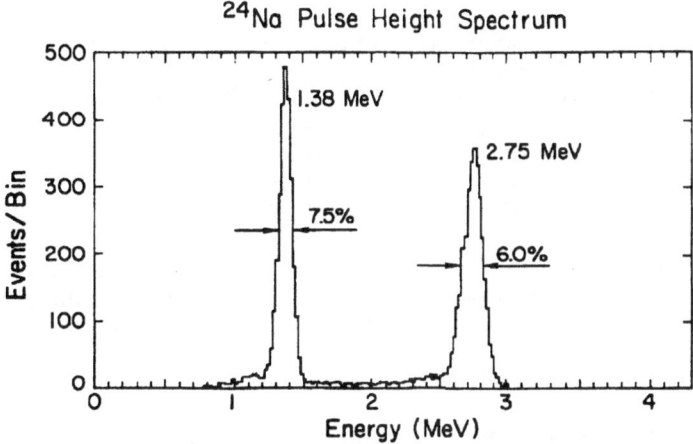

Figure 2.
The energy spectrum from a monoenergetic γ-ray source produced under the constraint that the γ-ray scatter angle ϕ be consistent with the source direction.

attenuating the solar γ-ray flux. The capabilities of COMPTEL for burst detection have been discussed by Winkler *et al.* (1986). When the burst system on COMPTEL receives a signal from the Burst and Transient Source Experiment (BATSE) (Fishman *et al.* 1989) indicating a burst of any origin, it starts accumulating spectra in these two modules at programmed time intervals, nominally every 2 s, for a total of six spectra, after which it switches to a so-called tail mode integration time of ~ 10 s. This fast accumulation rate provides information about the initial solar burst of γ-rays. These integration times are adjustable on-orbit, but only before the event. The evolution of both the bremsstrahlung and nuclear emission spectra is important in understanding the nature of the particle acceleration and transport processes. The earliest moments of a flare often carry the greatest information in terms of the relative timing of the two forms of γ-ray emission. Longer duration flares, such as those of 1982 December 7 and 1981 April 24 will be covered by the tail mode of the BSA in addition to the individual photon data in the telescope mode.

COMPTEL OPERATION DURING SOLAR FLARES

Within 5 s after the burst onset, the BATSE (Burst and Transient Source Experiment) instrument on the Gamma Ray Observatory sends a second signal to the On-Board Computer (OBC) if the burst originated from the general direction of the Sun (Fishman *et al.* 1989). Within the next two minutes, depending on the timing of the BATSE signal relative to the OBC telemetry frame, the OBC commands COMPTEL into a solar neutron mode for a time interval of 90 minutes or one orbit. COMPTEL still accumulates spectra in the two D2 burst detector modules, which collect burst spectra independent of instrument mode.

Double scatter telescope events are assigned a telemetry priority depending upon the measured event characteristics. During a solar flare, the event priority is determined primarily by the internal TOF, which separates γ-rays from non-relativistic neutrons. The priority γ1 is the highest priority, reserved for cosmic or solar γ-ray events, with other events, γ2 priority, (i.e. background γ-ray and neutron events) largely filling the remainder of the telemetry stream. After the onset of the γ-ray flare and before the instrument switches into the solar neutron mode, it remains in its primary cosmic γ-ray observing mode, collecting solar γ-rays in addition to extrasolar photons. In intense flares the solar flux will dominate all others. Upon the delayed command of the OBC, COMPTEL switches into the solar neutron mode. While the instrument is in the solar neutron mode γ1 (priority) events (solar and cosmic γ-rays) are still in the data stream, while the γ2 channel is modified so as to selectively record solar neutron events. Solar neutrons, γ2 events, are separated from the γ-rays by TOF. Pulse height requirements in D1 and D2 and Pulse Shape Discrimination (PSD) criteria in the D1 detectors are also employed to further enhance the neutron signature in the γ2 data channel. The delay in commanding COMPTEL into the solar neutron mode has no impact on the scientific content of the data. Solar neutrons < 150 MeV lag behind the initial γ-rays by up to 7 minutes, in which time COMPTEL is ready to measure and record the events.

COMPTEL NEUTRON RESPONSE

The ideal type of neutron interaction in COMPTEL occurs when the incoming neutron elastically scatters off a hydrogen nucleus in the D1 detector. The scattered neutron then proceeds to the D2 detector where it may interact, depositing some of its energy to produce a trigger signal as indicated in Figure 1. The energy of the incident neutron is computed by summing the proton recoil energy E_1 in the D1 detector with the energy of the scattered neutron E_s deduced from the TOF from the D1 to the D2 detector. The scatter angle for non-relativistic neutrons (< 150 MeV) can be computed by the formula:

$$\tan^2 \phi = E_1/E_s. \tag{2}$$

As with γ-rays, neutrons can be traced backwards from D2 to D1 through the angle ϕ to a cone mantle restricting the incident direction to include the Sun. This is a geometrical constraint identical to that of the γ-ray measurements. The pulse shape from recoil protons is sufficient to reject more than 95% of electron-recoil events greater than about 1 MeV, the energy threshold in D1 for neutron detection. This method of detecting and measuring the neutrons is clean, in that a delayed coincident scatter with the correct pulse shape in D1 is required, yielding a large signal-to-noise ratio. Inelastic neutron reactions with carbon also occur in the liquid scintillator, particularly at energies greater than about 50 MeV. The carbon interactions in D1 often produce γ-rays, deuterons or alphas which can be identified. Typical problem reactions are

$$^{12}C(n,np)$$
$$^{12}C(n,2n)$$
$$^{12}C(n,n'\gamma)$$
$$^{12}C(n,n'3\alpha)$$
$$^{12}C(n,n'p\gamma).$$

These reaction channels can be included to further increase the instrument response to neutrons. These interactions, however, are difficult to interpret because the lost energy (nuclear binding energy or escaping γ-rays) results in an inaccurate measures of E and ϕ. In the case of a solar neutron flare, identified by time, these events can be used to supplement the information obtained from clean elastic scatters.

With COMPTEL in the solar neutron mode, neutron interactions appear in the γ2 channel, covering the TOF interval from about 8 to 40 ns. The PSD and TOF criteria in this channel are such that solar neutrons incident on D1 in the energy range from about 10 MeV to 150 MeV are recorded. In this energy interval COMPTEL can observe neutrons from about 14.5 to 55 minutes after release from the Sun. This corresponds to a minimum observed delay time of 6 to 47 minutes after the onset of the γ-ray flash (assuming neutrons are not produced without accompanying γ-rays).

A prototype of COMPTEL (Science Model 3) consisting of two D1 and three D2 modules was exposed at the Indiana University Cyclotron Facility to calibrated pulsed neutron beams from 20 to 200 MeV incident at various angles with respect to the telescope axis. The resulting data were inspected to select events obeying the proper kinematic relationship for elastic scatters. The measured efficiencies of single D1/D2 module pairs (minitelescopes) are listed in Table 1 for various incident angles and energies. The resulting effective area of COMPTEL for neutrons incident at 29.2° is 12.1, 16.1 and 3.1 cm^2, where the calculated area is scaled up to the full COMPTEL instrument from the single minitelescope efficiencies in Table 1.

Table 1

Efficiencies for neutron detection with the constraint that the inffered scatter angle ϕ be consistent with neutron source direction.

Angle	Energy (MeV) 18.5	35.7	77.0
5.5°			6.7×10^{-5}
19.6°	2.9×10^{4}	4.4×10^{4}	
25.4°			7.6×10^{-5}
29.0°	3.1×10^{-4}	5.0×10^{-4}	
29.2°	2.1×10^{-4}	2.8×10^{-4}	5.4×10^{-5}
38.0°		1.2×10^{-4}	

Energy resolution figures are available for the analyzed Science Model 3 data. In Figure 3 are the measured energy spectra from monoenergetic neutron beams as indicated. Note the low energy tail in the 77 MeV data resulting from inelastic carbon reactions in D1. The prototype Science Model 3 was not tuned for precise location of neutron interactions within the individual detectors, so the acceptance angle windows are unusually wide (±10°), determined here by the physical size of the detectors rather than the spatial resolution of the detectors. The energy resolution is largely determined by the TOF measurement between D1 and D2. The pulse height or energy resolution in the scintillation measurement in D1 is generally negligible. There are three contributions to the energy resolution: the finite thickness and diameter of the two detectors creating an uncertainty in the path length over which the TOF is measured, the energy spread in the neutron beam due to the thickness of the target and the electronic TOF resolution, which for COMPTEL is 1.5 ns (FWHM). For the minitelescope D1/D2 pair at 29.2°, *for which there was no interaction location performed*, we expect from an error analysis energy resolutions (FWHM) of 17%, 15% and 18% for 18.5, 35.7 and 77.0 MeV, respectively. We measure 24%, 24% and 25% for those energies. For COMPTEL, in which the spatial resolutions are 2 and 1 cm for D1 and D2, respectively, we expect a greater rejection of spurious events by narrowing the scatter angle acceptance window from ± 10° to ± 2°. The discrepancy between the *a priori* estimate of the neutron energy uncertainty and that measured is not resolved pending re-examination of several instrumental gains and offsets.

An uncertainty (1 σ) of 8% (or 18% FWHM) in the measured energy of a 50 MeV solar neutron translates into an uncertainty of < 3 minutes in the production time of the neutron at the Sun. For solar flares which exhibit prolonged acceleration of protons on the order of 10 minutes (Ryan and Lee 1991) such the 1982 June 3 event (Forrest *et al.* 1985), this time resolution will aid in constructing a time dependent neutron production and, consequently, primary proton acceleration/precipitation spectrum.

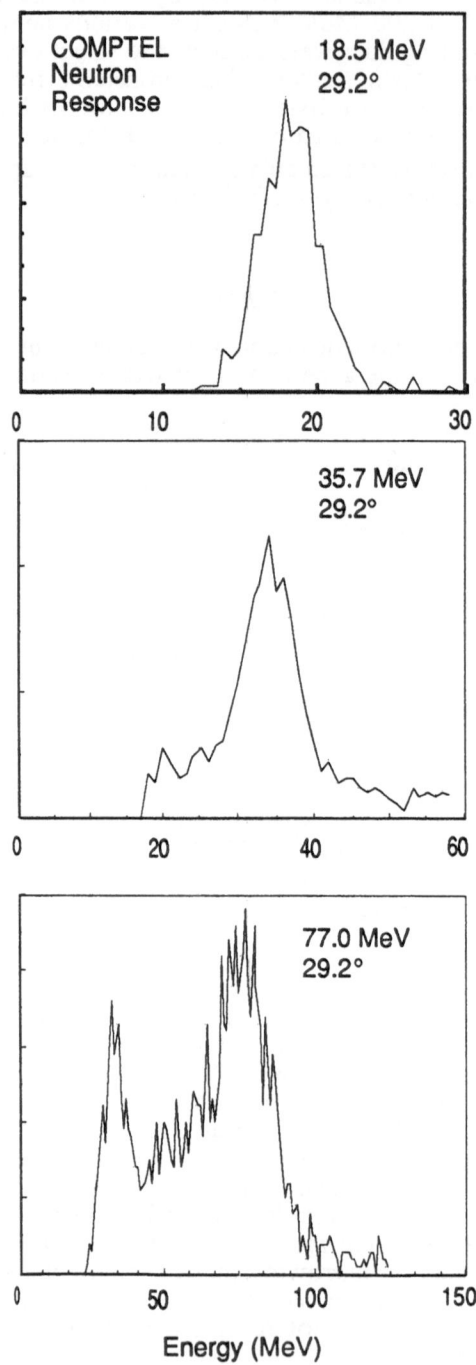

Figure 3
The energy spectrum from monoenergetic neutrons as indicated.
Data are selected based on scattering kinematics consistent with the
source direction.

COMPTEL RESPONSE TO THE 1982 JUNE 3 SOLAR γ-RAY FLARE

We can use these extrapolated efficiencies to estimate the response of COMPTEL to the solar neutrons from a flare event such as that of 1982 June 3. The calculated neutron energy spectrum in total neutrons produced per MeV at the Sun as a function of energy according to Murphy *et al.* (1987), assuming that the neutrons are impulsively produced, is plotted in Figure 4. A power law in rigidity times an exponential in energy is assumed for the solar neutron production spectrum, i.e.

$$dN/dE = A\,p^{-5}\exp(-E/1000),\tag{2}$$

where p is in Mv and E in MeV. The resulting neutron energy spectrum at the earth corrected for the neutron lifetime is shown in Figure 4. The energy range of COMPTEL for solar neutrons, as indicated by the heavy line from about 10 MeV to 200 MeV, covers the maximum in the neutron energy spectrum at earth. An event such as that of 1982 June 3 would have produced about 3300 clean neutron events below 100 MeV in COMPTEL over a time interval of about 40 minutes or an average event rate of $\leq 2\ \mathrm{s}^{-1}$, within the telemetry bandwidth of γ_2 events. The count rate spectrum peaks in the range of 40 MeV where the neutron spectrum peaks as well as the instrument effective area. A statistically significant spectrum can be produced with such a large number of events. The background neutron event rate is difficult to estimate until it can be measured directly. However, we can generously take the event rate > 1 MeV in the D1 detector to be $1000\ \mathrm{s}^{-1}$ and that in D2 to be $2000\ \mathrm{s}^{-1}$ and with a coincidence window of 30 ns this yields an accidental event rate of $0.1\ \mathrm{s}^{-1}$, a large fraction of which will be rejected when the scatter angle and pulse shape restrictions are applied.

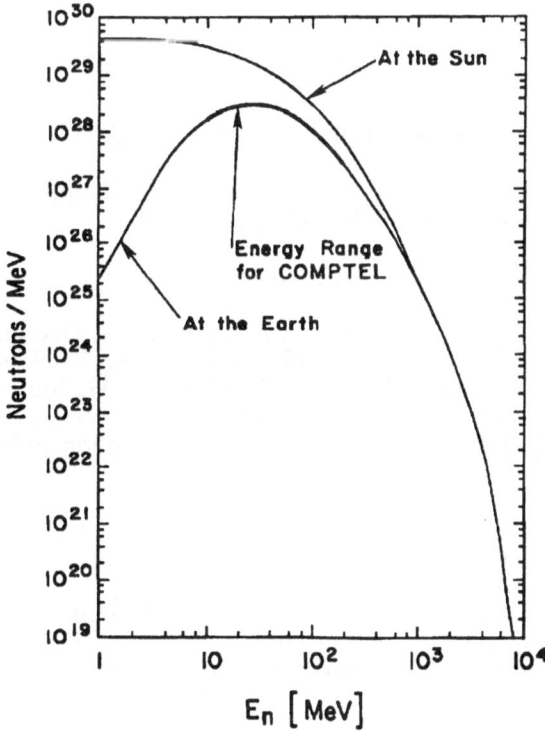

Figure 4
Neutron Production Spectrum from the 1982 June 3 solar
flare (Murphy *et al.* 1987).

CONCLUSIONS

After April 1991, the projected launch date for GRO, while still on the elevated part of the solar cycle, we should expect a number of solar γ-ray events with emission of photons > 0.8 MeV. From February 1980 to December 1982 Rieger *et al.* (1983) found ~ 130 solar flares with emission > 0.3 MeV, of which 8 emitted > 10 MeV γ-rays. Two of these events emitted measurable neutrons > 50 MeV. Consequently, we should expect to find solar flares with both high energy γ-rays (> 1 MeV) and neutrons (> 10 MeV) present, since the energy threshold of COMPTEL for neutrons is much lower than that of the SMM detector and since the neutron response curve of COMPTEL is well-matched to the solar neutron energy spectrum at earth.

By combining the normal imaging capabilities of COMPTEL with its burst mode operation and solar neutron spectroscopic abilities, definitive information should be obtained about the spectral evolution of nucleonic and electronic processes in solar flares. Neutron energy measurements provide more precise and extensive data about proton acceleration. For the unusual solar flare occurring on the west limb of the Sun in which prompt, energetic solar protons (> 500 MeV), γ-rays and neutrons are observed and measured, we should be in a position to understand better the relative roles of stochastic and diffusive shock acceleration processes at the Sun.

REFERENCES

Fishman, G.J., C.A. Meegan, R.B. Wilson, W.S. Paciesas, T.A. Parnell, R.W. Austin, J.R. Rehage, J.L. Matteson, B.J. Teegarden, T.L. Cline, B.E. Schaefer, G.N. Pendleton, Jr. F.A. Berry, J.M. Horack, S.D. Storey, M.N. Brock, and J.P. Lastrade 1989, BATSE: The Burst and Transient Source Experiment on the Gamma Ray Observatory, Goddard Space Flight Center, Greenbelt, MD: NASA.

Forrest, D. J., W. T. Vestrand, E. L. Chupp, E. Rieger, J. Cooper, and G. Share 1985, *Proc. 19th Internat. Cosmic Ray Conf.*, **4** : 146-149.

Murphy, R. J., C.D. Dermer, and R. Ramaty 1987, *Ap. J.*, **316** : L41-56.

Rieger, E., C. Reppin, G. Kanbach, D. J. Forrest, E. L. Chupp, and G. H. Share 1983, *Proc. 18th Internat. Cosmic Ray Conf.*, **10** : 338.

Ryan, J.M., and M.A. Lee 1991, *Ap. J.*, **368** : 316-324.

Schoenfelder, V., K. Bennett, W. Collmar, A. Connors, A. Deerenberg, R. Diehl, J.W. den Herder, W. Hermsen, G.G. Lichti, J.A. Lockwood, J. Macri, M. McConnell, D. Morris, J. Ryan, G. Simpson, H. Steinle, A. Strong, B.N. Swanenburg, B.G. Taylor, M. Varendorff, C. de Vries, and C. Winkler 1991, The GRO-COMPTEL Mission: Instrument Description and Scientific Objectives, Erice, Sicily.

Vestrand, W.T., D.J. Forrest, E.L. Chupp, E. Rieger, and G.H. Share 1987, *Ap. J.*, **322** (2) : 1010-1027.

Winkler, C., V. Schönfelder, R. Diehl, G. Lichti, H. Steinle, B.N. Swanenburg, H. Aarts, A. Deerenberg, W. Hermsen, L. Lockwood, J. Ryan, G. Simpson, W.R. Webber, K. Bennett, A.V. Dordrecht, and B.G. Taylor 1986, *Adv. Space Res.*, **6** (4) : 113-117.

NEURAL NET APPROACHES

FOR EVENT LOCATION IN THE DETECTOR MODULES

A. Connors[1], H. Aarts[2], K. Bennett[3], A. Deerenberg[2] *, R. Diehl[4], J.W. den Herder[2], W. Hermsen[2], G. Lichti[4], J. Lockwood[1], M. Loomis[1], J. Macri[1], M. McConnell[1], D. Morris[1], K. Reinhard[1], J. Ryan[1], V. Schönfelder[4], G. Simpson[1], B.N. Swanenberg[2], H. Steinle[4], A. Strong[4], M. Varendorff[4], C. de Vries[2], W.E. Webber[1], C. Winkler[3], V. Zeitlmeyer[4]

1. University of New Hampshire, Durham, NH, USA

2. ROL – Laboratory for Space Research, Leiden, The Netherlands

3. Space Science Dept. of ESA, ESTEC, Noordwijk, The Netherlands

4. Max-Planck Institut für Extraterrestrische Physik, Garching, Germany

1 INTRODUCTION

From the description of the Compton telescope given previously (Schönfelder *et al.*, this volume), one can see that the accuracy with which one determines the position of a cosmic gamma-ray source depends not only on the measurements of the energy deposited in the upper (D1) and lower (D2) detectors, but also on how accurately one estimates the (X, Y, Z) positions of each gamma-ray or neutron interaction (an *event*). If nothing were known about the position of each event except in which module it occured, it would increase the uncertainty in the position of a source by on the order of 10°. Within each COMPTEL module, one extracts position information from comparisons of relative intensities of signals in the photomultipier tubes. This technique was introduced in the 1950's for medical imaging by Anger (1958),

---

* deceased

and later was adapted to astrophysical applications (Zych *et al.* 1983; Schönfelder *et al.* 1984; Stacy 1985).

In practice, one cannot obtain exact (X, Y, Z) position information from the measured PMT signals. First, there are processes which "smear" the signal, such as statistical fluctuations within the PMTs, or multiple scatterings and partially absorbed events within the detector modules. Second, as is illustrated in the next section, the function describing event interaction position, (X, Y, Z), as a function of the relative intensities of PMT signals, is sometimes multi-valued. Also, this function is not easily expressed analytically. The problem then becomes to reliably and rapidly calculate a best estimate of a non–linear, non–analytic, vector–valued function. This is an ideal application for a neural net. For COMPTEL, we chose to implement CMAC, a readily available software neural net. This technique gave us roughly a factor of 20 to 50 improvement in the speed of processing events, compared to methods that had been tried earlier (McConnell 1990; R. Freuder 1991).

In the next section, we present an overview of the responses of the D1 and D2 detectors. In the third section, we review function approximation and interpolation as it applies to neural nets. In the final section, we display some results of COMPTEL D1 and D2 event location.

2 CELL RESPONSE OVERVIEW

2.A (X, Y) *RESPONSE*

As part of the overall COMPTEL calibration procedure, a source–plus–collimator was moved across the top of each COMPTEL module, and the signals from each PMT were recorded, for about 10^4 events per position from $\sim 10^3$ (X, Y) positions per module. In Figures 1 and 2, we display contour plots of the D2 and D1 relative PMT responses as a function of (X, Y) position within each module.

If PMT_j is the overall signal from the jth photomultiplier tube, let the PMT ratio R_j be defined by

$$R_j \equiv N \times \frac{PMT_j}{\sum_{k=1}^{N} PMT_k},\tag{1}$$

where N is the total number of PMTs in the cell (8 for D1 and 7 for D2).

Figure 1.A and 1.B show contour plots of R_1 and R_2, the mean PMT ratios in PMT 1 and PMT 2, calculated from the D2 mapping data. In this figure, PMT 1 is located in the center of the module, and PMT 2 at an angle of $\frac{\pi}{6}$ radians from the +X axis. The D2 R_1 response resembles a constant plus a Gaussian in r, the distance from its center, with width on the order of 5 cm; while the D2 R_2 response is more asymmetric.

Figure 2.A shows a contour plot of R_1, the mean PMT ratio for PMT 1, taken from D1 mapping data. In this mapping data coordinate system, the center of the face of PMT 1 is located at a radius of 13.5 cm from the center of the module, and at an angle of $\frac{5}{8}\pi$ radians measured counterclockwise from the +X axis ($\frac{\pi}{8}$ from the +Y axis). The D1 PMT response can be described roughly as the sum of a Gaussian

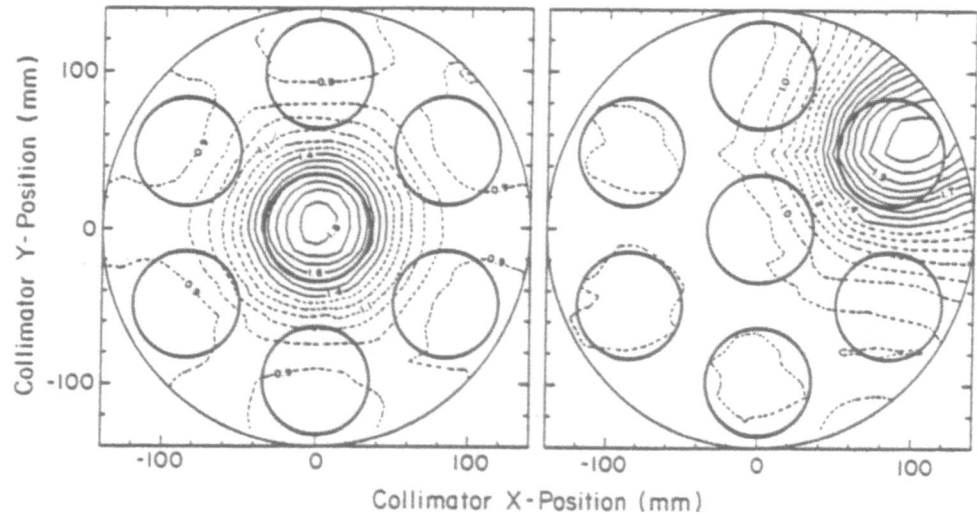

Figure 1.A (left). The relative intensity of the signal received in PMT1 (mean PMT 1 ratio, or R_1); and Figure 1.B (right): the relative intensity of the signal received in PMT2 (mean PMT 2 ratio, or R_2), as a calibration source is moved across a D2 module. Contours range from 0.80 to 1.91 with a contour interval of 0.08. Dashed lines indicate the intensity is less than the median value. Thicker lines indicate the module edge and the photomultiplier tube faces.

of width ~ 3 cm in r, the distance from the center of the PMT, plus an exponential in r with decay rate of ~ 30 cm. In Figure 2.B we display a cross–section of the D1 PMT1 response around $X = 0$. (For this figure the mean ratio for PMT 1, R_1, was calculated from mapping data points within $\pm 7°$ of the Y–axis.)

2.B Z RESPONSE

Although there was good (X, Y) data, no measurements were available on Z, the height of each interaction within a module. To indirectly infer Z, COMP-TEL used a principle components analysis (Finger 1987; Varendorff 1987). Several thousand events from each (X, Y) collimator position were used to calculate the covariance matrix of the PMT signals. Then an eigenvector decomposition was performed. For that (X, Y), the direction of the eigenvector with the greatest eigenvalue was designated the λ direction. In many cases this direction of greatest variation was associated with Z, the interaction depth (Varendorff 1987; Loomis 1991). In practice, therefore, COMPTEL event location estimates (X, Y, λ) rather than (X, Y, Z) from the PMT signature.

Since the D1 PMTs are positioned symmetrically on the walls of each D1 module, it is clear there is limited depth information. One expects the PMT response to be symmetric in Z about the middle of the cell. For D2 cells, using a light propagation model originally developed by Zeitlmeir (1988), Loomis (1991) has demonstrated that interaction height is also not a monotonic function of λ and cannot be unambiguously inferred from the PMT signature. This is displayed in Figure 3.

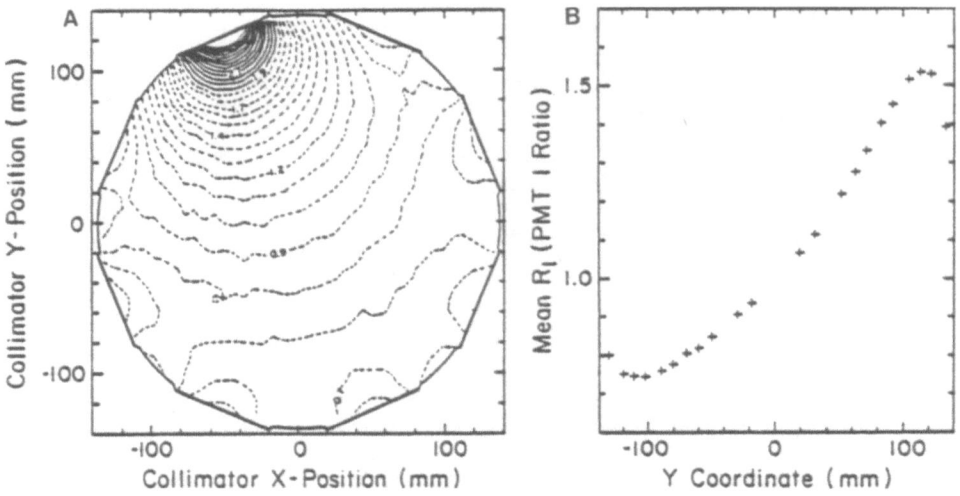

Figure 2.A and 2.B. The relative intensity of the signal received in PMT1 (mean PMT 1 ratio, or R_1) as a calibration source is moved across a D1 module. Thicker lines indicate the module edge and the photomultiplier tube faces. Contours range from 0.56 to 2.86 with a contour interval of 0.08; dashed lines indicate the intensity is less than the median value. The plot on the right, 2.B, is a cross section of the contour plot to the left, taken at $X = 0$.

2.C INCOMPLETE POSITION INFORMATION

In Figure 2.A, notice that the inverse function — the (X, Y) position as a function of the D1 PMT ratios — will be double–valued near the edges of the cell, betwen the phototubes. This is illustrated further in Figure 2.B, the cross–section of Figure 2.A, taken at $X = 0$ mm. In other words, even for the case of infinite signal–to–noise, the inverse functions would be ill–defined, so one has incomplete position information for (X, Y) as well as for Z.

Further, in practice one does not have infinite signal–to–noise. Even when physical effects such as multiple scatterings and escaping photons and electrons are ignored, one expects statistical fluctuations in the PMT signals (Engstrom 1980). At best, one infers a certain probability that an event occurred at (X, Y, λ), given the measured PMT signals. This full probability distribution was deemed too cumbersome for COMPTEL event processing. One therefore summarizes this information with a single position estimator, (X^*, Y^*, λ^*). Typical estimators can be: the most probable position, given the PMT signature; the expectation value, which gives the minimum rms error; the value which maximizes the information entropy, which makes the fewest assumptions about missing information; or some completely different kind of estimator. Recall that for non–Gaussian probability distributions, these can give very different numerical results. A minimum χ^2 search method, such as the one used previously (McConnell 1991; Varendorff 1987), looks for the most probable position; while as the neural net tries to minimize the RMS error, it comes closer to approximating the expectation value of (X, Y, λ). Note that for COMPTEL, since the PMT signature is not a unique indicator of position, one does not expect the distribution of (X^*, Y^*, λ^*), calculated for a large number of events, to trace out the true probability of finding an event at (X, Y, λ), no matter which estimator is chosen.

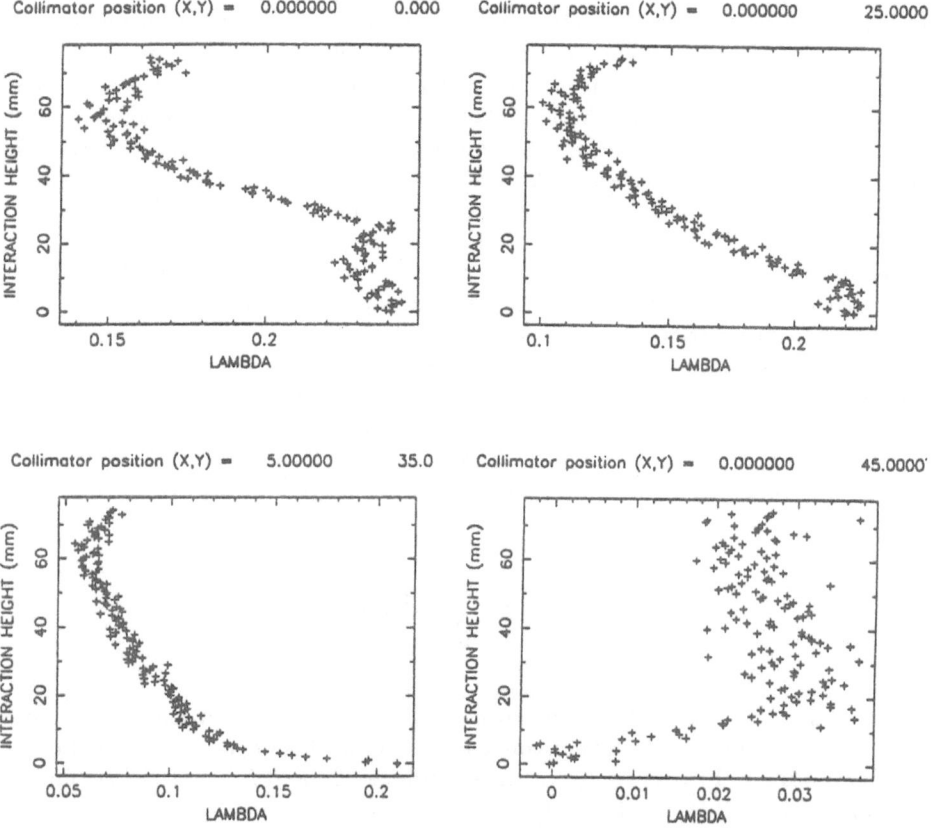

Figure 3. For D2, interaction height as a function of λ is displayed from Monte Carlo simulations for four different collimator positions. From Loomis (1991).

3 FUNCTION APPROXIMATION AND INTERPOLATION

3.A OVERVIEW

Once a position estimator is chosen, to be practical, one needs to find a fast way of approximating (X^*, Y^*, λ^*) as a function of the PMT signals, and of interpolating between the $\sim 10^3$ discrete mapping data points. In general, function approximation and interpolation methods can be classified as either local or non-local. Non-local methods include expansion in eigenfunctions, such as Fourier series, or Legendre polynomials. Sometimes one can cleverly choose a basis function so that the series converges very fast and the approximation is accurate using only a few terms. Local methods can be more flexible, especially when there is no analytic expression for the function of interest (or when an analytic expression would take too long to compute). Local methods include piecewise continuous polynomials, such as cubic splines. For example, most contour plots of unevenly spaced data (such as the contour plots in Figures 1, 2, 4, and 5) are produced using a standard two-dimensional Akima spline algorithm (Clare, Kennison, and Lackman 1987; IMSL 1987; Tennant 1989; and references therein). It requires a look-up table roughly of dimension twice the number of the data points. Another popular local method is to sum Gaussians of varying widths. (See, for example, Bendillini 1991.) This approximation is infinitely continous and differentiable, which is sometimes an advantage. However in some cases it can require storage space for a great many components. An even simpler algorithm, but one that requires even more storage space, is the sum of many constant components. For example, if one were to approximate the one-dimensional function in Figure 2.B by this method, one would divide the Y-axis into discrete bins and associate a weight A_j with the jth bin, so that

$$R_1(Y) \approx \sum_{j=0}^{M-1} A_{S(Y)+j}, \qquad (2)$$

with S(Y) the first bin associated with an input value Y; and the number of bins M is chosen by the user to give a convenient overlap between adjacent points. Note that if two inputs, Y_1 and Y_2, are assigned indexes S_1 and S_2 that are H bins apart, the number of overlapping bins is:

$$OVERLAP = \max(0, M - H); \quad H = ||S_1 - S_2|| \qquad (3)$$

Once the weights A_j are assigned, this method can be extremely fast.

3.B NEURAL NETS

Of course, all the local methods sketched above can be formulated as Neural Nets. (See, for example, Poggio and Girosi 1990.) Rather than relying on algorithmic complexity, the neural network (or connectionist) philosophy emphasizes using many extremely simple elements, connected together, to represent complicated, non-linear, functions. They can be particularly useful when one has many examples but no analytic form for a function. The local methods described in the previous section are all classified as single layer localized receptive field networks, with no hidden layer. Equations 2 and 3 describes CMAC, or Cerebellar Model Arithmetic Computer, which is the neural net currently used in COMPTEL event location. It gives a real vector output in response to discrete state (i.e. integer) vector input, from a look-up table (Miller, Glanz, and Kraft 1990; 1987).

CMAC is categorized as having *fixed receptive field centers*. In practice, this means that the mapping from the real-valued PMT ratios to integer look-up table indexes is set ahead of time by the user, independent of the data. For historical reasons these look-up table indexes are called *state vectors* $\vec{S}$ (see Albus 1972, 1975). COMPTEL currently uses a roughly logarithmic function with the spacing between bins set by a *Scale Parameter*. A smaller scale parameter means a smaller bin size, or finer mesh, is used to map the real-valued input into indexes for the look-up table of weights A_j; a larger scale parameter implies a larger bin size, or coarser mesh. For example see Reinhard (1989), Figure 7.1, which uses a scale parameter of .05.

CMAC also uses *fixed receptive field widths*. This means the number of cells summed to form the approximation, M, is also set by the user when the net is made, independent of the data. Since one uses a very large number of cells (typically 50,000), as one is *training* the neural net, one is solving for a large number of weights, A_j. By fixing both the widths and the centers of the receptive fields, one greatly simplifies the optimization procedure. If all three were free to vary, there is no guarantee that any training method would converge.

To reduce the large storage space required one maps the table of weights A_j to a compressed version A_j'. However CMAC retains the convenient property that, even with multi-dimensional input, two inputs $\vec{Y}_1$ and $\vec{Y}_2$ that are close together, having look-up table indexes $\vec{S}_1$ and $\vec{S}_2$, will have an overlap of:

$$OVERLAP = \max(0, M - H); \quad H = ||\vec{S}_1 - \vec{S}_2||. \tag{4}$$

This allows some interpolation between adjacent data points. Note that the amount of overlap depends on both the number of cells to be summed, M, and the mapping from the real inputs $\vec{Y}$ to the integer indexes $\vec{S}$.

With CMAC the weights A_j' are determined by supervised incremental learning. All weights in the net are initialized to zero; then pairs of input and output vectors (termed *exemplars*) are presented to the net. For COMPTEL, the inputs are PMT ratios from the mapping data, and the outputs are the corresponding collimator locations. The weights are adjusted to minimize the square of the distance between the example output vector and its neural net approximation, using a standard Widrow-Hoff (gradient descent) technique.

Let $\vec{S}_i$ represent the ith vector-valued input; $\mathbf{W}$ the array of weights; $\vec{X}_i$ the ith vector-valued output; $\vec{f}(\vec{S}_i, \mathbf{W})$ the estimator of the output; and E_T the total

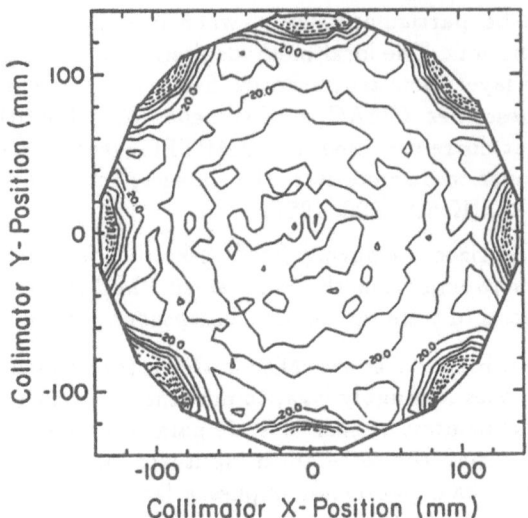

Figure 4. Contour plot of (X, Y) position resolution, in mm, across a D1 module. Contours range from 5 mm to 65 mm with a contour interval of 5 mm. Dashed lines indicate regions with resolution worse than the median. Thicker lines indicate the module edge and the photomultiplier tube faces.

square error. Then

$$E_T = \sum_{i=1}^{M} (\vec{X}_i - \vec{f}(\vec{S}_i, \mathbf{W}))^2. \tag{5}$$

If one were to minimize the total error, one would find an extremum by setting all components of the gradient of the error E_T with respect to the weights $\mathbf{W}$ to zero, and solving the simultaneous equations. In a neural nets context, this is called *batch learning*. This involves inverting a very large, sparse, matrix — a numerically awkward procedure. (COMPTEL currently uses an array of 50000 weights.)

Instead, CMAC uses *incremental learning*. In contrast with batch learning, in which the gradient of the total error is set to zero to find an extremum, in standard Widrow-Hoff incremental learning, one adjusts the weights $\mathbf{W}$ based on the incremental error E_i due to the ith exemplar $(\vec{S}_i, \vec{X}_i)$:

$$E_i = (\vec{X}_i - \vec{f}(\vec{S}_i, \mathbf{W}))^2,$$

and

$$\mathbf{W_{jk}(i)} \equiv \mathbf{W_{jk}(i-1)} - \frac{1}{2}\beta \frac{\partial E_i}{\partial \mathbf{W_{jk}}}. \tag{6}$$

The parameter β is called the learning rate, and $0 < \beta < 1$. For the neural nets used in COMPTEL, β is typically 0.05.

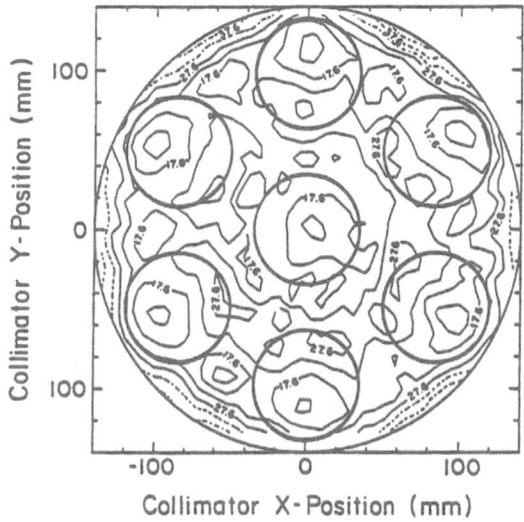

Figure 5. Contour plot of (X, Y) position resolution, in mm, across a D2 module. Contours range from 10.0 to 47.6 with a contour interval of 5. Dashed lines indicate regions with resolution worse than the median. Thicker lines indicate the module edge and the photomultiplier tube faces.

4 RESULTS

Using the neural net method, the root mean square error over the whole module in (X, Y) location was 25 mm for D1 and 23 mm for D2, while the root mean square error in λ estimation was 0.190 for D1 and less than half that, or 0.085, for D2, where λ ranges from -1 to $+1$. In Figures 4 and 5 we display contour plots of the (X, Y) resolution (in mm) across a D1 and a D2 module. Notice that the resolution tends to be best near a photomultiplier tube, and worst around the edges, where the inverse function (i.e. event position as a function of relative PMT signals) is ill-defined. (These two figures were made from mapping data, using neural nets trained on $\sim 10^6$ exemplars, with $M = 50$ cells summmed to form the approximations, and with Scale parameters of 0.05 and 0.02 for D2 and D1, respectively.)

This is also visible in the scatter plots of event locations displayed in Figures 6 and 7. (These two figures were made using data from a 6 MeV CaF source 10 meters from the telescope, at an angle of 10° from the vertical.) Also, there is some slight clustering around the centers of each of the PMTs in both the D1 and D2 modules. This is thought to be due to a tendency of CMAC to broaden sharp peaks (i.e. *overlearning* on regions with steep gradients).

Our studies so far indicate that non-uniformities at this level do not seem to affect the overall COMPTEL angular resolution. In Figure 8, we display a plot of the Angular Resolution Measure (A.R.M.) distribution, for the 6 MeV data described above. (The A.R.M is defined as the angular distance between the inferred photon arrival direction and the true source position.) We show the A.R.M. distribution for

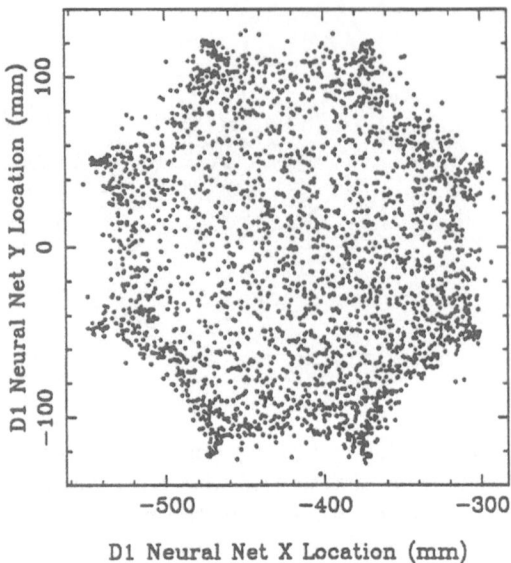

Figure 6. Scatter plot of (X, Y) event locations in a D1 module, from a 6 MeV full telescope run.

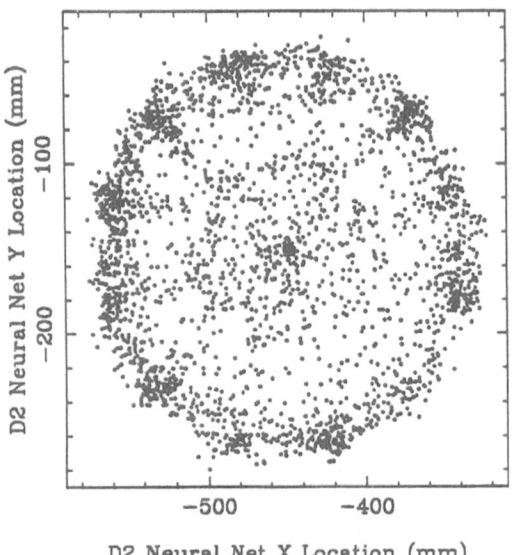

Figure 7. Scatter plot of (X, Y) event locations in a D2 module, from the same 6 MeV full telescope run.

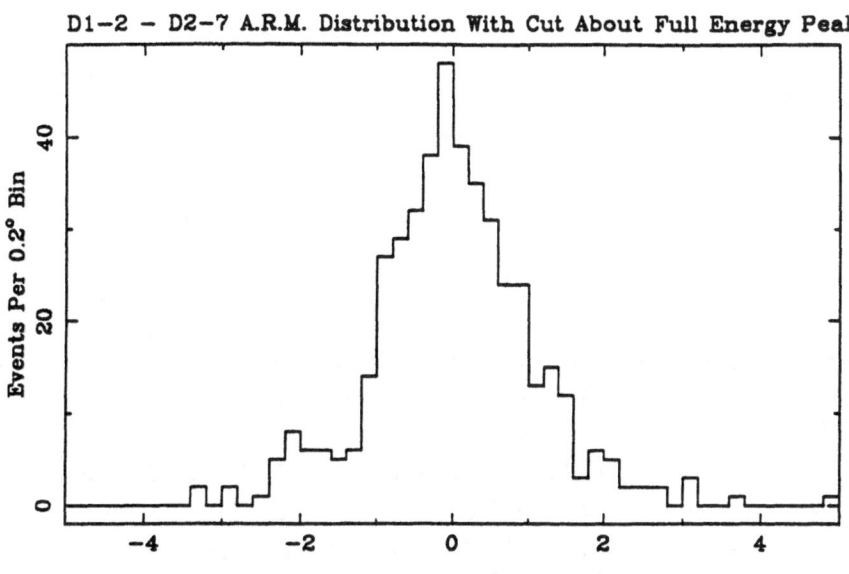

D1-2 - D2-7 A.R.M. Distribution With Cut About Full Energy Peak

Events Per 0.2° Bin

Angular Distance From True Source Position (Degrees)

Figure 8. 6 MeV photopeak A.R.M. distribution, in events per 0.2° bin.

events with total energy of $6 \pm 0.1 \text{MeV}$. This effectively includes only gamma–rays that have deposited their full energy in the D1 and D2 detectors, and is a reasonable way to observe the effect of just the event location on COMPTEL's angular resolution. The one σ width is about 0.9°.

Advances in electronic computation can change not only the speed with which we calculate numeric answers, but also the ways in which we formulate the questions, and even what kinds of problems are considered interesting to ask. In the use of "connectionist" techniques such as neural nets, or in Bayesian and Entropic analyses of the information content of data, are we beginning to see qualitative changes in how we approach astrophysical data analysis? The coming years may be very interesting.

5 ACKNOWLEDGEMENTS

Much of the summary of neural nets relies on notes from *Advanced Topics in Systems Engineering: Artificial Neural Networks*, as taught by M. J. Carter at the University of New Hampshire.

6 REFERENCES

Albus, J. S., 1972, Ph.D. Thesis, University of Maryland.

Albus, J. S., 1975, *J. Dyn. Sys., Meas., Contr.,* **97,** 220.

Anger, H. O., 1958, *Rev. Sci. Instr.,,* **29,** 27.

Bendellini, O., 1991 *Ap. J.,* **366,** 599.

Clare, F., Kennison, D., and Lackman, R., 1987, *NCAR Graphics User's Guide,* (National Center for Atmospheric Research: Boulder, Co).

Engstrom, R. W., 1980, *RCA Photomultiplier Handbook,* (RCA Corporation Electro Optics and Devices: Lancaster, PA).

Finger, M., 1987, Ph.D. Thesis, California Institute of Technology.

IMSL 1987, *User's Manual: Math/Library Fortran Subroutines for Mathematical Applications.*

Freuder, R., 1991, private communication.

Lampton, M., Margon, B.,and Bowyer, S., 1976, *Astrophysical Journal,* **208,** 177.

Loomis, M., 1991, U.R.O.P. Report, University of New Hampshire.

McConnell, M., 1990, *Comparisons of Neural Net vs Conventional Location Techniques,* COM-TN-UNH-MMG-074.

Miller, W. Thomas, Glanz, Filson H., and Kraft, L. Gordon, 1987, *International Journal of Robotics Research,* **Vol. 6, No. 2,** 84.

Miller, W. Thomas, Glanz, Filson H., and Kraft, L. Gordon, 1990, *Proc. of the IEEE,* **Vol. 78, No. 10,** 1561.

Poggio, T., and Girosi, F., 1990, *Proc. of the IEEE,* **Vol. 78, No. 9,** 1481.

Schönfelder, V., it et al., 1984, *ISEE Trans. Nucl. Sci.,* **NS-31,** 766.

Shafer, R. A., Haberl, F., Arnaud, K. A., 1990, *XSPEC: An X-Ray Spectral Fitting Package User's Guide,* (EXOSAT Observatory: European Space Agency, Noordwijk).

Stacy, J. G., 1985, Ph.D. Thesis, University of Maryland.

Tennant, A. F., 1989 *The QDP/PLT User's Guide,* (Space Science Laboratory: Marshall Space Flight Center).

Reinhard, Kent, 1989, M.S. Thesis, University of New Hampshire.

Varendorff, Martin, 1987, Diplomarbeit, MPE.

Zeitlmeir, Valentine, 1988, Diplomarbeit, Augsburg.

Zych, A. D., Tumer, O. T., and Dayton, B., 1983, *IEEE Trans. Nucl. Sci.,* **NS-30,** 383.

THE OSSE DATA ANALYSIS SYSTEM

M.S. Strickman, J.E. Grove[1] , W.N. Johnson, R.L. Kinzer,
R.A. Kroeger, J.D. Kurfess, and M.D. Leising

E. O. Hulburt Center for Space Research, Naval Research
Lab., Washington DC

D.A. Grabelsky, S.M. Matz, W.R. Purcell, and M.P. Ulmer

Department of Physics and Astronomy, Northwestern University,
Evanston, IL

R.A. Cameron and G.V. Jung

Universities Space Research Association

K.M. Brown and C.M. Jensen

Physics Department, George Mason University, Fairfax, VA

INTRODUCTION

The Oriented Scintillation Spectrometer Experiment (OSSE) on board the Gamma Ray Observatory (GRO) spacecraft has a wide variety of operating modes, each of which produces its own type of data. Each operating mode has various adjustable parameters that effect the availability and interpretation of the various data types. OSSE, its modes and data products, have been described in detail elsewhere [1].

The OSSE data analysis system has the primary task of interpreting the data stream from the instrument. The system must be able to produce properly sequenced streams of clean, error-free data from any of the OSSE operating modes with any combination of instrument parameters. In addition, anomalies must be identified, isolated, logged and eliminated from the data stream. Data must then be archived in formats that give convenient, reliable access to all information required to perform meaningful scientific analyses. The scientific analysis portion of the system must provide the tools required to access the data archives, manipulate and select the various types of data, remove instrumental effects, model the data with empirical and/or as-

---

[1]NRC/NRL Postdoctoral Fellow

Data Analysis in Astronomy IV, Edited by V. Di Gesù *et al.*
Plenum Press, New York, 1992

trophysically meaningful models, visualize the results, and determine the significance of observed phenomena, both statistically and as possible instrumental artifacts.

OSSE data analysis software runs on DEC VAX computers using the VMS operating system. Although it is not currently configured to be easily portable to other operating systems, this is a long-term goal. Portions of the data analysis system are written in IDL, a proprietary data analysis language and visualization tool widely used in astronomical research.

Other portions are written around the INGRES database management system. However, the bulk of the system is written in VAX FORTRAN. VAX extensions to the ANSI FORTRAN-77 standard have been used where appropriate and calls to VMS system service and i/o routines have been used where needed.

The first section of this paper will discuss the major data formats used in the data analysis system. These formats allow a wide variety of OSSE data to be stored in such a way that all information required for scientific analysis is conveniently available to the scientist.

The second section will discuss Production Data Analysis, the set of processes routinely applied to all incoming data. Production Data Analysis receives incoming telemetry, and outputs data archives usable by the scientist.

The final section of the paper will discuss Scientific Data Analysis, a set of processes and tools that allow the scientist to access the archived data, manipulate it into an interpretable form, and model the results. This section will deal primarily with the analysis of spectra from discrete gamma-ray sources. It will also mention temporal analysis of periodic signals from pulsars.

DATA FORMATS

The overall philosophy of OSSE data storage is to keep the data themselves together with the auxiliary information required to analyze the data. In some cases, this means that the two are actually stored together, in some cases it means that the data storage format includes pointers to auxiliary data, and in some cases it means that the data storage format includes information required to call tools to produce auxiliary data.

Data formats must include data stored on long-term archival media, data stored on-line, typically on magnetic disk, and data stored in the Interactive GRO/OSSE Reduction Environment (IGORE). IGORE is discussed in more detail in [2]. This paper will discuss some of the major IGORE data storage formats.

In general, data storage in archive, disk and IGORE environments are parallel to one another as much as possible. While the same software cannot always be used to access all three types of stored data, the same information is generally present in more or less the same format. Hence, analysis applications that require a given set of data can expect that data to be present no matter what the source.

A summary of major OSSE science data products together with the formats in which they are typically stored can be found in table 1.

SPECTRUM DATA

The primary medium for all OSSE spectrum storage is the Spectrum Database (SDB) format. SDB has been implemented as a random access file for disk media,

Table 1. Science Data Products

Data Type	Integration Time (s)	2-wk Volume (Mbyte)	Storage Format
"TSAC" Spectra (TSAC = Time of spectrum acquis.)	4-32	~640 for 8 sec TSAC	SDB (multispectrum)
Pointing Interval Spectra	120	~100	SDB
Burst Time Histories	.004	~0.01/burst	SDB
Pulsar Data	0.000125	~1000	PDF

as a sequential access file for tape media, and as a record type within IGORE. While the primary purpose of SDB is the storage of spectra, it is capable of storing a large variety of other data types as well. These auxiliary data types generally contain data required for analysis of the spectra. In addition, each spectrum is accompanied by an extensive header containing spacecraft and instrument housekeeping information and information pertaining to the in-orbit environment such as magnetic field and charged particle dosage data. Spectrum manipulation applications are required to keep this header up-to-date when performing operations on spectra.

The random access implementation of SDB, the SDB disk file, is a VAX/VMS indexed sequential access method (ISAM) file, each record of which contains a structured key field encoding a spectrum tag and datatype identifier. The SDB format has very few restrictions in terms of data that must be present. Tags may have auxiliary data with no spectrum data or vice versa. Each data type is accompanied by a header describing the data. Headers may be present with no accompanying data records.

A single "spectrum" normally contains, for a single detector, data for all three energy ranges (0.05-1.5 MeV or Low; 0.05-10.0 MeV or Medium; and 0.05-200 MeV or High) concatenated into a single array. The header that accompanies a spectrum (referred to as the SDB header to distinguish it from auxiliary data type headers) contains a wide variety of spacecraft and instrument pointing, configuration, housekeeping rate, and environment information. The intent of the SDB header design is to supply all information (or pointers to information) required to analyze the accompanying spectrum.

Auxiliary data types may contain arrays of integers or real numbers or may be arrays of structured variables. The currently defined SDB data types include spectra, uncertainties, calibration and instrument response data, and analysis strategy information.

Some of the OSSE data products that are stored and used in SDB format are the pointing interval spectra produced and archived by Production Data Analysis, "TSAC" (Time of Spectrum Acquisition) spectra accumulated at the basic OSSE spectrum mode time resolution and produced as an optional product, diagnostic spectra produced by the Calibration PHA (CALPHA) and the Roving PHA (RPHA), and various derived spectra produced by Scientific Analysis. In addition, OSSE burst mode time histories are stored as pseudo-spectra in SDB.

Spectrum data are stored within IGORE as an IGORE spectrum data record (SDR). The SDR has a pointer to the spectrum on disk and pointers to the SDB header, spectrum data and up to eight auxiliary data types in virtual memory. The virtual memory arrays may or may not be populated with data. In the case of large arrays of SDR's (for example 20,000 SDR's associated with two-minute pointing interval spectra over a 2-week observation period), memory is generally insufficient to store the required volume of spectra, or even the headers. Only the disk file pointers are populated in this case. When the array of SDR's is passed to an analysis application, the application will read the data in from disk.

On sequential media, spectrum data are stored in a format identical to SDB on disk, but with the indices removed. Records are stored sequentially in key order. This format is used for spectrum archives.

PULSAR DATA

OSSE pulsar mode data can be collected in one of either event- by-event mode or rate mode. Event-by-event mode contains time- and energy- tagged individual photon events with time resolutions down to 0.125 msec. Rate mode, with a best time resolution of 4 msec, consists of binned counts in up to eight energy bands.

The Pulsar Data File (PDF) format is defined to include rate or event-by-event mode data together with pulsar mode information, detector configuration information and solar system barycenter vectors. The PDF is a sequential, variable record length file containing a variety of record types, each of which contains one of the aforementioned data types. It is designed to be identical in implementation on both random access and sequential media. Due to the large volumes of pulsar data generated over a two-week observation period, PDF is designed to be used from magnetic tape media as well as from random access disks. Also due to the volume involved, there is no corresponding IGORE record to represent pulsar data in memory.

FIT DATABASE FORMAT

The Fit Database Format (FDB) stores the results of parameter estimation problems. Past experience has indicated that, even if data are stored in an organized manner with enough resident documentation to identify them unambiguously, the results of applying models to the data are often stored in such a way that they cannot be easily identified or related to the data to which they apply. The FDB format attempts to address both these problems.

FDB stores not only the best-fit parameters resulting from an estimation procedure, but also stores a model identifier, all setup information for the procedure and the model, and a pointer to the data. Any number of parameters can be stored for each fit and any number of fits can be stored for each spectrum.

In memory, fit data is represented by the Fit Data Record (FDR), an IGORE record with a structure similar to FDB. Each FDR contains the information for a single fit to part or all of a single spectrum. IGORE utilities exist to read and write FDR's from and to FDB files on disk.

FDB files are represented on sequential media in a manner similar to SDB files, that is, as a sequential, variable record length file with the records written in key order.

Table 2. Archive Structure

TAPE	CONTENTS
1	Telemetry
2	Primary Data
	(Science Spectra, Diagnostic Spectra, Database Updates)
3	Pulsar Data Files (PDF)
4	Telemetry Scalar Files (TSF)

TELEMETRY SCALAR FILE

The Telemetry Scalar File (TSF) contains virtually all of the scalar (i.e. other than spectra or pulsar data) variables in the telemetry stream at or near the best time resolution available. TSF contains instrument rates, pointing information, etc. In addition, TSF contains rates in sixteen spectrum bands summed from the TSAC spectra. TSF is useful for temporal analysis, especially of background variations and transient events, and for manual and automatic data screening.

PRODUCTION DATA ANALYSIS

Production Data Analysis (PDA) consists of the set of all processes that are run routinely on all data received from the instrument. The goal of PDA is to produce a consistent database of on-line and archived data products. It must decommutate the telemetry, divide the various OSSE data products into individual time-aligned streams, produce summed spectra and calibration data corresponding to detector pointing intervals, perform routine pulse height analyzer calibrations and archive the resulting data products. In addition, both automatic and manual data screening operations take place during PDA, the results of which are included in a data quality database.

PDA tasks are discussed briefly below:

Data Capture and Archiving: Telemetry data are received daily over a 56 kbps telephone line to Naval Research Laboratory 48 hours after the data are collected on-board the spacecraft. PDA will be run on a daily basis as data are received, one day's worth of processing taking approximately six hours plus the time required to receive the data over the phone line. DCA handles the data reception protocol, writing the data to magnetic disk as it is received. The telemetry packets are not opened at this stage, but are augmented with solar system barycenter vectors for later use.

Telemetry and Science Archive Processes (TAP and SAP): TAP is responsible for writing the telemetry files to archive media. At the beginning of the mission, these will be 4mm DAT's. By the end of the first year of the mission, an additional deep archive on optical disk will also be written. The volume of archived telemetry data is about 1 Gbyte per two-week observation. SAP archives science data products produced by the processes described below. The volume of some SAP products are listed in table 1. Table 2 lists the 4mm DAT archive structure for a two-week observation. Each two week observation consists of a self-contained set of 4mm tapes.

Data Stream Formation (DSF): DSF's major tasks include decommutation and subdivision of telemetry packets; time alignment of housekeeping, spectrum, and pulsar data; summation of spectra from basic time resolution to detector pointing intervals (typically two minutes); updates of instrument configuration and data quality databases; and output of data streams to disk, for the most part in SDB format. Special purpose runs of DSF can output PDF's and spectra with TSAC time resolutions (typically 8-16 seconds).

Automatic Calibration Processes: The automatic calibration processes fit models to pulse height analyzer and pulse shape analyzer data and place the resulting best-fit model parameters into data streams for archiving.

Routine Data Plotting Processes: Instrument rates, orbit-integrated gamma-ray spectra, instrument orbital environment data and housekeeping data are routinely plotted to microfiche for archival and data screening purposes.

Automatic and Manual Data Screening Processes: Data screening, both automatic and manual, attempts to identify known types of data events and detect previously unknown events. Some of the known events screened for include electron precipitation events, orbiting reactor events, gamma ray bursts and solar flares. Results of screening are logged to the Data Quality File.

Science Data Production (Post DSF Merge): Post DSF Merge processing takes the results of the calibration and screening processes and merges them into the science data streams. The resulting files contain virtually all the information (or pointers to information) required to analyze the data.

Database Updates: PDA maintains a number of database tables that keep track of archive media, archived files, commanded instrument state, data quality, etc. Some of these tables are in file format while others are managed as tables by the INGRES database management system. These are updated on a daily basis using database update files produced by the PDA run. As indicated in table 2, the database update files are archived along with the science data products.

SCIENCE ANALYSIS

Science analysis is defined as the collection of processes and tools used by the scientist to produce an interpretable result from the OSSE data. These processes apply to analysis of both spectra and time series. We will discuss spectrum analysis in detail and will touch on the analysis of periodically varying sources (i.e. pulsar analysis).

The science analysis processes are built around IGORE, the Interactive GRO/OSSE Reduction Environment. IGORE is an interactive and programmable data analysis language based on the commercial product IDL. Further discussion of IGORE can be found in [2]. Virtually all of the science analysis applications discussed below are linked to IGORE and pass data and parameters to one another via either IDL variables or IGORE data constructs.

Spectrum Analysis Concepts

Spectrum analysis, by which pulse height spectra from the OSSE detectors are converted into estimated incident photon spectra, is the primary OSSE analysis mode.

The features of spectrum analysis are largely driven by two factors concerning space-borne scintillator experiments. First, the signal-to-background ratio is extremely small. Based on on-orbit background estimates, for a source at the OSSE limiting sensitivity of 5 millicrabs, the signal-to-background ratio is ~0.001. Consequently, much of the effort that goes into spectrum analysis is involved with estimation of background. Another source of "background" that requires treatment is the presence of strong gamma ray sources not in the nominal field of view of the OSSE detectors. At energies greater than a few hundred keV, the modulation of sources by the OSSE collimator is incomplete. Hence, the effects of sources out of the field of view must be considered.

Normal OSSE operations are driven by concern over background. Recalling that the OSSE detectors can rotate about a single axis relative to the spacecraft, yielding various detector position angles, the normal operational mode consists of a sequence of on-source and off-source pointing observations by each detector. Each observation lasts two minutes. By comparing on-source points to off-source, the background contribution can be estimated and removed. Each of the four detectors produces its own stream of on- and off-source points which is analyzed separately. Results are combined at the end of the process.

Second, the detector does not yield the same pulse height for all photons incident at the same energy. The incident spectrum is subject to both Gaussian smearing ("energy resolution") and partial energy deposition in the detector. These effects must be removed from the data in order to estimate the incident photon spectrum. Problems involved in this inversion or deconvolution process are well known but solutions involve various compromises to be discussed below.

Background Estimation

The OSSE background is a function of many parameters including detector pointing direction in the sky, pointing direction relative to the spacecraft, pointing direction relative to the atmosphere, cosmic ray environment, time from last SAA passage. etc. The background estimation process takes as input some number of spectra measured with the source out of the nominal field-of-view, together with the values of the relevant parameters. Given the values of these parameters during the source-pointing observation, the process interpolates in some or all of the parameters to produce an estimated background spectrum during the source point.

Deriving a detailed background model that depends explicitly on the aforementioned parameters (and possibly others as well) is a time consuming task requiring literally years of on-orbit data. For example, after 10 years of Solar Maximum Mission Gamma Ray Spectrometer (SMM/GRS) observations, the background is still not completely understood despite intensive efforts to do so [3].

As an alternative to a rigorous solution to the problem, several simplifying assumptions can be made that yield an approximate solution early in the mission. The simplest case is interpolation in detector position relative to the spacecraft. If the stream of pointing positions consists of a cyclic repetition of a background observation above the source, a source observation, and a background observation below the source, the source point background may be estimated by linearly interpolating in detector position angle (DPA) between the two adjacent background points to the source point. The justification for this approximation is that particles and photons generated by cosmic rays interacting with the local mass of the spacecraft and leaking into the detector vary with detector position angle relative to the spacecraft.

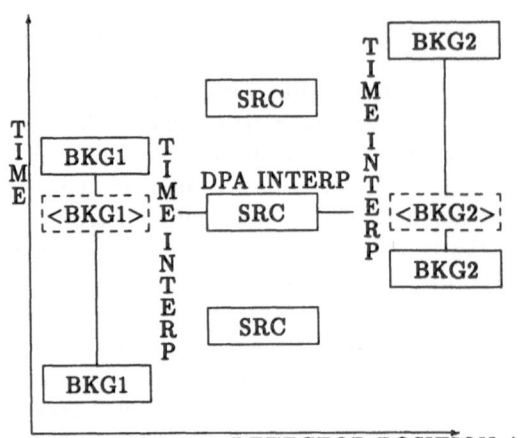

DETECTOR POSITION ANGLE (DPA)

$$BKGESTIM = (<BKG2> - <BKG1>)\frac{DPASRC-DPA1}{DPA2-DPA1} + <BKG1>$$

Figure 1. Background linear interpolation in time and detector position angle (DPA) (<BKG> is time- interpolated)

Testing this technique on data from the SMM GRS has shown that it does not produce background estimates sufficient for OSSE limiting sensitivity observations. A variety of background sources do not depend on detector position angle but do depend on the other parameters mentioned above. These have the common characteristic that they can be parameterized as functions of time. Hence, the next level of approximation, shown schematically in figure 1, involves a two dimensional interpolation in which multiple observations at the same detector position angle but at different times are interpolated in time to the time of the source observation. This is done for each background position angle in the observation sequence, typically alternating source and background points as described above. Finally, the time-interpolated background spectra at each position are interpolated in detector position angle to the source position as before. Analysis of SMM-based simulations indicates that this technique will produce sufficient results in most cases. The only variables in this technique are the number of measurements used at each position, the order of interpolation in time used, the number of off-source position angles used (generally limited to two or three per source), and the order of interpolation in position angle. For any given source spectrum, the background estimation scheme to be used requires that a given set of backgrounds be available. In the simplest schemes only the adjacent backgrounds (one on either side of the source in detector position angle) are required. In the more complex scheme shown in figure 1, a larger set of backgrounds is necessary. In order to assure that all backgrounds are not only available but also satisfy all data selection criteria, the concept of an analyzable unit (AU) has been defined. An AU is a collection of a single source spectrum and all the background spectra required to produce an estimated background for that source. As such, the AU is the minimum set of spectra required for analysis given a background estimation model. The AU may consist of any number of background spectra, but it generally will not span an SAA passage or a change of pointing strategy. Since any one background observation may be used to estimate the background for several source spectra, AU's can and usually do overlap one another.

Each background estimation technique (such as those outlined above) carries with it an AU definition, stored in a specific format in an SDB auxiliary data type. The definition is applied to the data in order to identify specific AU's in the data stream.

Background estimation techniques considered for use are all linear in the measured background spectra. In other words, they can be written:

$$B_{\text{estimate}} = \sum_i A_i B_i \tag{1}$$

where i ranges over the background pointings in the AU, B_{estimate} is the estimated background spectrum for one AU and the B_i are the background spectra specified in the AU identification. The A_i are the coefficients of the linear model.

Spectrum Analysis Tasks

In general, three factors should be noted concerning the sequence of operations required to fully analyze a spectral observation. First, as mentioned above, the analysis proceeds in parallel for each of the four OSSE detectors. Generally, results are combined during or after deconvolution and modeling of the spectra. Second, backgrounds are estimated one analyzable unit at a time, producing a set of difference spectra that may then be summed. Finally, spectrum analysis is a highly iterative process. The main variables in the process are background estimation technique and hence AU selection and data selection criteria. Only experience with flight data will indicate what combination of techniques and criteria will maximize sensitivity.

The tasks required to perform a spectrum analysis are as follows:

Archive Access: This process consists of determining which files are required to perform the desired analysis, reading the files from the archive media and writing them to a user-owned "database" on magnetic disk. Typically, the user will require the pointing interval (usually two-minute) spectra and access to shared databases such as the command state and data quality databases. In addition, this process builds a database of SDB header quantities (mostly rates and magnetic field and charged particle environment quantities) useful for AU identification and data selection. Each entry in this database contains a pointer to the spectrum it represents.

AU Definition and Identification: Next comes selection of a background estimation model and AU definition. Once selected, the AU definition is applied to the data stream using the SDB header database. The result is an array of IGORE Spectrum Data Records (SDR's), one element per AU. Referred to as DIFF records, these will ultimately contain the estimated backgrounds and difference (i.e. background subtracted) spectra. At this point, they contain AU identifications – pointers to the source and background spectra for each AU. During the AU identification process, a certain amount of preselection of the data is allowed. This preselection is generally used to include only spectra from the desired observation and instrument configuration.

Background Estimation and Subtraction: The DIFF SDR's are read in one at a time. The AU identifications in each SDR are used to read in the spectra for the AU and the background estimation model is applied. The estimated background is then subtracted from the source, and both the estimated background and difference spectrum are written to the SDR (or to a disk file pointed to by the SDR).

Data Selection and Summation: The systematic uncertainties inherent in background estimation will, in many cases, determine the limiting sensitivity of an OSSE observation. SMM/GRS experience has shown that background estimation systematics are heavily dependent on data selection [3]. In particular, selection criteria are applied to the parameters on which background levels depend, such as magnetic cut-off rigidity, SAA passage history and the angle between the field-of-view and the horizon. Simple linear background estimation techniques work best when the background is not varying in an extremely nonlinear fashion, such as it would near the maximum or minimum of rigidity or with the field-of-view near the horizon. Setting selection criteria requires striking a balance between restrictive selection to control background estimation systematics on the one hand and broad selection to improve counting statistics on the other. The learning curve for this process will depend on experience with on-orbit data.

Data selection is performed interactively by the scientist using either an IGORE record-based array of subsets of the SDB header called the selection record or an INGRES DBMS-based SDB header table containing the same information. The former executes selections faster but has a less friendly user interface. Queries may include simple ranges of parameters or more complicated criteria such as selecting orbits or other periods of time during which certain parameters have dynamic ranges below a cutoff value. In general, all spectra in an AU must meet the selection criteria for that AU to be selected. However, the user has the option of turning this feature off if he so desires.

Selected difference spectra are summed over the observation for each detector. The summed spectra are the input to the deconvolution or modeling stage of the analysis. If the analysis is accounting for confusing sources either in or out of the field of view, a separate sum must be accumulated for each confusing source visibility configuration (i.e. each combination of visible and earth-occulted confusing sources).

Instrument Response Generation: The instrument response is derived from a Monte Carlo simulation of the instrument that has been verified by comparison to extensive calibration data. An instrument response matrix describes how incident photons at an array of energies are distributed among the pulse height channels output by the detector. It takes into account interaction probabilities, energy deposition in various elements of the detector, event veto probabilities and instrument energy resolution. Each response matrix is computed for a given instrument configuration and incident beam angle. These are referred to as simple responses.

The detector produces a pulse height spectrum C_i for pointing position i according to the the following equation:

$$C_i = \sum_k R(\theta_{ik}) P_k + B_i \tag{2}$$

where k ranges through all sources (both the primary target and confusing sources), $R(\theta_{ik})$ is the response matrix for a source at angle θ_{ik} in the collimator, P_k is the incident photon spectrum for source k and B_i is the background spectrum for pointing position i. Note that, due to the incomplete modulation afforded by the OSSE collimators, especially above a few hundred keV, the response at angles outside of the nominal field-of-view is nonzero.

Estimating the background for a given AU according to equation 1, we get

$$\sum_i A_i C_i = \sum_i A_i \sum_k R(\theta_{ik}) P_k + \sum_i A_i B_i \tag{3}$$

$$= \sum_i A_i \sum_k R(\theta_{ik}) P_k$$
$$+ B_{\text{estimate}}$$

where i ranges over the background pointings within an AU. Taking the difference between the source-pointing pulse height spectrum C_{src} and $\sum_i A_i C_i$ yields difference spectrum D as follows:

$$
\begin{aligned}
D &= C_{\text{src}} - \sum_i A_i C_i \qquad\qquad (4)\\
&= \sum_k R(\theta_{\text{src}\,k}) P_k - \sum_i A_i \sum_k R(\theta_{ik}) P_k \\
&= \sum_k \left[R(\theta_{\text{src}\,k}) - \sum_i A_i R(\theta_{ik}) \right] P_k \\
&= \sum_k R_{k\,\text{compound}} P_k
\end{aligned}
$$

where we have assumed that $B\_\text{estimate} = B\_\text{src}$. Equation 4 is referred to as the "deconvolution equation". The salient feature of the deconvolution equation is that, if the response matrices are combined into a compound response, the formulation of which is governed by the background estimation model (i.e. the A_i coefficients from equation 1), the equation represents the well-known inversion problem.

Deconvolution and Spectrum Modeling: Deconvolution refers to the solution of the inversion problem (equation 4) in order to remove instrument effects from the data. This process would be trivial except that the response matrices are numerically ill conditioned and, hence, are impossible to invert in a straightforward manner.

The standard approach to such problems in high energy astrophysics, and the approach adopted by OSSE for our initial system, has been to use a technique we will refer to as forward folding. In forward folding, the user selects a parameterized model for the incident photon spectrum. An automatic parameter estimation routine then varies the parameters of the model, folding the model through the response for comparison to the pulse height spectrum at each iteration, until a best fit is achieved.

The advantages of this method are straightforward implementation and interpretation of the resulting best-fit model. The disadvantage is that the technique does not directly result in an estimated incident photon spectra. A variety of techniques are available to produce such an estimate, but the results can be misleading since the shape of the estimate tends to conform to the shape of the model used in the forward fold [4]. In addition, there is some ambiguity in the production and interpretation of uncertainties for the incident spectrum. In general, the estimated incident spectrum produced from a forward fold model is unsuitable for use by a scientist who wishes to perform further modeling on the spectrum. To do this, he requires access to the instrument response in order to perform the forward fold with the model of his choice.

Other so-called "model-independent" techniques such as Maximum Entropy [5] or Backus-Gilbert [4] will be investigated by the OSSE team as follow on techniques for deconvolution. These have the potential to produce estimated incident spectra directly, without assuming a parameterized model. Hence, the resulting spectra can be modeled at will without access to the instrument response.

In either case, the OSSE system will supply a standard set of both empirical and physical models in common use for predicting gamma-ray emission and the capability for users to add their own models as desired.

Pulsar Analysis

OSSE pulsar mode data can supply data at a variety of time resolutions and over a limited number of energy bands. In general, analysis of pulsar data for periodically varying sources is performed via summed epoch analysis, in which a test period (and, if required, period variation) is assumed and the data are summed in phase with this period over some interval of time. The result is an estimated mean light curve for the source assuming the period. By examining the astatistical modulation of the light curve as a function of assumed period, the actual period can be estimated. The OSSE data analysis system supplies a summed epoch analysis package. Starting with pulsar data, periodically calculated solar system barycenter vectors and a range of test periods and period derivatives, the package can determine whether a signal is present and, if so, supply a period and light curve. In it's initial configuration, it cannot deal with period variations due to motion in a binary system, but this will be a follow-on capability.

CONCLUSIONS

Analysis of OSSE data can be a complex procedure due to both the variety of data types and the difficulties involved in background estimation and response deconvolution. Experience indicates that a long learning curve will result, during which time sensitivity is pushed to progressively better levels. The OSSE team anticipates that both team members and guest investigators will contribute heavily to the understanding of the processes involved and the tuning of the system.

References

[1] W.N. Johnson, J.D. Kurfess, R.L. Kinzer, W.R. Purcell, M.S. Strickman, G.V. Jung, M.P. Ulmer, C.M. Jensen, G.H. Share, D.D. Clayton, C.S. Dyer, and R.A. Cameron, in Proceedings of the GRO Science Workshop, W.N. Johnson, ed., 2-22,(1989).

[2] D.A. Grabelsky, S.M. Matz, J. Pendleton, W.R. Purcell, M.P. Ulmer, W.N. Johnson, and M.S. Strickman, IGORE, Interactive GRO/OSSE Reduction Environment, these proceedings (1991).

[3] M.D. Leising, private communication (1991).

[4] T.J. Loredo and R.I. Epstein, Ap. J., 336:896, (1989).

[5] J. Skilling and R.K. Bryan, MNRAS, 211:111, (1984).

IGORE: INTERACTIVE GRO/OSSE REDUCTION ENVIRONMENT

D.A. Grabelsky, S.M. Matz, J. Pendleton, W.R. Purcell, and M.P. Ulmer

Northwestern University, Department of Physics and Astronomy
Evanston, IL 60208

W.N. Johnson, and M.S. Strickman
Naval Research Laboratory
Washington, DC 20375

Abstract

The interactive scientific data analysis system developed for the Oriented Scintillation Spectrometer Experiment onboard the Gamma Ray Observatory is described. The Interactive GRO/OSSE Reduction Environment, or IGORE, integrates a high-level, interactive programming language, analysis tools, specialized data structures, and a standard interface all into a command-driven environment. The basic system requirements and design are summarized, followed by a description of OSSE data analysis in IGORE.

1 Introduction

The Oriented Scintillation Spectrometer Experiment (OSSE) onboard NASA's Gamma Ray Observatory (GRO) consists of four identical NaI detectors, optimized for gamma-ray observations in the 0.05-10 MeV energy range. Detector pairs are co-axially mounted on two parallel axes, and each detector can be independently oriented about its axis. The numerous instrument operating configurations and orbital systematic background effects make OSSE scientific data analysis quite complex. While certain initial tasks will be carried out routinely and automatically in a production mode, generation of scientifically meaningful results will require interaction of the scientist in the analysis process.

The interactive analysis of OSSE science data will involve several steps, and utilize numerous data structures for storing raw science data, intermediate, and final results. A single observation will consist of roughly 50 megabytes of raw science data. One of the most difficult tasks from the scientific point of view is expected to be the removal of systematic background effects. All of this, coupled with the complexity of the instrument and its operating modes, present a data analysis challenge. The scientist must be able to develop a variety analysis strategies through the imposition of various selection criteria on the data, the use standard analysis routines, and the ability to create customized analysis tools. The entire processes will be highly iterative, requiring the examination and evaluation of intermediate results, possible revision of data selection criteria and alogrithms, and recomputation of results.

Data Analysis in Astronomy IV, Edited by V. Di Gesù *et al.*
Plenum Press, New York, 1992

These general considerations, plus experience gained from data analysis of previous gamma-ray missions (e.g., SMM), convinced us of the need for an integrated data analysis environment. A variety of options existed for the skeleton of such a system, including IRAF, TAE, and IDL. (Starting from scratch was ruled out fairly quickly.) An important factor in the evaluation of these options was an OSSE project-wide descision to adopt the VAX/VMS computing enviroment. This decision, plus the general requirements described below, led us to choose IDL, a commercially-available package. In this paper we describe the scientific data analysis system developed for OSSE: Interactive GRO/OSSE Reduction Environment, or IGORE.

IGORE is the primary user interface to the scientific data products of the initial routine processing of OSSE data. Besides providing access to the science data, IGORE provides a library of standard analysis routines in a high-level, interactive programming environment. Users may extend the IGORE environment with programs written either in the native language (IDL), or in Fortran. Integration of external modules into IGORE is made simple by the use of a standard interface. IGORE also includes a set of novel data structures designed to accomodate OSSE science data types; additional data structures may be easily added as the need arises.

In the following sections, we summarize the general and specific requirements which drove the IGORE design, and give a brief overview of the system. OSSE's primary mission is gamma-ray spectroscopy, so we focus here on those aspects of IGORE most closely associated with spectral analysis. It should be noted, however, that IGORE has been designed for use in the interactive analysis of OSSE pulsar and time series data, as well. In addition, IGORE has already been extensively used in the analysis of pre-launch calibration data. Following the requirements and design descriptions, we illustrate how IGORE will be used in the scientific analysis of OSSE spectral data. Finally, we mention possible future directions of IGORE.

2 Requirements

Translating the scientific requirements into software requirements leads to two categories of requirements: 1) general requirements, such as the capabilities of the interactive programming language, and extensibility of the environment; and 2) specific requirements which address scientific issues, e.g., definition of data structures used in background estimation.

2.1 General System Requirements

The most general (and challenging) requirement is to put a functional system in place that can the accomodate specific scientific software requirements which are still evolving. The following software requirements were developed to address the need for such a system:

- Programming/command language. The environment must integrate analysis tools into an interactive programming and command language which is capable of dynamic extension through the creation of variables and data structures, and the definition of new commands. The scientific programming capabilities must be equivalent to Fortran.

- Standard applications interface. A standard interface between the analysis system and external modules must be provided which makes integration of external application programs simple and quick. By far the most common language for OSSE scientific applications is Fortran, so the initial interface must accomodate Fortran applications. Inclusion of an applications interface for C programs should follow.

- Flexible data structures. Data flow during analysis will utilize a variety of data structures, many of which are still evolving. The analysis system must be capable of extending its inventory of data structure definitions. Structure definitions must support arrays which consist of a mixture of memory-resident and disk-resident data.

- Data plotting and fitting. The analysis system must include tools for interactive plotting and fitting of data. A library of standard models (e.g., Gaussian profiles) should be maintained, and the scientist should, with relative ease, be able to add new models.

Other general system requirements include error handling, journaling capability, batch-mode processing, and the possibility of making applications menu driven.

2.2 Specific OSSE Requirments for Spectral Analysis

An OSSE spectral observation is a two-week ($\sim 10^6 sec$) sequence of source and background pointings, obtained by alternately orienting the detectors on and off the source position in one to two minute intervals. OSSE science data will consist nominally of the two-week set of short-interval pointings (source and background) in the form of pulse-height spectra, calibration data, instrument state data, and orbital environment parameters. The scientist's objective is to produce a photon energy spectrum for the source which is free of instrument response and background effects.

Specific scientific requirements, provided by identifying the actual steps involved in the analysis, translate into specific required application routines, data-flow paths, and data structures. Here we summarize the fundamental software requirements which address the complicated nature of OSSE data and data analysis.

- Data base management system. Typical data analysis scenarios involve roughly 20,000 observations, requiring approximately 50 megabytes of storeage. Associated with the data acquired during this time are numerous auxilliary data describing the instrument state, the orbital environment, etc. All of this information may be relevant to a particular study, and therefore must be available to the scientist. A data base management system is required to facilitate selective access to the data.

- Spectrum data structure. This is the primary data structure for OSSE spectral data. All application programs which process spectral data will operate on spectral data structures. It must accomodate variable-length pulse-height spectra, variable- and fixed-length auxilliary (header) data, and must include data and information pertinent to background estimation for the associated spectral data. In addition, it must handle storeage of all the included data either directly in memory, or on disk.

- Analyzable unit data structure. For each source pointing, this defines the spatial and temporal off-source pointing pattern, relative to the source pointing, required to produce an estimate of the background at the source. The required pattern is determined by the particular background model chosen by the scientist.

- Model-fitting data structures. Various stages of OSSE data analysis require fitting models to data. Data structures are required to store the fit model definitions, best-fit model parameters, fitting ranges, fit statistics, etc.

- Background estimation. Routines must be available which use off-source spectra to estimate the background at the source position during a given pointing interval. Background estimation may be as simple as using the closest sequential background point-ing, or as complex as fitting an energy-dependent model to a series of background pointings. The specific model used imposes requirements on the relative positions and times of the background observations, with respect to the source observation. These requirements define an *analyzable unit*.

- Instrument response model. A model of the instrument response as a function of energy and source position within the detector must be available for convolution of incident model spectra.

- Incident spectrum estimation. Routines must be available which convolve model source photon spectra with the instrument response to produce model pulse-height channel spectra. These are compared with the observed background-subtracted spectra in an iterative process, in order to determine the best parameters for the chosen incident model spectra. Methods for directly inverting the response matrix in order to de-convolve the observed spectrum are under review, and some form of deconvolution is expected to be available in the future.

A great deal of detail is obviously left out of the above summary, and some of the scien-tific software requirements are still under development. We anticipate that the scientific requirements will continue to evolve as the mission progresses.

3   IGORE System Overview

This section provides a brief description of IGORE, the system designed to address the above requirements. IGORE was developed in the VAX/VMS computing environment, as noted above, and its design and implementation make extensive use of VMS-specific features.

3.1   User Interface: IDL

IGORE uses a commercially-available package called IDL for the primary user interface and programming/command language. IDL, widely used in astronomical data analysis, is a powerful and versatile interactive programming language. Variables can be dynamically created, and the language can be extended by associating new programs with new commands. Native scientific programming capabilities are comparable to Fortran.

The version of IDL used during the initial requirements and design phase of IGORE was developed specifically for the VAX/VMS system. One particularly useful feature from the standpoint of IGORE system requirements is the ability to dynamically or statically link externally created modules directly into the address space of the IDL image. This feature has played an important part in the interface for external applications, and the ability of users to add new applications to the IGORE system.

The original version of IDL did not support structured data types. As described below, we designed our own system of structured data types which is layered on top of IDL. IDL has recently been rewritten in C, and now includes its own structured data types. We discuss the implications of IDL structures for IGORE in the last section.

3.2   General Application Interface

As noted in §2.1, most OSSE scientific application programs will be external programs written in Fortran. In order run external applications which are linked with the IDL image, an active interface was developed to handle the protocol of passing parameters between IDL and VMS Fortran data formats. Interface routines perform various checks and decide how the data will be passed (e.g., by moving the data or by passing a pointer to the data).

The implementation of the IGORE interface requires that every external application have an associated frontend program, called the application frontend. The core of the frontend program is a data structure, called the control record, that establishes a one-to-one pairing of IDL variables, passed as parameters, with correpsonding Fortran variables, which are the arguments of the application program. The control record contains basic information about both variables, and various flags which establish the action to be taken on any given call to the application. In addition to invoking the application itself, the frontend program passes the control record to the various interface routines which carry out the checking and data passing.

Each frontend routine is essentially identical to the next, except for the specific arguments, number of arguments, and a few tuning parameters which allow the programmer to customize certain aspects of the interface protocols. The design of the application frontend program isolates the application design from the interface. That is, with a few exceptions, applications can be designed and coded free from requirements imposed by the interface design.

3.3   Application Frontend Preprocessor

To further simplify the process, a standalone facility, called the Application Frontend Preprocessor, is provided which automatically generates the application frontend program code for any application program, given the code of the application as input. Only a few mild coding format conventions are required of the application code by the Preprocessor. These include explicit declaration of application arguments, one per line; and a selection of directives to the Preprocessor (these hide from the Fortran compiler behind comment characters). The output of the Preprocessor is a file containing the frontend code with the application code appended to the end.

The application may be integrated into IGORE by compiling the Preprocessor output file and linking it, either dynamically or statically, to the IGORE image. The static link feature is reserved for use by the system manager, since this procedure updates the IGORE image for all users. The dynamic link, which allows users to customize their individual environments, must be performed each time they log into an IGORE session. This dynamic feature requires that the modules be pre-linked into relocatable images; a standalone program is provided with IGORE for this purpose.

3.4   IGORE Data Structures

IGORE data structures are implemented by using IDL variables to contain pointers to data records, and structure definitions to provide maps of the data fields within records of a given structure type. An inventory of structure definitions is maintained by IGORE, and any number of records of the structure types in the inventory can be declared dynamically. A memory management system keeps track of all record creation and deletion.

Generally, an IGORE structure is defined only if some application program requires it and includes it as an argument. That is, IGORE structure definitions must be made to fit the compiled definitions within the application programs that use them. IGORE structure definitions are created by a standalone facility called the Structure Translator. This program takes the source Fortran source code defining the application's structure, and produces the IGORE structure definition. The output definition is stored either in a system or user library. System structure definitions are automatically loaded into IGORE at the start of every session; user structure definitions are loaded on command by the user.

The manipulation of records by their pointers is utilized also in the implementation of sub-structures, and in the interface protocols for passing structured data between IGORE and external applications. Substructures are created by embedding record pointers as data items in other records; such substructures are called indirect records. Indirection enables arrays of data records to be assemblies of non-contiguous record elements. Such arrays arise naturally through data selection (application of various criteria, e.g., orbital environment parameters), and allow subsets of the data to be created without proliferation of already existing data.

A set of IGORE record routines is provided for interactive manipulation IGORE records: reading and writing individual record fields, copying and duplicating records, etc. All of these routines are implemented as IDL commands, and take the IDL variables which contain record pointers as their arguments. The interactive routines which manipulate records make all indirection transparent to the user.

3.5 Data Base Management System: INGRES

To handle the various tasks of data base management, INGRES, a commercial relational data base management system was selected. For spectral analysis, INGRES will manage a data base of spectral header information; i.e., the auxilliary data associated with the on- and off-source pulse-height spectra. INGRES can be executed as a standalone system or accessed from within IGORE via a frontend program. In the latter mode, commands in the DBMS language (SQL) may be passed across the interface, or a query-by-form program may be executed. Return parameters to IGORE may be spectral header data from the data base, or file pointers for locating the pulse-height channel spectral data.

For certain data selection operations, it is possible to bypass INGRES and use IGORE directly. This can be done, for example, by populating appropriate IGORE records with auxilliary data containing the relevant selection parameters, and using standard IGORE functions to examine the fields of the records, and then make selections based on the values.

4 OSSE Spectral Data Analysis with IGORE

As noted above, one the primary scientific tasks of OSSE spectral analysis is to estimate the incident source photon spectrum from the on- and off-source observations. The two main scientific difficulties are that: 1) the background varies systematically with orbital parameters and detector position; and 2) the instrument response matrix cannot be uniquely inverted to estimate the incident photon spectrum. Both of these conditions make spectral analysis a highly iterative process requiring extensive interaction with the scientist. Added to this is the vast amount of data that must be processed and analyzed. In this section we outline the steps involved in the analysis of OSSE spectral data with IGORE. The purpose of this cookbook outline is to indicate how IGORE facilitates the complex and interactive analysis procedure.

1. The relevant archived data are loaded to disk. A special-format file, called SDB (Spectral Data Base), has been developed for OSSE spectral data. The SDB file includes the arrays of the observed on- and off-source spectral data (short-interval pointings; cf. §2.2), and the auxilliary data in a structured header.

IGORE provides routines for reading from and writing to SDB files, and includes a special data structure called the SDR (Spectral Data Record) for manipulating the data within

IGORE. The SDR is the primary data structure passed between IGORE and the spectral analysis tools, and its fields can be treated as IDL variables: any valid IDL operation may be performed on them (e.g., plotting).

In order to help alleviate memory requirements imposed by the large amount of data associated with the observations and analysis, the SDR can either represent data that is resident in memory, or simply contain a pointer to the appropriate SDB file. Application programs that process spectral data via SDRs can easily determine whether actual data or a file pointer has been passed; in the latter case, the application program reads the data directly from the file. In this way, arrays of several kilobyte-sized spectral data may be represented by arrays of much smaller SDRs.

2. The background at the source is estimated. This is likely to be an iterative procedure that depends on the background model chosen, and the instrument configuration at the time of the observations. The essential steps are:

    i. Chose a background model. The specific model will prescribe the spatial and temporal pattern of off-source pointings required for the background estimate, and weighting factors to be applied to the background spectra used in the computation of the estimate. This prescription defines the *analyzable unit* for the subsequent analysis steps.

    The background estimate at the source position at the epoch of observation may be as simple as a linear interpolation of the two nearest background pointings, or as complex as an energy-dependent fit to a series of background pointings. Generally, the simplest models will be attempted first. Progressively more complicated models, with more complicated analyzable unit definitions, may become justified as analysis and evaluation of the results warrants.

    ii. Identify all source spectra satisfying the definition. IGORE provides a utility that identifies all the on-source spectra for which off-source pointings satisfying the analyzable unit definition are available. The utility produces an array of spectral data records which contain file pointers both to the selected source spectra and the associated off-source spectra suitable for background estimation, and the weighting factors for the off-source points.

    Not all source spectra will satisfy the analyzable unit defintion. For example, the satellite's passage through the SAA, earth occultation of the source, or instrument configuration may cause would-be off-source points not to exist. The scientist must evaluate whether this identification procedure results in a sufficient number of source spectra to continue the analysis. If not, another, presumably less complicated, background model must be used, and the identification step re-applied.

    iii. Generate background estimate, and produce background-subtracted source spectra. For each source spectrum identified in the previous step (and stored in an SDR), a background is estimated from the weigthed sum of the associated background spectra. This estimate is subtracted from the source spectrum, and the result is stored in the appropriate spectral data record. The resulting spectral data record finally contains the original source spectrum, the background spectra satisfying the analyzable unit definition, the weighting factors, and the background-subtracted source spectrum.

The background-subtracted spectra that result from this procedure will not, in general, be entirely free of various systematic effects due to orbital enviroment, relative angle between the earth and the detectors, etc.

3. Screen the background-subtracted spectra for residual systematic background. Various

selection criteria (e.g., orbital environment parameters) are imposed on the background-subtracted data, and the surviving spectra are summed to produce a total spectrum. By iteratively examining the resultant total spectrum, and tuning of the selection criteria, a total spectrum which is reasonably free of celestial background and systematic effects is produced.

If no satisfactory total spectrum can be produced, or if data selection results in too few surviving spectra, this procedure may indicate that an inadequate background model was used. In this case, the scientist may need to back up to the previous step and try a different model.

4. Derive an incident photon spectrum from an assumed model photon spectrum. The model spectrum is convolved with the instrument response, using the on- and off-source viewing angles and the weighting factors defined with the analyable unit definition, to produce a model count-space spectrum. In another iterative procedure, the observed and model spectra are compared and the model parameters adjusted until the best fit is achieved. This step can be repeated for various assumed model photon spectra.

At each stage of the analysis outlined above, intermediate results may be examined interactively with IGORE. Depending on the scientific objectives, required sensitivity, and available data, the entire analysis may be repeated several times using different background models, data selection criteria, etc. In addition, the scientist may integrate customized analysis routines into IGORE, as described above (§3.2).

Similar analysis scenarios exist for OSSE pulsar observations. In this case, the background is not as much of an issue. The convolution step would be replaced by an epoch folding procedure, or an FFT routine.

5   Future Developments

Three general areas for IGORE's future can be identified: 1) system evolution; 2) sophistication of scientific analysis strategies; and 3) exporting to UNIX-based systems.

System evolution is mainly concerned with the recent release of IDL version 2. The new IDL system includes support for structured data types as native IDL variables. Migration of IGORE structures to IDL structures is not straightforward for two reasons: 1) IDL structures cannot be made to easily support the indirection which is central to IGORE records and used extensively by IGORE applications; and 2) IGORE structures start as Fortran application structures, and have limited use in the absence of applications which process them, while IDL structures are defined in isolation from any external applications. Possible migration paths are currently under study, and it is anticipated that some mixed mode of IDL and IGORE records will be developed.

Other aspects of system evolution address specific requirements corresponding to new scientific requirements, the second general area of ongoing IGORE development. As the GRO mission progresses and experience is gained in such areas as background estimation, new analysis strategies will undoubtedly be developed. This would be true, of course, for any analysis system. The flexiblity designed into IGORE should help accomodate the evolution of the scientific analysis of OSSE data.

Finally, the question of exporting IGORE is related somewhat to the demands of the GRO Guest Investigator Program. As the GRO mission moves ahead, IGORE's role beyond the OSSE team will become better defined.

THE EGRET HIGH ENERGY GAMMA RAY TELESCOPE ON GRO:
INSTRUMENT DESCRIPTION AND SCIENTIFIC MISSION

H.-D Radecke, and G. Kanbach

Max Planck Institut für Extraterrestrische Physik
8046 Garching, Germany

INTRODUCTION

The Energetic Gamma Ray Experiment Telescope (EGRET) is the high energy instrument among the four experiments aboard GRO and that with the highest angular resolution. Its primary photon energy range extends from 20 MeV to about 30 GeV, i.e. the energy range where pair-production dominates the interaction between cosmic photons and the detector material.

DESCRIPTION OF INSTRUMENTAL CHARACTERISTICS

The essential components of the instrument are (see fig. 1):

- an anticoincidence system to discriminate against the intense primary charged particle cosmic radiation. This is realized by a large dome on top of the instrument formed from a single piece of plastic scintillator of 2 cm in thickness which is viewed by an array of 24 photomultiplier tubes each with its own high voltage supply and preamplifier. For reliability these tubes produce two independent veto signals.

- a two-section spark chamber system which materializes the incident photons and determines the trajectory of the secondary electron-positron pair. The upper section, within which the gamma-ray is required to convert, consists of 28 spark chamber modules of an active area of 81 cm x 81 cm, interleaved with sheets of the conversion material tantalum. The lower section of the spark chamber system, located between the upper and lower scintillator tile arrays, contains six widely spaced spark chamber modules. Its function is to define the electron-positron trajectories until they intersect the lower tile array and impact on the NaI crystal.

- a triggering telescope that detects the presence of the charged pair particles with correct direction of motion and initiates the recording of the tracks in the spark chamber. It consists of two 4x4 arrays of square plastic scintillator tiles separated by the 60 cm long lower spark chamber section. The trigger telescope is able to configure "subtelescopes" out of the detector tiles in the two planes. It also carries out a time-of-flight measurement to determine whether a particle is traversing the detector in upward or downward direction and to immediately reject upward moving particles.

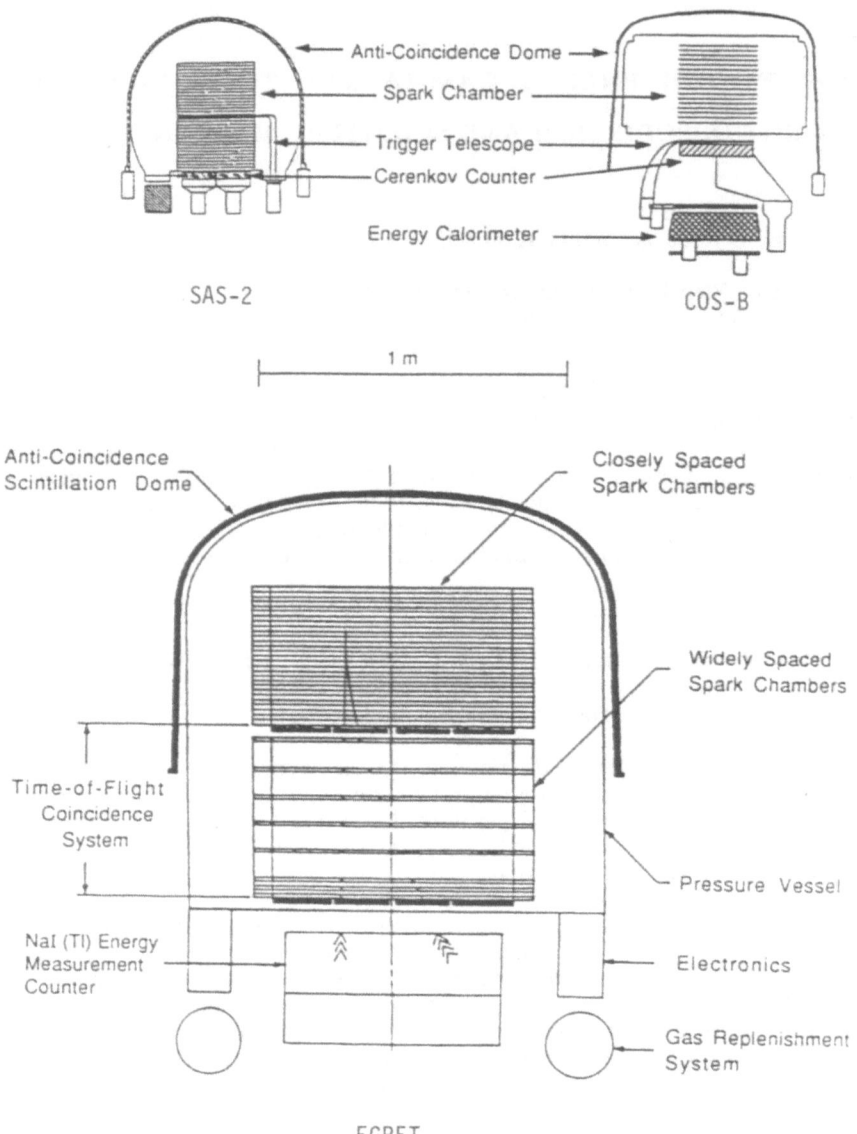

Fig. 1. Schematic view of the EGRET detector as compared to the earlier gamma-ray instruments SAS-2 and COS-B.

- a detector to determine the energy of the secondary charged particles released by the incident photon: the so-called Total Absorption Shower Counter (TASC) made of NaI which is located below the lower tile array. The crystal of 20 cm in thickness dissipates the energies of the incident particles in an electromagnetic cascade and is being viewed from below by an array of 16 photomultiplier tubes. The energy resolution is about 20% (FWHM) in the central portion of its energy range of 200 - 2000 MeV. The TASC has also the secondary purpose of analysing gamma-ray bursts. Following a burst trigger signal from BATSE, an energy spectrum in the 0.6 to 167 MeV interval is produced and recorded in four time intervals.

A gamma-ray entering the telescope within the acceptance angle has a known probability of converting into an electron-positron pair in one of the thin tantalum sheets of the closely spaced spark chamber modules. If at least one electron is detected by the time-of-flight coincidence system as a downward moving particle and there is no veto-signal in the anticoincidence scintillator, a high voltage pulse is applied to the spark chamber modules, and subsequently the event data is recorded in the form of a digital picture of the gamma-ray event. After that the analysis of the energy signal from the NaI crystal is initiated while the event time is given by an on-board clock.

EXPERIMENTAL CAPABILITIES

To address its scientific investigations EGRET provides special experimental capabilities which have been verified during the instrument calibration phase (see table 1).

The high energy telescope will be able to recognize point sources that are nearly two orders of magnitude fainter than the Crab pulsar. For strong sources, the position should be determined to about 10 arc min, the strongest sources even with a positional accuracy of 5 arc min. Spectra of the stronger sources should be measurable over the entire energy range.

Table 1 - EGRET Parameters (see also fig.2 and 3).

Energy Range:	20 MeV - 30 GeV
Energy Resolution:	about 20% FWHM (100-2000 MeV)
Effective Area (>200 MeV)	about 1500 cm^2
Sensitivity (>100 MeV):	about 5*10^{-8} cm^{-2} s^{-1}
Source Position Location:	10 to 5 arcmin for strong
	source with hard spectrum
Field of View:	about 40 deg FHWM
Timing:	0.1 ms

For the diffuse galactic plane emission, the spectrum will be measured with high accuracy and spatial variations in the spectrum should be measurable on a scale of a few degrees. Features which subtend more than about 0.5 degrees will be resolvable as extended sources. The diffuse radiation away from the galactic plane will be separable into galactic and extragalactic components on a scale of about 5 degrees. The extragalactic component will be studied for spatial variations in intensity and spectrum.

DATA COLLECTION AND TELEMETRY

The information from EGRET includes a variety of data types involving spark chamber event pictures of individual gamma-rays, burst and background spectra from the TASC, and housekeeping data showing the instrument operating mode and its state of health. All EGRET data are assembled into data packets that are partitioned regions of the GRO major telemetry frame by the GRO Command and Data Handling System. The spacecraft computer adds a primary and secondary header that contains the instrument identification, packet sequence count, time, position within 30 km, and aspect to within 2 arc minutes. The remaining portion

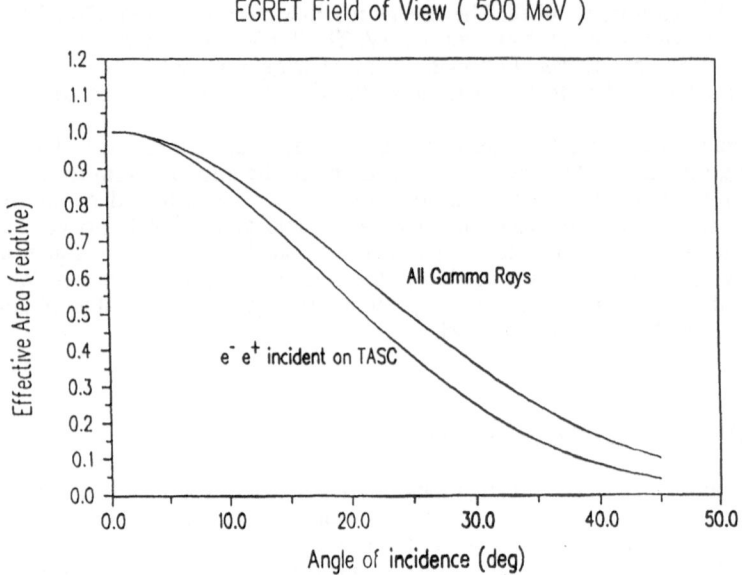

Fig.2. The EGRET field of view at 500 MeV.

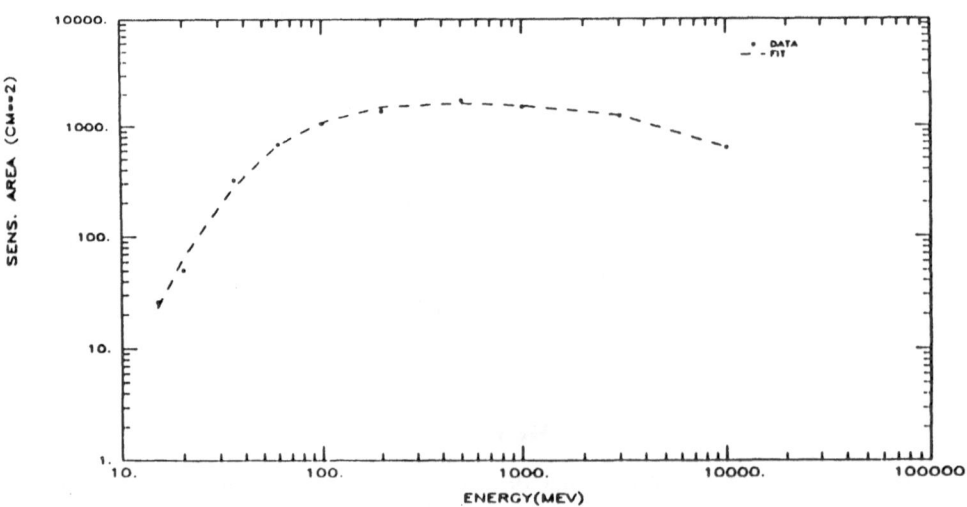

Fig.3. The energy dependence of the EGRET sensitive area.

of the packet is filled with instrument data: housekeeping data, counter rates, solar spectra, burst spectra and events. The major part of the packet is allocated to events: 95% of the packet space of 1768 bytes. Events occur asynchronous to the packet time, and they are of variable length. For this reason, events are stored in buffers as they occur, and at the appropriate time, the buffers are unloaded to the packet buffer. Each packet can hold on the average 5 to 6 events per second (see table 2).

Table 2 - The EGRET data packet structure.

ELEMENT	LENGTH BYTES	CONTENTS AND NOTES
Primary header	8	Identifiers and packet sequence count.
Secondary header	22	Time, position, attitude, direction of earth center.
Housekeeping	18	Instrument mode, temperatures, voltages, currents, command verification, etc.
Counter rates	40	Anticoincidence, coincidence, and TASC.
Solar spectra	32	32.8 sec accumulation. 16 channels read in each packet out of 256 channels total.
Burst spectra	2	One channel. Four spectra with 256 channels in each burst; i.e., 35 minutes readout time.
Events	1,664	Events are variable in length. Average capacity is about 10 - 12 events. Each event has a header of 22 bytes (instrument mode, fine time, TASC event data, etc.) plus spark information.
TOTAL	1,768	

NOTES:
1.   Packets are transmitted every 2.048 seconds (6.98 Kbits/sec.)
2.   Raw telemetry totals 75.3 Mbytes per day

A complete packet is produced and transmitted every 2.05 sec at a rate of nearly 7 Kbits per sec so that raw telemetry totals more than 75 Mbytes per day.

The spark chamber information of each gamma event is processed to define the secondary particle tracks from the pair production interaction. The event analysis involves two close particle tracks in each of two orthogonal projected views. A pattern recognition program identifies the structure of each event. Partly, however, a manual review using an interactive graphics system is necessary.

In its secondary mission the TASC is analysing the energy spectra of omnidirectional radiation in the energy range of 0.6 MeV to 167 MeV. This interval is divided into seven energy bands with varying resolution. The incident radiation (charged or neutral) is unrelated to spark chamber events. The analysis is being done in parallel with the energy analysis for spark chamber events. There are two accumulation modes: the so-called "solar mode" where the analysis is not triggered and a spectrum is taken about all 32.768 seconds and the "burst mode" in which four spectra within preset time-intervals are accumulated after being triggered by the BATSE - detector. The solar mode is used to establish a background profile as a function of orbit position and time, and to detect solar activity. The BATSE-triggered spectra will have to be corrected for background using the data obtained in the solar mode. EGRET may also record events in what is called the microsecond burst mode designed to record multiple gamma-rays that might be incident from a common source. In this mode a burst of counts in the anticoincidence counter that is detected within a commandable time window (typically less than one microsecond) will trigger the spark chamber. Processing software will analyse the direction and energy value of the events recorded during the burst.

The analysis system routinely processes each data type and stores it in specific databases with associated catalogs for use in subsequent analysis and mission planning. Skymaps of event counts and intensity will be constructed for each viewing interval of two weeks. To produce intensity maps, the instrument housekeeping database provides the operating modes as a function of time. Based on the map results, events from interesting region will be screened using software to test for significant increases indicative of a source.

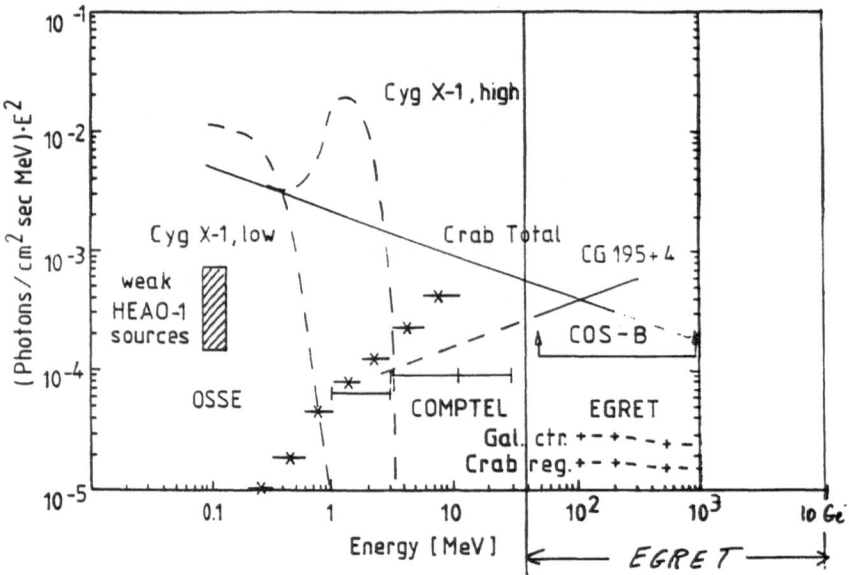

Fig.4. Detection limit for galactic sources.

SCIENTIFIC PERSPECTIVES OF EGRET

Most experimental characteristics and parameters of EGRET show a substantial improvement over previous gamma-ray telescopes (SAS-2, COS-B). Therefore it should be able to provide a deeper understanding of all segments of gamma-ray astrophysics. The main subjects of investigation will be:

- diffuse galactic gamma radiation, where EGRET should provide data about size and degree of the coupling of cosmic rays and matter. Within the interstellar medium, cosmic rays are interacting with matter, photons, and magnetic fields. Each process may result in gamma-rays. Cosmic ray nucleons may interact with the interstellar medium and subsequently produce many secondary particles, mostly pi-mesons (either neutral or charged). The neutral pions decay into two gamma-ray of about 68 MeV in the rest frame. This results in a unique spectrum not monotonically decreasing with increasing energy as in the other gamma producing processes like bremsstrahlung etc. but showing a maximum at the characteristic energy value. The SAS-2 and COS-B satellites revealed a strong correlation of the gamma-radiation in the galactic plane with the overall galactic structure. These data lack a high degree of sensitivity so that the conclusions drawn from the results are varying. With its greatly improved sensitivity, EGRET should bring about a better understanding of the physical processes in the interstellar medium.

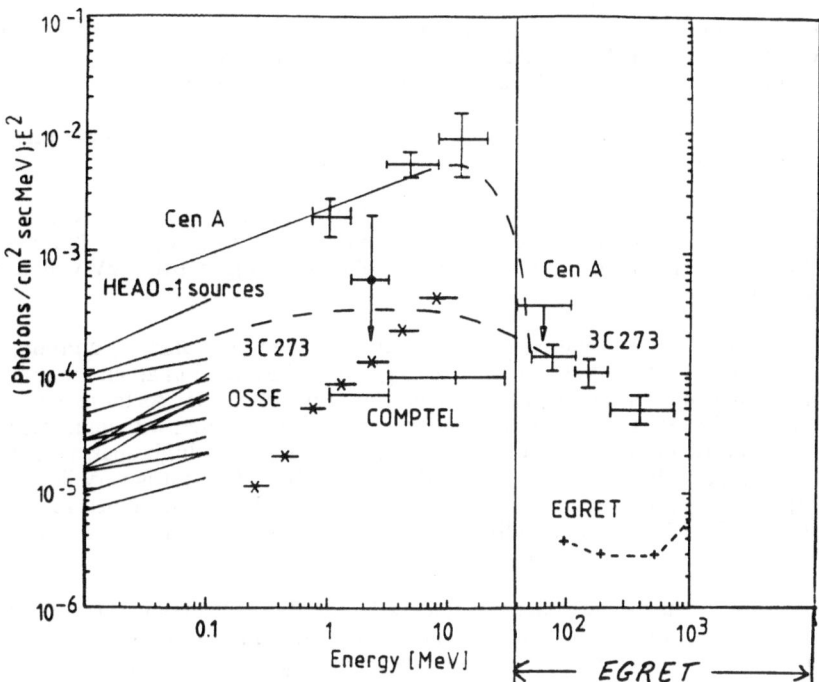

Fig.5. Detection limit for extragalactic sources.

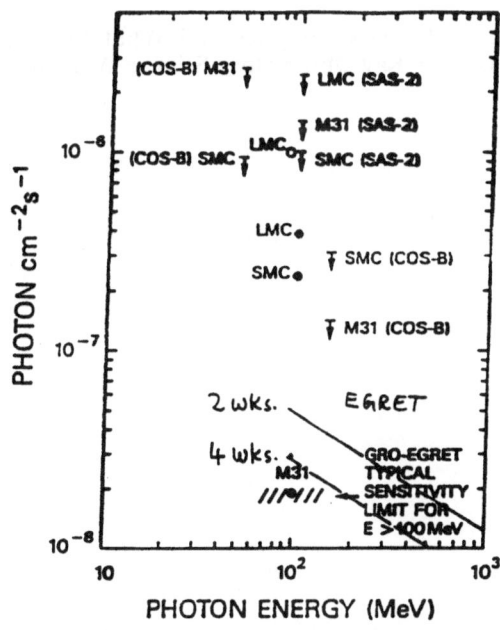

Fig.6. Detection possibilities of nearby galaxies by EGRET.

- galactic objects: COS-B point sources, new point sources, pulsars, X-ray binaries, black holes, molecular clouds, spiral arms. There are currently several galactic point sources on the record. The most famous are the Crab and Vela pulsars and a somewhat mysterious object called Geminga which has been observed by SAS-2 and COS-B. As for pulsars with the striking fact that the gamma and radio pulses are not in phase, EGRET will probably contribute to our understanding of the general pulsar phenomenon and will hopfully identify more gamma-ray pulsars (see fig.4).

- transient phenomena: gamma-ray bursts, solar flares, variability of sources. The EGRET spectrometer may record spectra of gamma-ray bursts (possibly neutron stars with high magnetic fields) and high energy solar flares up to 140 MeV with adjustable integration time.

- galaxies: local normal galaxies, active galaxies (Seyfert galaxies, quasars, and BL Lacertae objects); EGRET should be able to give new information about the structure and cosmic ray densities of the Magellanic clouds and possibly M 31. As far as active galaxies are concerned the assumed very high energy processes coupled with immense magnetic fields are surely a typical environment for the production of high energy gamma-rays. Gamma radiation has been detected in the quasar 3C273 and it is very likely that also other quasars as well as Seyfert and radio galaxies and BL Lacertae objects show gamma-ray activity. EGRET will hopefully detect new sources and improve the statistics of the known objects (see fig.5 and 6).

- the isotropic diffuse extragalactic gamma radiation. EGRET may well contribute a lot to our knowledge of this cosmological relevant radiation field. This may eventually lead to a decision between different theories about its origin, e.g. matter-antimatter annihilation at the boundaries of superclusters, sum of extragalactic sources gamma radiation, remains of primordial black hole radiation etc.

REFERENCES

Kanbach G. et al., "The EGRET Instrument", Proceedings of the Gamma Ray Observatory Science Workshop, Greenbelt, Maryland 1989

Fichtel C.E., "High Energy Ray Astrophysics and What EGRET Might Contribute", Proceedings of the Gamma Ray Observatory Science Workshop, Greenbelt, Maryland 1989

THE EGRET HIGH ENERGY GAMMA RAY TELESCOPE ON GRO:

DATA ANALYSIS SYSTEM AT MPE

T.L. Halaczek, G. Kanbach

Max-Planck-Institut für Physik und Astrophysik
Institut für Extraterrestrische Physik
8046 Garching, Germany

INTRODUCTION

The information from the EGRET instrument includes a variety of data types involving spark chamber event pictures of individual gamma rays, burst and background spectra from a Totale Absorption Shower Counter (TASC), and housekeeping data showing the instrument operating mode and its state of health. The data analysis system will routinely process each data type and store them in specific databases with associated catalogs for use in subsequent analysis and mission planning. According to these functions the EGRET software is generally classed in three categories [1]:
1. Instrument telemetry and performance monitoring
2. Production processing of event data
3. Scientific analysis
We describe the three groups subsequently.

INSTRUMENT TELEMETRY AND PERFORMANCE MONITORING

Data from EGRET (and the other instruments on GRO) are assembled into data packets by the GRO Command and Data Handling System. The spacecraft computer forms the data in packets of 2.048 sec. duration. The packet contains all important information to identify and analize the data, e.g., instrument data, the time, the spacecraft position, the packet sequence count etc. The biggest share of the packet of EGRET is occupied by the data of gamma ray events, which are recorded if they satisfy a set of commandable triggering conditions. They are tagged with their time of occurrence and are filled into 1664 bytes of storage per packet. The telemetry can accommodate a rate of about 5-6 events per second although an average rate of 1-2 sec is expected in orbit. Raw telemetry totals 75.3 MB per day. The data in this form are transmitted to the Eart and processed at the Goddard Space Flight Center facilities. This is the next level of data processing - production processing of the experimental data; it will be done centrally by GSFC for the EGRET collaboration.

RAW DATA BASIC PROCESSING

The raw experimental data sent to the Control Center are originally parts of the GRO major telemetry frame. They must be processed to a form easy to use in scientific analysis. The editing and derivation of physical event parameters is a routine process to transform the

Data Analysis in Astronomy IV, Edited by V. Di Gesù *et al.*
Plenum Press, New York, 1992

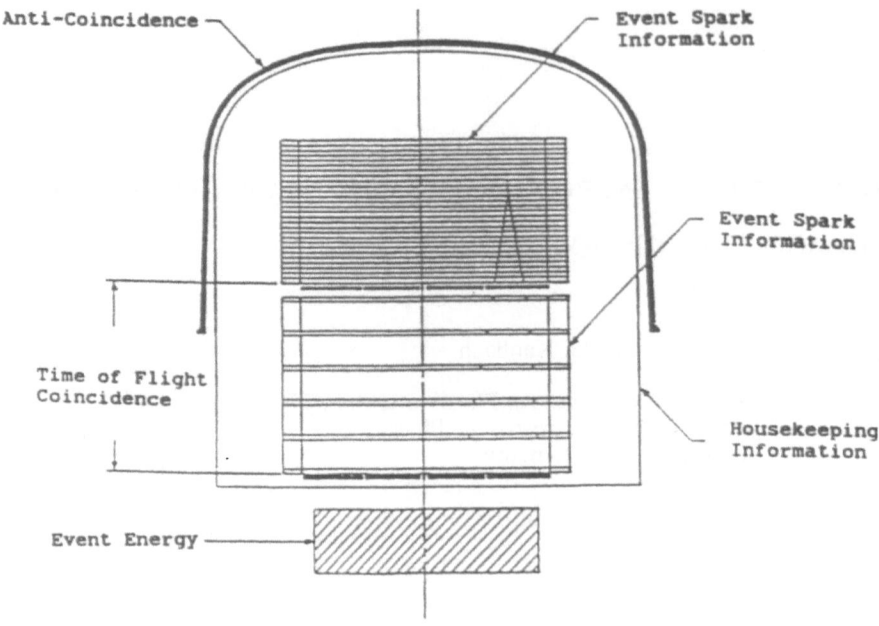

Fig.1. EGRET instrument and examples of output data types

Table 1. The Egret Packet Structure [1]

Element	Length/Bits	Contents and Notes
Primary Header	8	Identifiers and Packet Sequence Count
Secondary Header	22	Time, Position, Attitude, Direction of Earth Center
Housekeeping	18	Instrument mode, temperatures, voltages, currents, command verification, etc.
Counter Rates	40	Anti coincidence, Coincidence and TASC
Solar Spectra	32	32.8 sec accumulation, 16 channels read in, each packet out of 256 channels total
Burst Spectra	2	One channel, four spectra with 256 channels in each burst, i.e., 35 min readout time
Events	1664	events are variable in length. Average capacity is about 10-12 events. Each event has a header of 22 bytes (instr. mode, fine time, TASC event data, etc. plus spark information
Total	1768	

raw instrument data into data bases useful in subsequent analyses. The production processing is described in details elsewhere [1]. We mention only, that there are four flight data types: gamma ray events, TASC burst events, microsecond burst events and housekeeping data. Each has unique record structure and length. The raw instrument data are first packed into particular data type records. Each gamma ray event (the spark chamber readout), for example, is processed in the pattern recognition process to define the secondary particle tracks from the pair production interaction. The event analysis is a complex process involving generally two closely situated particle tracks in each of two orthogonally projected views. For a significant fraction of the events, still a manual review and perhaps editing is necessary using an interactive graphics system. So, the selected gamma events only, are subsequently written to the Primary Data Base (PDB). Similarly, the spectral information from the Total Absorption Shower Counter must be processed before packing to the PDB. It consists of spectra from 0.6 to 167 MeV that are routinely collected for 32 seconds integration times. These spectra will be used to develop a library of background as a function of orbit position and time. Count rates from this detector and from the anti coincidence detector will be screened for significant increases in counting rate. Processing software will analyze the time profile and energy spectra of these bursts.

The information covered in packets recorded within typically 6 hours of EGRET operation is assembled in PDB files and distributed (on tapes) to the centers participating in the experiment.

Table 2. The PDB's records format

Data Type	Record Type	Record Length
Gamma ray event	Variable	max 32756
Housekeeping	Fixed	5400
TASC burst	Fixed	650
Microsecond burst	Variable	max 32756

The PDB is transferred on IBM standard cartridge as variable-blocked records, with maximum logical record length 32756 bytes, and a physical block length 32760 bytes. The records of different data types are distributed randomly within a data file.

SCIENTIFIC ANALYSIS

Development of the Scientific Analysis Software

The software development tasks required for the scientific analysis have been distributed among the collaborating institutes under coordination by a central software group at the GSFC. The resulting software tools had to be adapted then, to the hardware and operating systems in particolar sites. The EGRET group in the Max-Planck-Institute for Extraterrestrial Physics uses IBM computers of the Computer Center in Garching, under the VM operating system. The data storage and maintenance system used by us is different from the one used by GSFC. For these two reasons we decided to maintain the functions of the GSFC's software tools but to change their structure substantially.

The Catalog

An integral part of the scientific analysis software at the MPE is the data acquisition and maintenance system, called the Catalog. It has been originally designed and tested for the EXOSAT experiment, but enlarged and updated for EGRET. The data reside in HADES (Heidelberger Automatisches Datenverwaltungs - & Editor System), and are of READ ONLY access for users. For safety reason, only the system operators have the READ/WRITE access. The particular data types are reselected from the PDB and then written to the Catalog as separate sets - the Distributed Data Bases. There are four of them with the flight data. During the PDB record reselection process another data set is created. It consists of 100 bytes-long records extracted from the PDB gamma ray event. It contains the most important information concerning the detected gamma event. This is the Summary Data Base, - a data set used by most analysis programs. The Catalog is also a place to store some results of analyses, especially the most time consuming applications like the sky maps, for example.

The Catalog Structure

A Log-File (LF) is the major unit in the EGRET archives. For the flight data the LF will generally correspond to one observation period. A sequence of the LFs is ordered with Log-File-Keys (4 byte integer). The LFK is the encoded start time of the observation period it has to be supplied to the catalog when generating a new LF. Each LF is labeled with Log-File Number, that carries a passport record (80 bytes) with the LFK and short description of the observation period. Furthermore the Log-File-Type - it is a 2 character code ("EG" for EGRET flight data) and the Data Type - it is a 2 character code to describe the data type, e.g., "EV" for event records (Gamma Event DB) are included to identify the data sets.

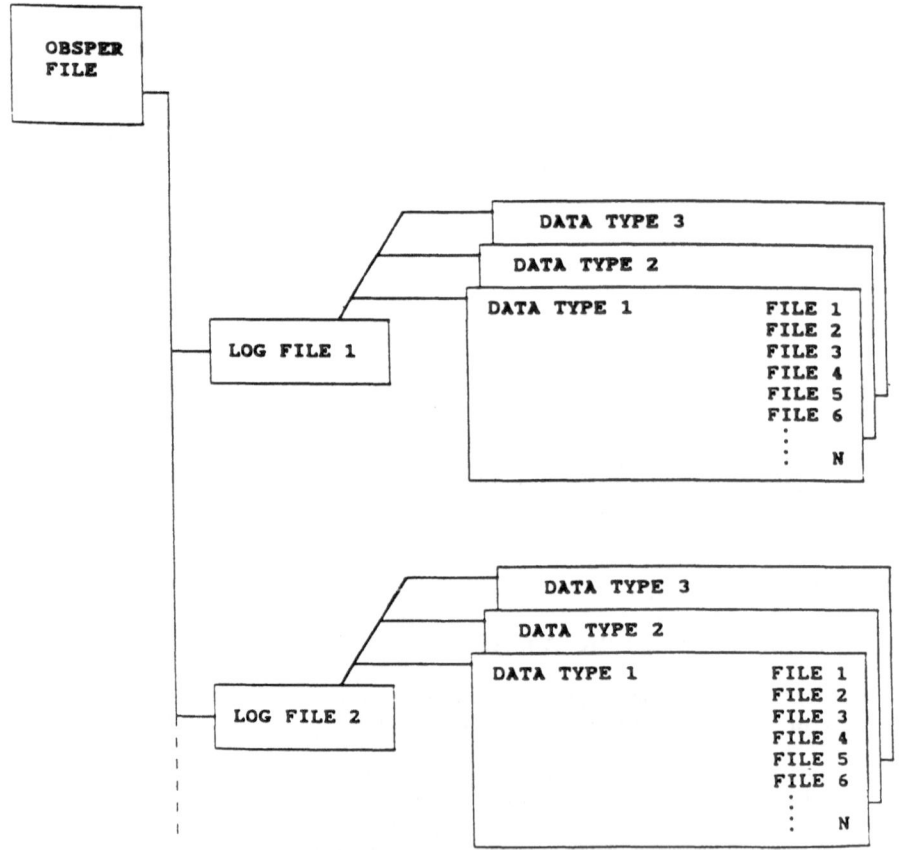

Fig.2. The EGRET archival system at MPE

The Calibration Data Catalog

The calibration data catalog, is the same kind of catalog as the above described. It holds one data type only - the calibration data. It is a separate data storage and maintenance system, that is not of common access. There is no analysis program that needs the original calibration data. The results of the analysis of the calibration data are stored in a set of calibration files. These files reside either on a CMS disk or also in the HADES storage and are not administrated by the Catalog.

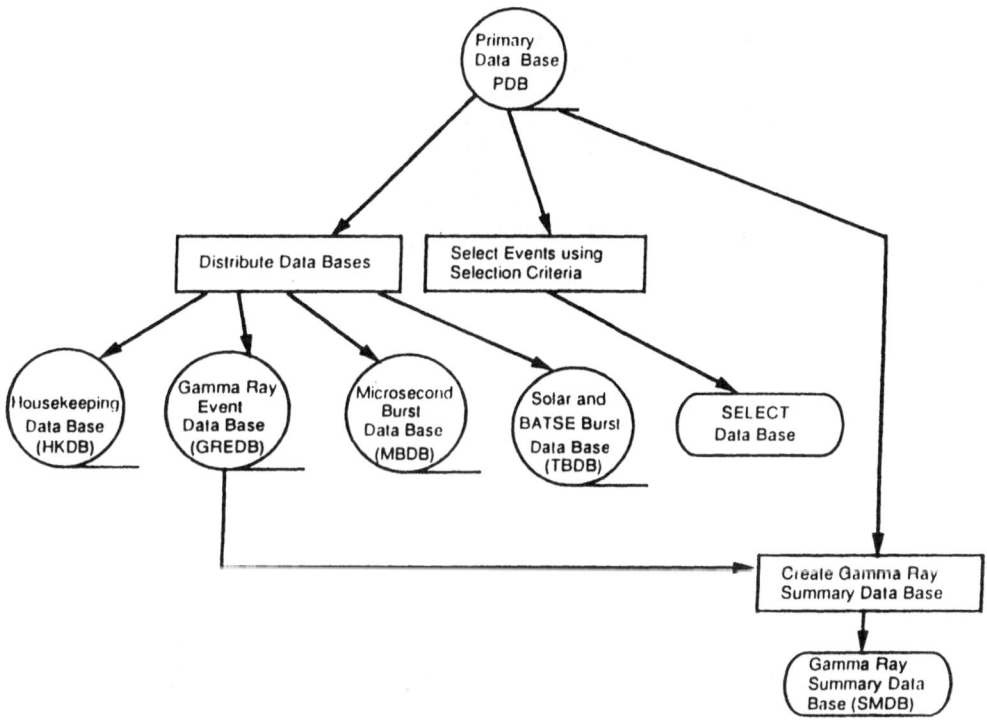

Fig.3. EGRET data bases

Table 3. EGRET data bases

DATA BASE	SIZE-ESTIMATE
Calibration Events	140 MB
Primary Database (PDB)	66.2 MB/day
Gamma Ray Event	8.5 MB/day
Solar and Burst	3.2 MB/day
Housekeeping	7.6 MB/day
Microsecond Burst	trivial
Summary (SMDB)	3.4 MB/day
Instrument Timeline	25 kB/day
TASC Spectrum Analysis and Background	0.9 MB/day
Calibration Files	
point-spread functions	16.3 MB total
energy dispersions	16.3 MB total
efficiencies	trivial

The Analysis Programs

The interpretive analyses will proceed only after the necessary editing and review steps have been carefully done. There are generally two groups of analysis programs: the standard application tools and the user's own tools. The standard tools set covers till now programs for:
1. Sky map creation
2. Pulsar analysis
3. Gamma source search/analysis
4. Spectral analysis.

As an example of the complexity of the tasks we can use the creation of sky maps:

Sky map of event counts and intensity will be constructed for each viewing interval of typically two weeks. To analyze data of one observation period, typically 14 days long, the user must address about 50 catalog data files plus some other files like the production involves three steps: The experimental data are first converted to FITS standard file by the Timeline file. Then the FITS file of event counts will be used by the program Intmap. It will recalculate the data using the calibration files and the instrument housekeeping data, and makes two outputs: the FITS file of exposure, and the FITS file of intensities. The latter one is used as an input by the Skymap program, that plots it in the form of maps.

Each task listed above involves similarly several programs to perform intermediate analysis, and to administrate huge amounts of data.

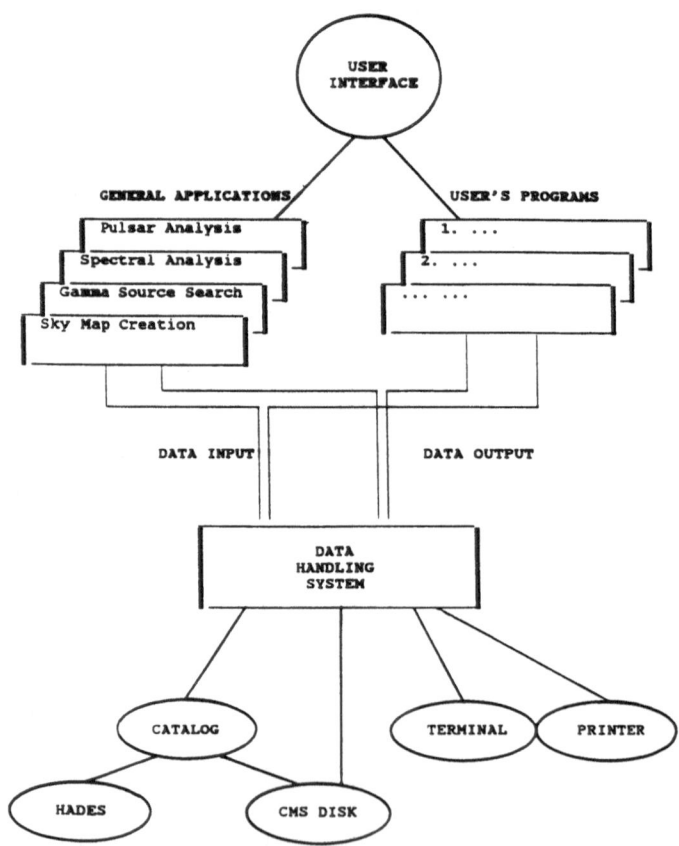

Fig.4. Data analysis system at MPE

The User Interface

The huge amount of data to access and handle on the one side and the big number of parameters required by the analysis programs demands automation. We have developed a set of programs and subroutines to perform this task. It is called the Data Handling System (DHS). The DHS has been written in Fortran and uses the X-Menu system (Kolinar Enterprises, Inc.) to design the User Interface. The user interface routines will supply the user with special menu screens that allow to input data much more clearly than typical Fortran "READ" input does. It simplifies data handling, and minimises the possibility of errors due to incorrect use of the catalog. Additionally a pre-analysis process will be available. It will include several common data tests to extract possible data errors e.g., the test of data recording time, and the energy filter, the time filter etc.

The data handling program displays a set of menus appropriate for selected analysis tasks. It allows the user to set up all necessary parameters, to call some standard filters and to access observation data according to specified time periods. For more convenience some parameters are available as defaults for the corresponding analysis programs. This process is called "configuring" an analysis program. It is possible to configure a program completely and then execute it as a batch job, but the user can work interactively, as well. Independent of operation mode the handling program writes a report in the form of a file. All data parameters are summarized, and a passed/rejected record statistics is written. Additionally all selected operation parameters are saved in a special file, and can be used as input values in subsequent analyses.

The problem of accessing the great number of big experimental data sets in the Catalog is solved by the DHS as follows: the user writes in a special menu the time or the LFN appropriate for the requested analysis data. Minimum one start time and one end time are required, but multiple periods are allowed. The Data Handling System searches for the appropriate datasets in the Catalog and suplies them to the program. The data files are accessed automatically without human intervention. The main control parameter is the time. There are several serving routines to complete special control functions, e.g., to test if the data do not overlap, to check the event's energy, etc. The system allows many such server tasks.

To automate the data access a special file, called OBSPER has been created. It correlates the observation period number, the observation's calendar date and time, julian date, and LFN in the Catalog. In this way all the user must know, is the required observation time and date. The DHS looks for this time in OBSPER file, opens the Catalog, and reads the data from appropriate data set. By each read, the record time is controlled until the required time is reached. Then the DHS starts to transmit data to the analysis program's input buffer.

To pass parameters to a program's input buffer the user can do in one of two ways:

1. The user can call the CONFIG program, which displays a set of menus and allow the user to choose appropriate analysis programs to configure, and subsequently the user is asked for all the requested program parameters. The set of parameters is saved then as so called Configuration Information (CI) (a named data set), in a special buffer file. The user interface system uses the file to save all the CIs. It the user starts now the analysis program with the CI's name as an option, e.g., "TOOL1 (CI1' where TOOL1 - it's an analysis program name, and CI1 - the name of an appropriate configuration set, the CI is imported automatically by the CONFIG subroutine covered in the TOOL1 program, and the program starts to work with the CI1 parameters. This way of configuring is requested for programs that must work as batch jobs, but it is possible for all our analysis programs.

2. The second possibility is to start an analysis program by typing in a terminal the program name without an option. It causes the CONFIG subroutine to call a default option for the program and then to start work interactively. The CONFIG subroutine displays for the user the same set of menues as the CONFIG program. There is only one difference: the CONFIG program is a general purpose tool, which allows to configure all the analysis programs. The CONFIG subroutine allows to configure the just started program only - so the configuring process starts at a lower level. During a program execution the DHS writes a job report (a log file), which lists the used CI set and the CI set name, supplies general information like user ID and date,and a data statistics as well. The data statistics includes e.g., the number of records read, or rejected and the reason for rejection.

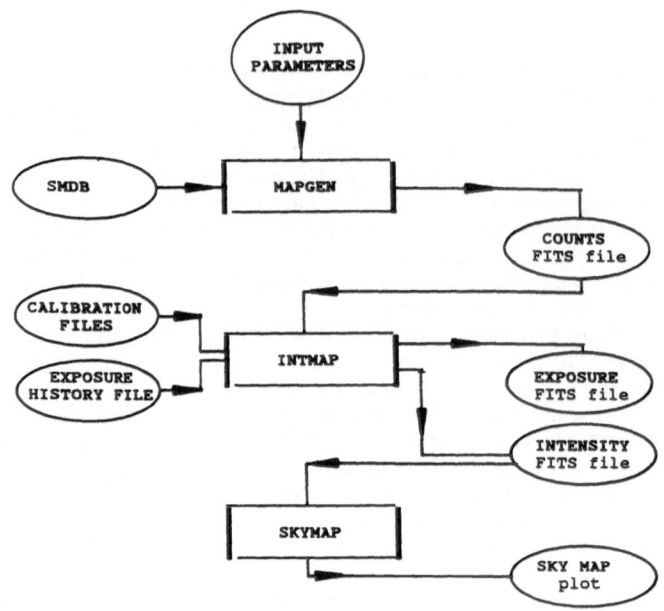

Fig.5. Typical program using the DHS facility

Because the standard analysis programs were created partly in other institutions of the EGRET collaboration, they must be subsequently adapted to MPE's computer operating system (VM) and our Catalog. While the changes due to the operating system are somewhat small, the difference in data archival system forced us to replace big parts of the source programs with our routines. The art of making a user input to the programs has been changed as well. It makes our system structurally different, though the functional compatibility is always maintained.

REFERENCES

1. The EGRET team, in: Proc. of the GRO Science Workshop, GSFC 10-12April 1989, ed. by W.N.Johnson, p.2-52
2. D. Bertsch, N. Laubenthal, EGRET Telemetry and Data Formats, Doc.No.82-3-5, Issue 6, Rev.:B, Internal EGRET
3. Egret, Instrument Data Analysis and Operations - Software Review, December 1, 1989, GSFC, Washington DC

FUTURE MISSIONS

THE EXTREME ULTRAVIOLET EXPLORER MISSSION:

SOFTWARE DEVELOPMENT

C.A. Christian

Center for EUV Astrophysics
Space Sciences Laboratory
University of California
Berkeley, CA 94720 USA

INTRODUCTION

The Extreme Ultraviolet Explorer (EUVE) is a NASA explorer-class satellite mission devoted entirely to observations in the wavelength range from 70 to 760 Å. The science payload incorporates four individual grazing incidence telescopes, three of which will be used to conduct a complete imaging sky survey in the EUV in four separate bandpasses. The fourth telescope, called the Deep Survey/Spectrometer (DS/S) has two coaligned instruments consisting of one "deep" imaging system and an EUV spectrometer with three overlapping bandpasses covering the entire 70–760 Å window. The spectrometer will be used by guest observers in a pointed observation program during the second part of the mission. The instruments were built by the Space Astrophysics Group, at the University of California, Berkeley (UCB).

The science payload will be integrated with a Multi-mission Modular Spacecraft (MMS), NASA's Explorer Platform, and launched into a low-inclination circular orbit on an expendable Delta II vehicle. Launch is scheduled for December 1991. A complete sky survey is scheduled for the 6-month period following the in-orbit checkout and calibration period. At the end of the survey, pointed observations will be conducted using primarily the EUV spectrometer. The mission lifetime is expected to extend to at least 3 years. The pointed observations will be awarded to selected guest astronomers who

Data Analysis in Astronomy IV, Edited by V. Di Gesù *et al.*
Plenum Press, New York, 1992

have responded to the NASA Research Announcement for EUVE, scheduled for distribution in the summer of 1991.

Additional information concerning the EUVE Mission can be found in Bowyer et al,[1] and a description of the instrumentation is contained in Welsh et al.[2].

MISSION OVERVIEW

Sky Survey

The first scientific goal of the mission is to make a complete sky survey in four bandpasses between 70 and 760 Å that will locate sources to better than 1 arc minute and obtain absolute flux measurements. This will be accomplished using the three coaligned imaging instruments which scan continuously along great circles as the spacecraft rotates about its anti-sun axis three times per orbit. The scanning survey is expected to achieve exposure times of over 500 seconds for the entire sky.

The three survey instruments, called scanners, each comprise a grazing incidence telescope, a two-dimensional imaging microchannel plate (MCP) detector, and the attendant controls and event-processing electronics. Two scanners are identical, redundant, short-wavelength imaging systems, each employing two filters between 70 and 290 Å. The third instrument is similar but operates in two longer-wavelength regions from 400 to 700 Å. The resulting sky map for fully scanned sources will be approximately 6x6 arc minutes in resolution and should have the sensitivity to detect objects about 200 times fainter than the brightest known EUV source.

The first "deep imaging survey" will also be made during the initial survey phase of the mission, using the fourth scanner, called the Deep Survey Scanner, which covers 2 bandpasses between 60 and 285 Å. This instrument will continually point along the anti-sun line. As the satellite orbits the earth, the deep survey detector will obtain long exposures on the order of 20,000 seconds/0.1° pixel over a 2° swath of sky along the ecliptic plane. This will produce a survey section with sensitivity greater than that of the other scanners alone, by about a factor of 3–30. The survey data will be proprietary for Berkeley scientists for a period of one year, upon completion of the survey data acquisition. The data will then be released to the community and may be accessed through the Astrophysics Data System (ADS).

Pointed Observations: Spectroscopy

The second scientific goal of the mission is to make pointed spectroscopic observations of selected EUV sources. The EUVE spectrometer disperses light from the DS/S telescope into three overlapping bandpasses from 70 to 760 Å with a resolution ($\lambda/\Delta\lambda$) of 200–400. For a 40,000 second observation, individual spectral lines from point sources can be detected with a 3-σ threshold of 7×10^{-4} photons/cm^2/second.

EUVE Guest Observer Program

Pointed observations will be offered to the scientific community through the NASA Research Announcement cycle. Proposers responding to the announcement for EUVE will receive a copy of the *EUVE Guest Observer's Handbook*, which contains a description of the EUVE mission and the instrumentation, as well as information pertinent to the preparation of proposals.

The EUVE Guest Observer (EGO) Program is intended to support all activities of the mission's pointed phase which involve the guest observer directly. The program will provide informative material and organize support for the proposal process, including technical evaluation of proposals prior to their submission to the Peer Review Panel. Additional assistance will be available via electronic correspondence as well as telephone and mail communications. Advice in scheduling observations will be available when choices must be made.

The pointed-mode phase will begin after the all-sky survey and continue through the end of the mission. The first observation cycle will last for 12 months, with as many subsequent periods of approximately one year as the mission's lifetime allows. Data taken for guest observers is proprietary to that scientist for a period of one year. The pointed phase observations are also released to the community through the ADS, once the proprietary period has expired.

IMAGING SURVEY SOFTWARE

Design

The software designed for science analysis of the imaging survey is part of an overall software system called the End-to-End System (EES). The EES provides a range of functionality for EUVE. During the prelaunch phase of operation the software has been particularly useful for generating telemetry similar to the data that will be obtained during flight. This simulated data are used for a variety of purposes: for example, to check analysis software modules, to create displayed data on operations consoles and check operations software, and to compare modeled data against instrument performance.

The operations part of the EES is used to receive formatted data from Goddard Space Flight Center, from the instrument during in-house testing (calibration) for the EUVE hardware simulator, and from software-generated telemetry streams. These modules display data on monitors, perform error checking, and in general monitor the health and safety of the instruments. The EES also stores incoming telemetry in a permanent storage device at UCB.

Finally, there exist modules to decommutate or reformat the incoming data for use in subsequent analysis modules.

Science Data Analysis

The part of the EUVE EES which provides automated reduction of incoming data is called "the pipeline". The primary output from the science data analysis modules are 1) a sky map of the accumulated survey data and 2) a catalog of detected sources. Ancillary information such as tabulated engineering values is produced as well.

In order to produce either sky maps or a list of detected sources, photons are mapped from detector coordinates to celestial coordinates using an "aspect module." A separate module computes exposure files as a function of position for subsequent use in other data analysis routines. Sky maps are produced by binning or tiling regions of the sky. Maps are normalized for exposure time.

Catalogs are compiled in two ways. The first method is to produce a list of sources using a deceptively simple concept. Sources are "detected" if the flux from the source is above a certain threshold or significance level. In practice such a detection scheme is not so straightforward for the scanning instruments. First, the detectors are continuously scanned across the sky so that photons from a given source fall on different parts of the detector and through different filter materials. This information must be preserved if color information is desired. Second, the point-spread-functions (PSF) of the detectors are variable basically because the curved focal plane of the telescopes impinges on the flat detector surfaces. The image quality is therefore poorer at the edges of the detectors, so that a variable PSF must be used to optimally detect sources. This effect is, of course, in addition to the usual distortions, pixel to pixel variations, and other effects characteristic of microchannel plate detectors. As well, the EUVE detectors have their own personality traits to be reckoned with during analysis.

A second method for handling the data has been devised. Lists of previously catalogued sources have been compiled into what have been coined "pigeon holes". These lists are used to select photons that occur only within some radius of these catalogued source positions. Clearly, the size of the pigeon hole could be varied according to exposure time, aspect accuracy, catalog position accuracy, or the whim of the user. This analysis strategy allows the pipeline to run quickly, sifting out only photons which fall within the pigeon holes. The scheme can be quite a timesaver considering that a vast majority of detected photons are expected to be background airglow photons. Pigeon hole lists can be extended using the positions of sources detected as described above which do not already appear in the catalogged lists. The information produced from the latter analysis method is particularly useful for determining upper detection limits in each bandpass for particularly interesting sources.

Considering that the survey data analysis takes place on computing facilities resident at UCB, and all EES routines should be optimized for timely processing of telemetry, the system has been developed on a UNIX platform. Many routines are developed in the C language, although modules in other languages exist. The operations modules are being developed so that health and safety monitors are accessible through a windowed interface. Consequently, the software is not designed to be highly portable or machine-independent.

<u>Subsequent Analysis</u>

Other data analysis tools to be developed are modules for timing analysis and deeper processing of the scanner data. Some processing tools may be developed in such environments as IRAF or IDL to take advantage of existing routines in those systems useful for the display and analysis of imaging data.

SPECTROSCOPIC ANALYSIS SOFTWARE

<u>Design</u>

The EUVE Guest Observer Data Analysis Software is being developed currently for analysis of EUVE spectrometer data to be acquired during the pointed phase of the mission. Software for use by Guest Observers to analyze scanner data will be developed in the longer term. Considering that the software will be used by a wide variety of scientists for a range of purposes, the software was designed to service many requirements. In order to be effective, a philosophy that the software be portable was adopted for Guest Observer utilities. While many software development environments exist, only one could be supported. The EUVE project chose the IRAF environment for the exportable packages since it is used in a significant number of institutions and offers a good basic set of analysis tools, particularly for the workstation environment. IRAF can be used to develop code within a structure that can be used to control code releases and documentation, a fairly important requirement for distributed code.

Implementation

Data from pointed mode observations will be collected in a temporary storage area on a tape carousel. In addition, data are stored in the normal fashion in the permanent archive in an optical disk jukebox. A given observer's data, i.e., the photon data and associated engineering values, normally also will be rewritten into binary tabular files compatible with the format designed for the Space Telescope SDAS packages, i.e., "ST Tables." The explicit purpose of repackaging is to bring the data into the IRAF environment. The repackaging is handled by software (called the EGO Server) which services requests for explicit data sets such as the data belonging to a given individual. The server allocates and fills the tables and creates a text database (an ascii file) that records various bits of information pertaining to the formatting of the tables and to the source of the location of the raw data. The raw data also may be written to FITS files for guest investigators who prefer to take the data, unprocessed, to their home institutions.

The arrangement of the data into tables allows specific portions of the data to be selected and processed. This strategy is especially useful as the sizes of the data sets are significant (200 MB), and processing of the photon data is facilitated by masking out particular photons (detector area, hot spots, etc.) or certain sequences of photon events based on particular values or ranges of the engineering quantities. For example, valid photon events might be selected on the basis of detector voltage or mirror temperature using particular ranges of these parameters to distinguish 'valid' data from suspect data.

The tables are designed so that entries may be time-tagged since the telemetry is time-ordered intrinsically. Each table contains engineering values that occur with a particular frequency. Note that the basic frequency of engineering data is one *major frame* in the telemetry, that is once every ~ 1.024 seconds. Some data is provided less frequently, for example, detector voltages are provided every major frame. Therefore one table contains values that are provided from the instrument every major frame, another table contains data provided every *other* major frame, and so forth. The photon data comprise the largest data set and is assembled in a separate table which lists the X, Y position of the photon as well as its time tag. At this stage, the user may elect to use standard IRAF and STSDAS tools for viewing, plotting, printing, etc., any of the tabulated data.

Once the valid, masked data has been selected, the photons are processed within the EUVE IRAF packages to correct for distortion in the instrument and pixel-to-pixel variations, i.e., flat field effects. The conversion of X, Y positions to aspect can be done using the information reported by the platform on-board computers, or, if need be, the source position from catalogged coordinates or the centroids from the Deep Survey images can be used. In order to map photons "back onto the sky", an assumption concerning the source position must be made for the spectrometer data. The simplest assumption is that the source is located along the boresight of the telescope, or that the angular distance between the boresight and the source is known. This assumption makes it possible to assign a wavelength and a spatial position to each photon, using the grating equation for the instrument and the calibration information which describes the variation of the spatial scale over the detector. For diffuse or extended sources, and therefore for the sky background, spectral features are smeared because the spectrometer does not contain a slit, but for point sources these assumptions are valid. The tasks for completing these steps are built upon a library of tools to handle the photon data specific to, but not restricted to, EUVE.

At the end of this processing, the data is in a format reminiscent of ground based spectroscopy, with the X axis parallel to the dispersion direction and the Y axis corresponding to a spatial dimension; however, the data is still organized as a list of time-tagged photons. The IRAF supports this data format as a "QPOE" (Quick Position Oriented Event). Therefore, the processed photons are written in an IRAF QPOE file. Once in the QPOE file, the full complement of utilities for manipulating, viewing, graphing and writing the data is available as well as the packages specifically designed for the EUVE spectrometer. Note that all image utilities can be used with QPOE files, including writing the file as a FITS image.

Additional processing packages that will be available include utilities for treating the data for the significant higher order throughput which is a feature of the EUVE spectrometer. Modeling and synthesis routines will be built upon the existing tasks available in the IRAF *artdata* package. More subtle effects and deeper analysis tools, including polarization analysis, timing analysis, and deconvolution of instrumental resolution as a function of wavelength and spatial postion, are being planned. Other tasks for processing user input models to analyze the data will be included in the longer term. Note, however, that once the data is in the QPOE file, other spectral analysis routines in IRAF and STSDAS may be used for examining the data.

All of the IRAF tasks will be exported as layered packages to Guest Observers for installation in the IRAF environment at their home institutions, and each scientist will be encouraged to visit the EUVE project after their data has been acquired to become familiar with the use of the routines. By facilitating the interaction between observers and the Guest Observer Program at EUVE, it is hoped that a versatile suite of tools can be provided to the community in a timely fashion commensurate with the anticipated short duration of the mission.

ACKNOWLEDGEMENT

This work would not be progressing without the efforts of E. Olson, chief programmer for the EUVE Guest Observer Program. Much thanks goes to A. Miller who has provided much of the mission overview material. I also wish to thank the EUVE principal investigator, Stuart Bowyer, Roger Malina, and the EUVE science team for their advice and support. This research has been supported by NASA contract NAS530180, which is administered by the Space Sciences Laboratory of the University of California, Berkeley.

REFERENCES

S. Bowyer and R. F. Malina, in *Extreme Ultraviolet Astronomy*, R. F. Malina and S. Bowyer (eds.), (New York: Pergamon Press), pp. 397-408, 1991

B. Welsh, J. V. Vallerga, P. Jelinsky, P. W. Vedder, S. Bowyer, and R. F. Malina, *Opt. Eng.*, **29**, 752-758, 1990.

THE JET-X EXPERIMENT ON-BOARD SPECTRUM-XΓ MISSION

T. Mineo and G. Cusumano

Istituto di Fisica Cosmica ed Applicazioni dell'Informatica, CNR
Via M.Stabile 172
90139 Palermo, Italy

ABSTRACT

The Joint European X-ray Telescope JET-X is one of the experiment on-board SPECTRUM-XΓ. It consists of two identical co-aligned X-rays imaging telescopes, with a cooled CCD detector on the focal plane, and an optical monitor. A description of the experiment with some informations about the satellite and the ground segment are given.

INTRODUCTION

JET-X (Joint European X-ray Telescope) is designed for detailed studies of the X-ray emission from sources and cosmic background in the energy band 0.2-10 KeV. The experiment is being developed by a consortium of Italian groups from Istituto di Fisica Cosmica ed Applicazioni della Informatica del CNR in Palermo, Istituto di Fisica Cosmica e Tecnologie Relative del CNR in Milano, Istituto dell'Osservatorio Astronomico dell'Università "La Sapienza" di Roma, Dipartimento di Fisica dell' Università di Milano, Osservatorio Astronomico di Brera-Merate; British groups from University of Birmingham, University of Leicester, Rutherford Appleton Laboratory, Mullard Space Science Laboratory; a German group from Max Plank Institut fuer Extraterrestrische Physik, Garching bei Munchen; and the Space Science Department of ESTEC, ESA.

JET-X consists of two identical co-aligned X-rays imaging telescopes with a cooled CCD detector on the focal plane. This will provide an high spatial resolution combined with a good energy resolution, in particular, around the 7 KeV Fe-line complex E/ΔE is less than 50. An optical monitor is co-aligned with the X-ray telescopes for simultaneous observations of the optical counterparts of the X-ray target sources.

JET-X is one of the components of the overall payload of the USSR's SPECTRUM-XΓ, an international mission whose combined response extends from 20 eV to 100 KeV. The satellite will be placed in orbit by PROTON launch vehicle. The SPECTRUM-XΓ scientific payload includes two groups of instruments:

1)Narrow field instruments employed for detailed investigations of quasistationary sources.

2)Burst and Survey instruments with large field of view intended for locating X and Γ rays burst sources.

JET-X, a stand alone instrument of the first group, is chosen as reference experiment for the alignment, and all the other instruments of the payload are aligned relative to this. In fact, it has the best spatial resolution, and is equipped with an optical sensor with an angular reso-lution up to 5 arcsec.

Data Analysis in Astronomy IV, Edited by V. Di Gesù *et al.*
Plenum Press, New York, 1992

The main requirement for the orbit is for observing sources beyond the earth magnetosphere (above 80000 Km). To ensure a minimum time of passage in the magnetosphere, an high elliptical earth satellite orbit is needed. The principal initial parameters of the orbit are:

perigee altitude	500 km
apogee altitude	200 000 km
eccentricity	0.935
inclination	51.5°
period	96.25 h

The launch is scheduled for the end of 1994, well before the world class missions AXAF and XMM. In comparison with contemporary missions, JET-X will have a lower throughput, but a better limiting sensitivity, spectral and imaging resolution than SODARD; a comparable throughput, but better spatial resolution than ASTRO-D the Japanese mission and a larger throughput and a higher resolution than the imaging spectrometer on SAX.

INSTRUMENT DESCRIPTION

Telescope

Each of the two X-ray telescope is composed by a nested array of twelve Wolter I mirrors (paraboloid+hyperboloid). The focal length is 3.5 meters and the field of view 20 arcmin. The instrument effective collecting area is 360 cm^2 at 1.5 KeV and 140 cm^2 at 8 KeV. Fig. 1 shows the effective area as function of energy at different incident angles.

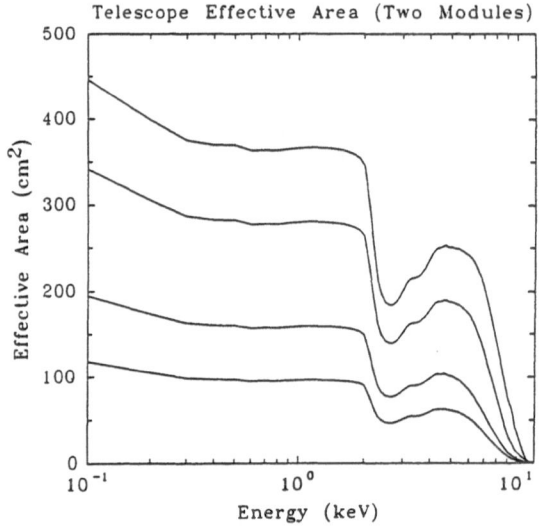

Figure 1. Total collecting area of one JET-X telescope on axis and at off-axis angles of 7.5, 15.0, 22.0 arcmin.

The angular resolution is strongly dependent on the microroughness of the mirror surface and on the manufacturing tolerances (Aschenbach et al. 1980). Using the same construction technique as the Italian satellite SAX, based on the electroforming replica, it will be better than 30 arcsec. A mandrel is machined to the required figure, and then superpolished in such a way to have a surface microroughness less than 5 Å at 8 KeV. On each mandrel a gold layer is evaporated and a nickel layer is electroformed on top of this. The gold layer gives the reflecting surface for X-rays, while the nickel, a few millimeters thick, is for supporting the gold (Citterio et al. 1987).

CCD Detector

The CCDs on the focal plane provide a spatial resolution compatible with the optics together with a broad band spectroscopy, covering the energy range 0.2-10 KeV. The type of CCD chosen for JET-X is based on an enhanced version of the P88200 CCD, with a lower noise amplifier included. It is a deep depletion CCD, 65 μm thick, with a 40 μm depletion depth and 25 μm undepleted silicon underneath (Lumb and Holland 1989). The device is fabricated on high resistivity silicon to obtain good quantum detection efficiency at energies above 3 KeV (see fig. 2). It is front illuminated with a substantially reduced electrode thickness to improve the sub-KeV response, compared with the standard CCD.

The field of view is covered by a set of two CCDs closely butted; each of them is fabricated in two identical sections: one collects the images and the other is shielded and used to read out the signals. The coverage of the total focal plane, considering both telescopes, is 98% (Wells and Lumb 1989). The CCDs are cooled at temperature around 170-180 K in order to reduce the dark current. The FWMH energy resolution is 150 eV at 8 KeV and is determined by the equation:

$$\delta E = 8.58 \ \sqrt{(n^2 + f*E/3.65)}$$

where n is the electronic noise, E the energy of the incident photon, and f the Fano factor. Fig. 3 shows the energy resolution versus the energy.

There will be two operating modes:

framestore: the image data are collected in one half of the array and with a rapid shift, the stored signal changes to the shielded store section. Data are readout whilst the next image accumulates in the image section. The time necessary to complete the readout operation is almost 2.5 sec.

timing mode: the readout mode comprises a row shift followed by a pixel shift and read, repeated indefinitely. The time resolution increases to 0.3 msec, that is the time for a single pixel to across the photon impact area.

The high thickness of the device will improve the charge particle background rejection. In fact, the mean energy deposited from each particle in such thick device will be almost 25 KeV, well out the working range. By a pixel-to-pixel anti-coincidence discrimination a 99.95% rejection efficiency is reached.

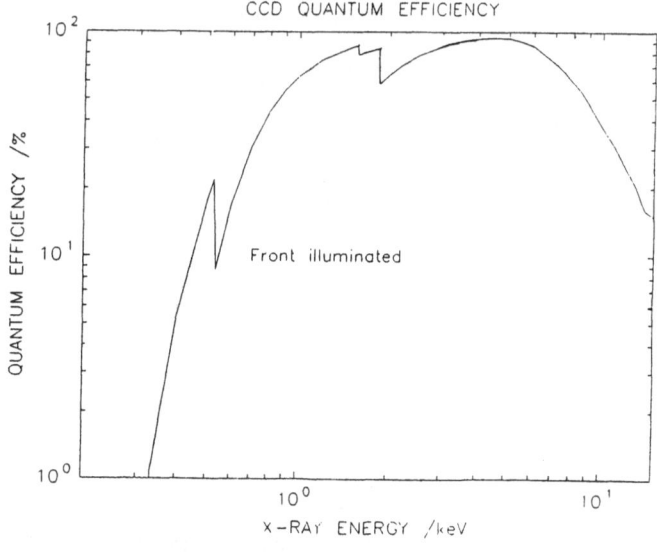

Figure 2. JET-X CCD quantum efficiency versus energy.

Optical Monitor

The presence of an optical monitor as part of an X-ray experiment derives from the scientific need of having data covering various wavelengths, to better understand the phenomenology of different objects. JET-X optical monitor consists of a small Ritchey-Chretien optical reflector of 26 cm, with a field of view of 30 arcmin equipped with two optical CCDs as detector. The first CCD is devoted to the observation of a small field of view (8 arcmin). The angular resolution, defined by the optics quality, is better than 1.67 arcsec. The other CCD has a wider field of view (30 arcmin), but lower angular resolution (better than 6.27 arcsec). The readout noise in the CCD leads to a limiting magnitude of mv=22 (Antonello et al.1990).

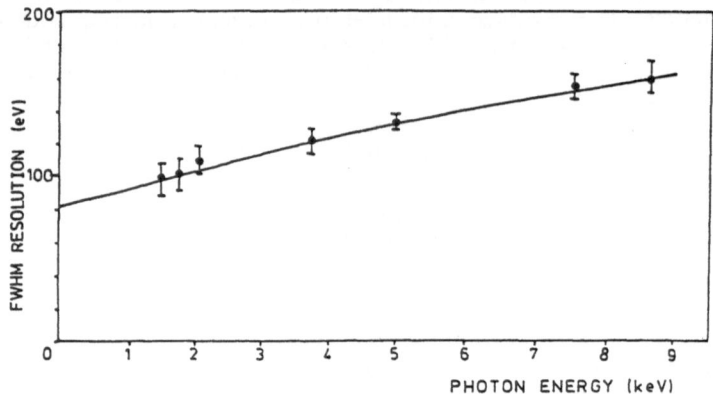

Figure 3: Predicted CCD energy resolution versus energy. The points represent the measured values.

SCIENTIFIC CAPABILITY

The two characteristics that determine the scientific strength of JET-X are the high spatial resolution that gives a potential high point source sensitivity, and the spectral performance. The sensitivity to X-ray sources depends on the flux level of the source and on the source spectrum. Fig. 4 compares the source sensitivity in three different energy bands. The source is assumed to have a power law spectrum with photon index 1.7 and absorbing column density of $3*10^{20}$ cm^{-2}. The detection limits is 5σ over the background. The kind of background considered is represented by three components: the cosmic X-ray background, the galactique diffuse X-ray background and the charge particle background, for altitude greater than 40000 Km, after rejection. The sensitivity does not depend strongly on the

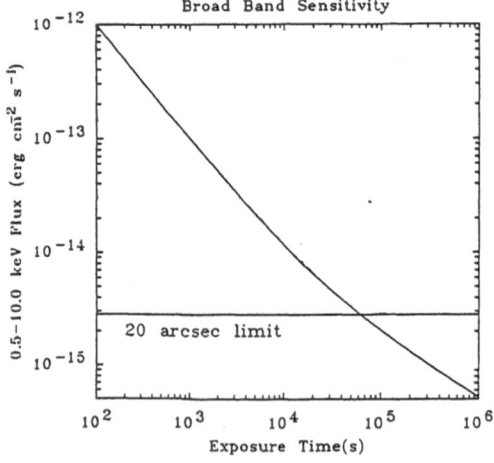

Figure 4. Time required for a 5σ detection of a point source for JET-X. Also shown is the confusion limit assuming a Euclidean extrapolation of the EINSTEIN logN-logS.

assumed source spectrum. Changing the photon index by 0.5, changes the observed count rate by less than 25%, while increasing the assumed column density to 10^{21} cm^{-2}, would decreases the count rate by 30%.

The limiting sensitivity, computed using one source per 30 beamwidth criterion, for a detector with 20 arcsec beamwidth is almost $3*10^{-15}$ erg/cm^2/sec in the range 0.5 - 10 KeV. The confusion limit has been extrapolated considering the EINSTEIN logN-logS population of sources. The time necessary to reach this limit is of the order of 10^5 sec. Fig. 5 shows the minimum flux detected with a 5σ level for a line at 7 KeV with an equivalent width of 2 KeV, 1 KeV, 100 eV over a power law continuum with photon index 1.7.

GROUND SEGMENT

The SPECTRUM-XΓ spacecraft will be commanded to dump stored memory data to ground, nominally once per day for about 1 and 1/2 hours. A low speed telemetry channel (65 kbit/sec) will be used for commanding and housekeeping. Normal science data will be telemetered using 1 Mbit/sec link. The prime ground station is at Eupatoria in Crimea. All data from SPECTRUM-XΓ will be transmitted when possible by 1 Mbit/sec analogue data link from the Ground Station to Moscow (IKI).

As back-up to this link, data may be sent by magnetic tapes with small sub-sets being transmitted by 4.8 kbit/sec public data link. The back-up ground station for SPECTRUM-XΓ is at Ussuriysk near Vladivostok. At IKI, the data will be filtered in order to pass to West-European share of JET-X data, together with the necessary spacecraft housekeeping data, to the JET-X Quick Look Facility (QLF) in IKI, where a brief analysis on the data, before the transmission to the United Kingdom, will be performed. The nominal volume of data for JET-X is 40 Mbyte per day, when spacecraft housekeeping, parity and the other administrative data are added the total volume will be 60 Mbyte /day.

From the QLF at IKI data will be transmitted by dedicated commercial data link to Rutherford Appleton Laboratory in UK, where all data will be checked for completeness, in order to inform the QLF that they have arrived safely. At Leicester University there will be a Instrument Analysis Centre responsible for the off-line functions of the instrument calibration and detailed science analysis. JET-X data will be distributed to the consortium groups in data sets corresponding to the observations requested.

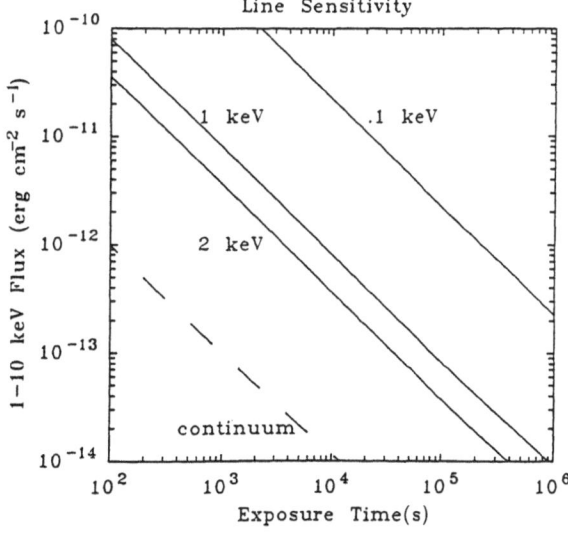

Figure 5: Time required by JET-X for a 5σ detection of lines of varying equivalent widths as a function of the continuum intensity.

REFERENCES

Aschenbach, B., Brauninger, B., Hasinger, G., Trumper, J., 1980, Proc. SPIE **257**, 223.

Citterio, O., Bonelli, G., Conti, G., Mattaini, E., Santambrogio, E., Sacco, B., Lanzara, E., Brauninger, H., Burkert, W., 1987, Proc. SPIE **830**, 139.

Lumb, D.H., Holland, A.D., 1989, Proc. SPIE **830**, 116.

Wells, A., Lumb, D.H., 1989, Proc. SPIE **1159**, 372.

Antonello, E., Citterio, O., Mazzoleni, F., Mariani, A., Pili, P., Lombardi, P., 1990, Proc. SPIE on 'Astronomical Telescopes & Instrumentation for the 21st Century', Tucson, Arizona.

THE SAX MISSION

R.C. Butler [1], L. Scarsi [2]

[1] Agenzia Spaziale Italiana - Roma - Italy
[2] Istituto di Fisica Cosmica e Informatica - CNR - Palermo
and
Dipartimento di Energetica e Applicazioni della Fisica
University of Palermo - 90128 Palermo - Italy

ABSTRACT

SAX (X-Ray Astronomy Satellite) is a programme jointly developed by the Italian Space Agency (A.S.I.) and the Netherland Agency for Aerospace Programmes (NIVR) devoted to systematic, integrated and comprehensive studies of galactic and extragalactic sources in the energy band 0.1 - 200 KeV.

Scientific objectives are:
- Imaging (with moderate angular resolution of 1 arcmin) and broad band spectroscopy over the energy range from 0.1 to 10 KeV.
- Spectral measurements, spectroscopy and timing on sources from 3 to 200 KeV.
- All sky monitoring (2-30 KeV) for the investigation of long time variability and localisation and study of transients.

The payload complement includes: a low energy (0.1-10 KeV) concentrator/spectrometer (LECS), a medium energy (1-10 KeV) concentrator/spectrometer (MECS) consisting of three units, a high pressure gas scintillation proportional counter (3-120 KeV) (HPGSPC) and a phoswich detector system (15-200 KeV) (PDS), all of which have narrow fields of view and have the optical axis coaligned to the same pointing direction. Two wide field cameras (2-30 KeV), field of view 20°x20° (WFC) which point in diametrically opposed directions perpendicular to the narrow field instrument axis, complete the payload.

The SAX mission is, for the Payload and Science, under the responsability of a Consortium of Italian Institutes with the partecipation of the Space Research Institute of Utrecht/SRON-Holland and the Space Science Department of the European Space Agency (E.S.A.).

The spacecraft has a total mass of 1200 Kg, is three axis stabilised and it will be placed into a circular orbit at 600 Km with an inclination of two degrees, by an Atlas G-Centaur.

SAX, to be launched at the end of 1993, will have a minimum mission life time of two years, extendable up to four years.

INTRODUCTION

The X-ray Astronomy missions ongoing and planned aim at specific objectives in order to obtain deeper insight in the phenomena of interest to Astrophysics; this approach drives the choice of the typical characteristics of payload, orbit and mission prophile.

Data Analysis in Astronomy IV, Edited by V. Di Gesù *et al.*
Plenum Press, New York, 1992

Along this line, ROSAT devotes the first part of its operational life to an all sky survey below 2 KeV which should provide a systematic catalogue an order of magnitude richer than what is presently available; Astro-D will introduce, together with Spectrum X, a definite step forward in spectroscopy in the band below 10 KeV adopting CCD based detectors; XTE will be devoted essentially to timing; the last generation observatory-missions AXAF and XMM will push at the extreme the grazing incidence telescope technology, for imaging in the first case and throughput level in the second, for the low-medium energy range, extending the upper limit possibly up to 20 KeV.

The results so far have on the other hand demonstrated the importance of simultaneous source investigation over a broadened energy range; the choice of SAX has been to carry out systematic and comprehensive observations in the 0.1-200 KeV energy band with special emphasis on spectroscopic, spectral and timing measurements.

MISSION OVERVIEW

The acronym SAX stands for "Satellite for Astronomy in X-rays" and it refers to a major joint programme of the Italian Space Agency (ASI) and the Netherlands Agency for Space Programmes (NIVR). The mission is, for the payload and science, under the responsability of a Consortium of Italian Institutes (See Appendix I) with the participation of the Space Research Institute of Utrecht/SRON-Holland and the Space Science Department of the European Space Agency (ESA).

SAX finished its phase B activities in 1988 and entered phase C/D (a continuous phase to completion) in April 1989; the planned launch date is at the end of 1993.

Observational goal

The observational goal to be addressed by SAX, is to continue and expand upon previous observations of celestial sources in those areas for which the existing information is missing or inadeguate and will remain uncovered in the mission planned in the forseable future.

Specifically, the scientific objectives are:

- Imaging (with moderate angular resolution of 1 arcmin) and broad band spectroscopy ($\lambda/\Delta\lambda \sim 10$) over the energy interval from 1 to 10 KeV, extended down to 0.1 KeV for spectroscopy.
- Continuum and line Spectroscopy ($\lambda/\Delta\lambda$ from 5 to 20) over the energy interval 3-200 KeV.
- Time variability studies of bright source energy spectra both on short-term (ms) and long-term (days to months) time scales.
- Systematic studies of the long-term variability of X-Ray sources through periodic surveys of preselected regions of the sky (minimum intensity equivalent to a 1mCrab source).
- Gamma-Ray burst studies.

Instrumental approach

The payload adopted in SAX to provide the observational capability required to pursue the scientific programme outlined above is schematized in Table I. The chief characteristics have been described also in (1, 2, 3, 10).

It comprises a package of coaligned Narrow Field of View Instruments (NFI): a medium energy (1-10 KeV) concentrator/spectrometer, MECS, consisting of three units; a low energy (0.15-10) KeV concentrator/spectrometer, LECS; a high pressure gas scintillation proportional counter (3-120 KeV), HPGSPC; a phoswich detector system (15-290 KeV), PDS.

All the NFI point in the same direction.

The set of Wide Field of view instruments (WFI) is based on two Wide Field Cameras (2-30 KeV), which point in diametrically opposed directions perpendicular to the NFI common axis.

Fig.1 shows the SAX payload accomodation. Fig.2 gives the relationships between the instruments with respect to the energy coverage.

Table I. Principal characteristics of SAX Instrumentation

INSTRUMENT	Energy Range (KeV)	Field of View (degree)	Angular Resol. (arcmin)	Effective Area (cm²)	Energy Resolution
Concentrator/ Spectrometer (C/S) 1 unit-LEC/S 3 units-MEC/S	0.1-10 1-10	0.5	1	56 at 0.25 KeV 200 at 7 KeV	28% (0.27 KeV) 8% (6 KeV)
High Pressure Gas Scintillation Proportional Counter (HPGSPC)	3-120	1		300 at 6 KeV 280 at 60 KeV	10% (6 KeV) 3% (60 KeV)
Phoswich Detector System (PDS)	15-200	1.5		680 at 20 KeV 140 at 200 KeV	17% (60 KeV)
Wide Field Cameras (WFC) (2 units in complement)	2-30	20x20	5	250 (one unit) through mask)	20% (6 KeV)

THE NARROW FIELD INSTRUMENT PACKAGE

The concentrator spectrometer (C/S)

The instrumentation consists of four separate concentrator Mirror assemblies each with a focal lenght of 185 cm and a position sensitive, Xenon filled Gas Scintillation Proportional Counter (GSPC) in the focal plane.

The mirror sistems are schematically described in Table IIa. They are produced by electroforming Nickel (0.2-0.4 mm) onto the gold coated surface of superpolished mandrels. X-Ray tests on prototype models at the Panther X-ray beam facility at the MPE/Garching have shown that the design goal of less than 10 Angstroms of surface roughness has been reached, and that an angular resolution of better than one arc-minute MPR will be achieved [4], [5].

The optics are designed to maximise effective area in the iron line region (Fig.3); the comparison with AXAF is particularly notable in this respect.

Three of the GSPC'S will have 50 microns beryllium windows which are essentially opaque below about 1.2 KeV and are read out by crossed wire anode Hamamatsu photomultipliers; they are described in Table IIb.

The fourth GSPC, developed by SSD/Estec and described in Table IIc extends the energy range down to 0.1 KeV and be viewed by a nine anode readout Hamamatsu photomultiplier. This detector is described in detail in (6).

The source confusion limit of the concentrator/spectrometers reached in 4×10^4s will be approximatively three times below the lower limit of the ROSAT all-sky survey and thus they will be fully capable of exploiting the survey in the selection of representative samples of objects for detailed studies up to 10 KeV (Fig. 4).

The High Pressure Gas Scintillation Proportional Counter (HPGSPC)

The characteristics of the detector are summarized in Table III.

The Xenon filled gas cell of the HPGSPC is viewed by an Anger Camera arrangement of seven photomultiplier tubes, and surrounded at the sides and from below by a graded lead/tin, shield.

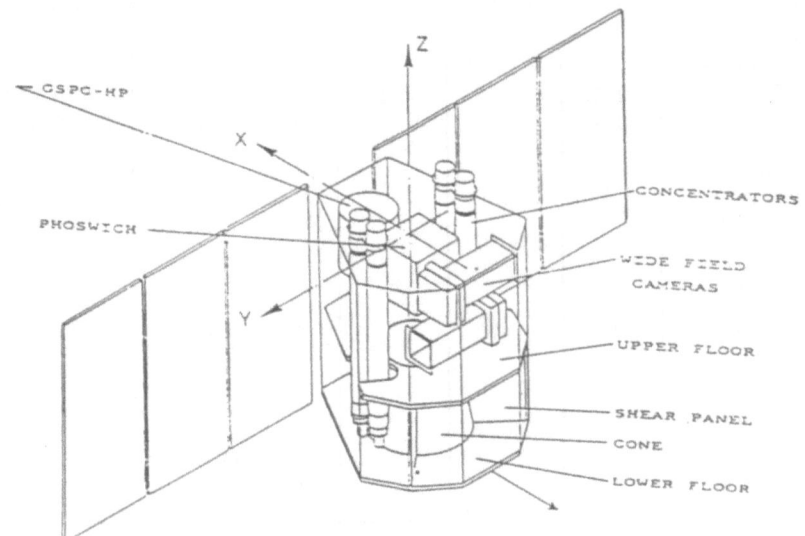

Fig.1. SAX payload accomodation.

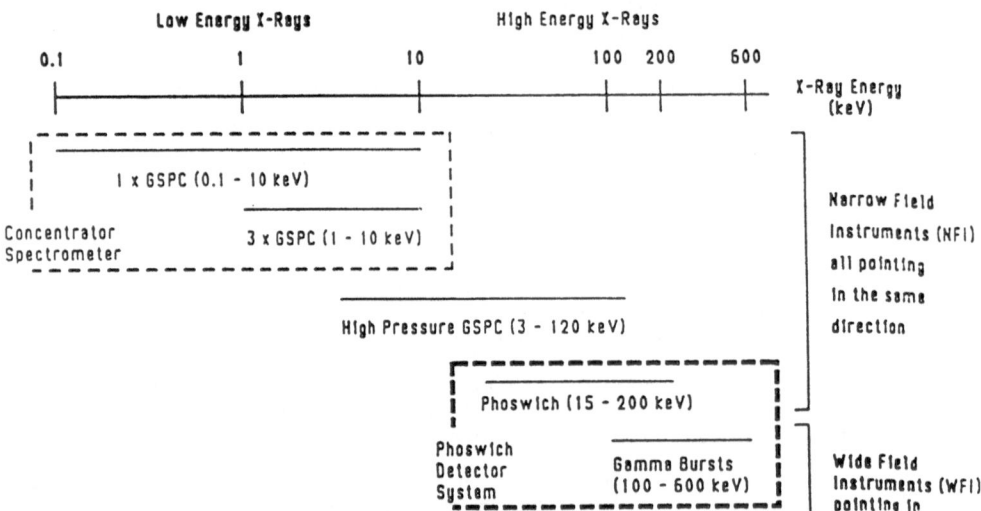

Fig.2. SAX payload energy coverage.

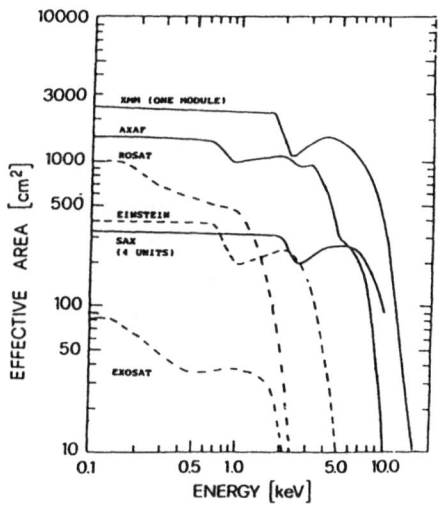

Fig.3. The area of the SAX mirrors compared to other missions.

Table IIa - SAX C/S Mirror Sistems: (4 Identical Units)

- Double cone approximations to Woltjer I configuration
- Number of nested mirrors/Unit is 30
- Focal length   185 cm
- Half power radius   1 arcmin
- Aperture (1 KeV)   30 arcmin
- Effective collecting area on axis per unit:

0.1 KeV	88 cm^2
1 KeV	86 cm^2
7 KeV	56 cm^2
10 KeV	23 cm^2

Table IIb - SAX C/S - Medium Energy (MEC/S): (3 Identical Units)

- Energy Range   1-10 KeV
- Focal Plane Detector: GSPC with 50 µm Beryllium Window

- Effective Area/Unit (Including Mirrors):
    3 KeV   48cm^2:   7 KeV   50 cm^2:   10 KeV   15 cm^2

- Energy Resolution   0.08/ $\sqrt{\text{E(KeV)}}$/6 FWHM

- Position Resolution of GSPC   0.75/ $\sqrt{\text{E(KeV)}}$/6 mm FWHM
                                  =>   0.76 arcmin HPR at 6 KeV

Table IIc - SAX C/S. Low Energy (LEC/S): (one unit)

- Energy Range   0.1 - 10 KeV
- Focal Plane Detector: GSPC with 1.5 μm Polypropylene windw

- Effective Area/Unit (Including Mirrors):
    0.25 KeV   56cm^2;   0.6 KeV   28 cm^2;   1 KeV   61 cm^2
    Above 3 KeV approx as MEC/S

- Energy Resolution   28% at 0.27 KeV (Measured)
                                                 and as for MEC/S above 1 KeV

- Position resolution expected to be as for MEC/S

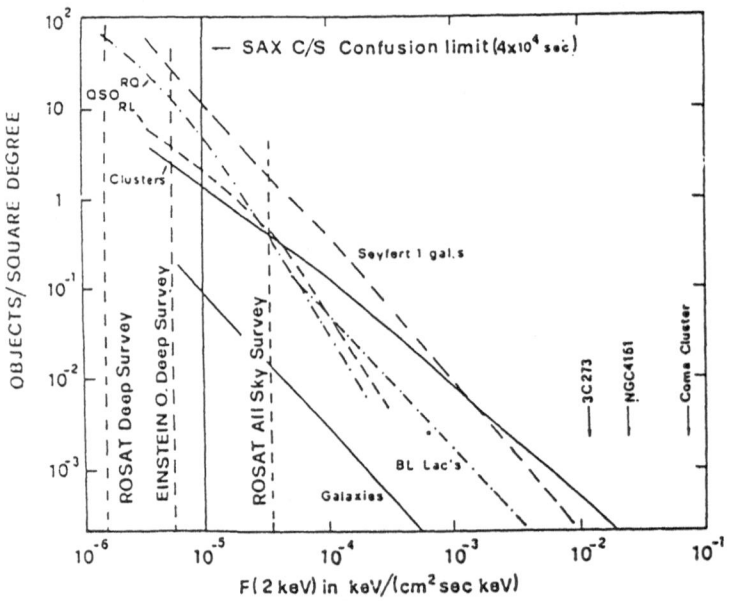

Fig. 4. Sensitivity-confusion limit of the C/S compared with Einstein and ROSAT

The X-ray event arrival position is used to correct the event energy with an overall improvement of about a factor of two when compared with proportional counters.

A rocking collimator is used for alternatively sampling the flux in the source direction and the background.

The prototype instrument is described in detail in (7).

The very good energy resolution (about 5 times better than that of NaI/CsI detectors), will be particularly important in the detailed study of narrow cyclotron lines, both in emission and absorption, as illustrated in Fig. 5.

Table III - SAX High Pressure Gas Scintillation Proportional Counter (HPGSPC)

- Energy range	3-120 KeV	Effective area:	6 KeV	300 cm^2
- Field of view	1 Deg FWHM		60 KeV	280 cm^2
- Geometrical area	450 cm^2		100 KeV	160 cm^2
- Window thickness	1300 µm/Be	Energy Resol.:	6 KeV	10% FWHM
			60 KeV	3% FWHM
			$\Delta E/E$	$0.25 \sqrt{E}$ (KeV)
- Pressure of Xenon	5 atm	Residual Background:		
		E<35 KeV		10^{-3} ct/cm^2 s KeV
- Depth of drift Region	10 cm	E>35 KeV		10^{-4} ct/cm^2 s KeV
		(Via K-Fluorescence Escape Gate)		

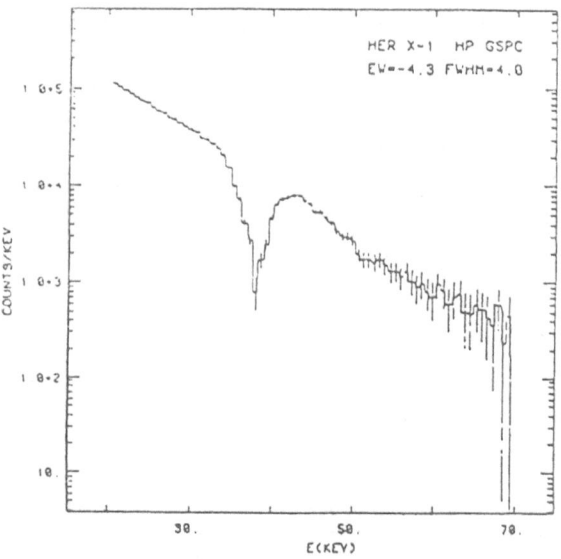

Fig.5. Simulation of count spectra (on minus off-source) from the HPGSPC obtained in 10^5 s on Her X-1 in the phase interval 0.0-0.1 around the pulse peak, with cyclotron absorption on an exponential continuum.

The Phoswich Detector System (PDS)

The PDS characteristics are summarized in Table IV.

The instrument is based on a central group of four phoswich units (3 mm of NaI(T1) above 50 mm of CsI (NaI), surrounded at the sides by an anticoincidence shield of CsI(Na) and with a plastic scintillator particle shield over the aperture. The source direction is viewed through two collimators that can be offset to sample the background.

The instrument, developed by ITESRE and IAS using the Lapex balloon borne telescope as test bed is described in Frontera et al. [9]. Extrapolating the background from the Lapex results to the SAX orbit has confirmed the sensitivity of the PDS.

The energy resolution, demonstrated on the prototype to be better than 17% FWHM at 60 KeV, combined with its high source flux sensitivity over a broad energy range make the instrument particularly useful in the detailed studies of the continuum in galactic and extragalactic sources and their time variability. It is also well suited to the study of cyclotron line features in known sources and the search for these features in other binary pulsars where it will extend the possible measurements of the HPGSPC particularly for broad line features. Additionally, the active shielding around its sides will be used for all-sky monitoring of gamma-ray bursts with a limiting sensitivity of 10^{-6} erg/cm s and a timing capability of 0.5 to 10 ms.

Table IV - SAX Phoswich Detector System (PDS)

- Energy range	15-200 KeV	Energy resolution 60 KeV 17% FWHM
		$\Delta E/E$ $1.4 \sqrt{E}$ (KeV)
- Field of view	1.4 Deg FWHM	Residual Background:
- Geometrical area	800 cm^2	30-40 KeV 2.3×10^{-4}ct/cm^2 s KeV
		40-80 KeV 1.7×10^{-4}ct/cm^2 s KeV
		80-200 KeV 1.1×10^{-4}ct/cm^2 s KeV
- Effective area:	20 KeV 600 cm^2	As gamma ray burst monitor:
	60 KeV 500 cm^2	Energy range 100-600 KeV
	100 KeV 500 cm^2	Sensitivity 10^{-6} erg/cm^2 s
	200 KeV 140 cm^2	

The NFI System

The overall broad band sensitivity of the NFI is shown in Fig. 6, where the minimum detectable source strength is given as a function of energy (1-200 KeV) for a 10^4 s observation.

The potentiality can be best illustrated by the simulated results for a typical source shown in Fig. 7, 8 and 9. The spectral index can be determined with an uncertainty approximatively one order of magnitudine smaller than in previous experiments (EXOSAT, HEAO 1). Deviations from the power law (soft excess and breaks or hard components at high energies), if present, should be clearly visible; Fe line spectroscopy is well accessible.

The narrow field instruments will also be used for time variability studies on time scales from milliseconds to days and months.

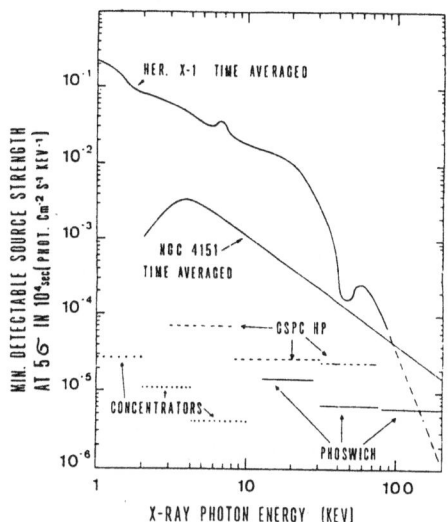

Fig.6. Broad band sensitivity of the three NFI for an exposure of 10^4 s.

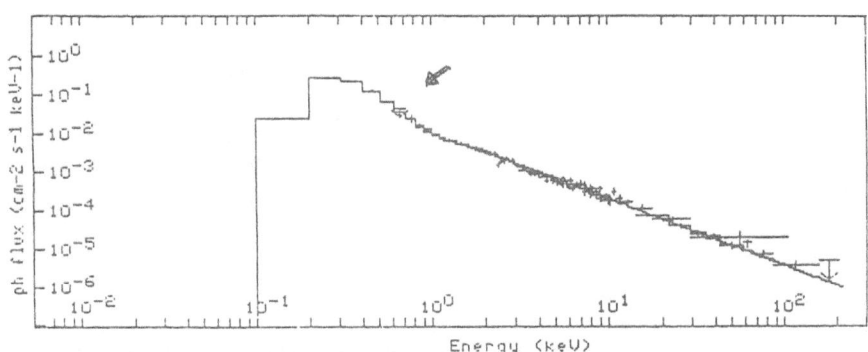

Fig.7. SAX simulation Exposure time 10^4 s. AGN (F=1.5 mCrab) with soft X-ray excess. Spectral parameters: $N_H = 4 \times 10^{20}$ cm^{-2}; $\alpha_1 = 0.7$ (energy index); $\alpha_2 = 2.5$; $E_i = 1$ KeV

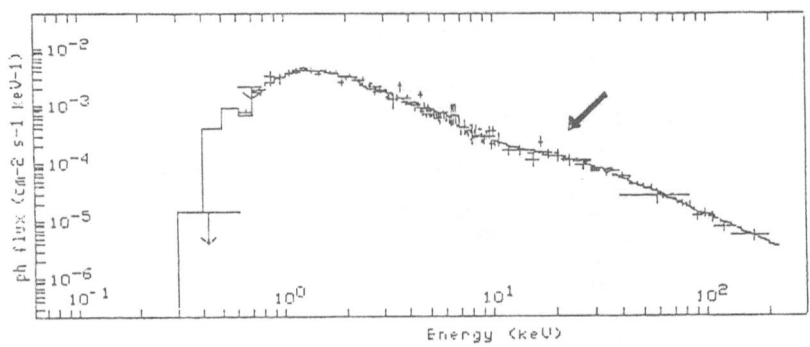

Fig.8. SAX simulation. Exposure time 10^4s. AGN (F=2 mCrab) with high energy bump described by a partial covering model. Low energy abs. $N_H = 6 \times 10^{21}$ cm^{-2}; α=0.7 $f_{cov} = 0.6$; $N_{Hcov} = 6 \times 10^{24}$ cm^{-2}

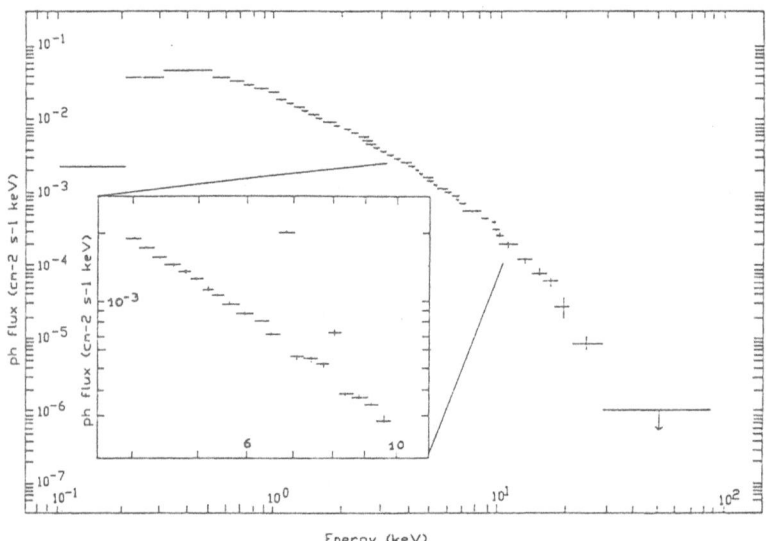

Fig.9. SAX simulation. Exposure time 10^5 s. Deconvolved spectrum of a typical cluster of Galaxies. Flux = 3 mCrab at 3 KeV; T = 8 KeV; $N_H = 4 \times 10^{20}$ cm^{-2}

The Wide Field Cameras

Each WFC contains a position sensitive proportional counter filled with two atmosphere of Xenon, which view the sky through a random mask aperture. These instruments, described in (6) are under construction by SRU/SRON and are a development of their COMIS experiment operated on the Soviet MIR Space Station. Table V summarizes the essential information. The instruments are described in (8).

Table V - SAX Wide Field Cameras (WFC)

Detector	Position sensitive proportional counter (2 atmos of Xenon) viewing sky through a random mask	Energy range Full Field 2-10 KeV Centre of Field 2-30 KeV Timing (no imaging) 2-35 KeV Down to 2 millisec
Field of Voew 20x20 Deg FWHM		Energy resolution 20% at 6 KeV
Effective area 250 cm^2 (one unit through mask)		Sensitivity in 10^{4}s 3 milliCrab
Intrinsic angular resolution 5 Arcmin		

The observation programme will include the long term monitoring of both galactic and extragalactic variable sources and the detection/localisation of transients on timescales from 2 ms upwards.

The WFC sensitivity as a function of observation time is illustrated in Fig. 10, while Fig. 11 shows the capability to detect fractional changes in source intensity.

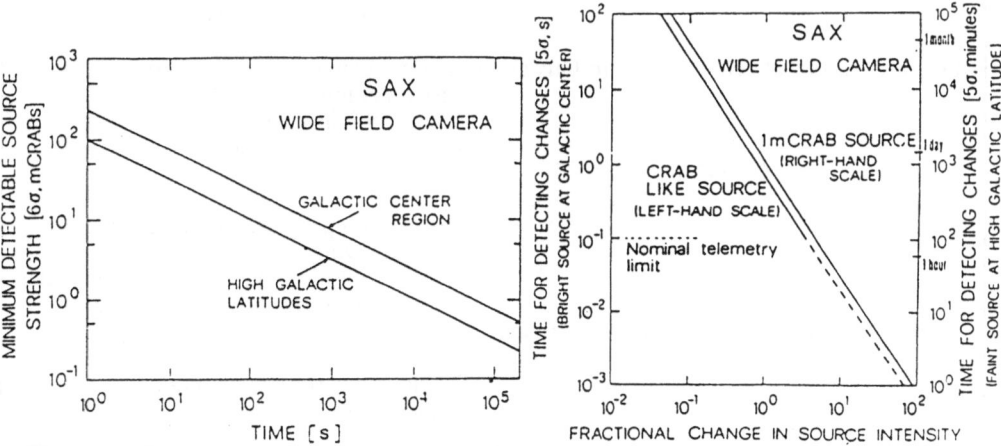

Fig.10. WFC sensitivity as a function of observation time.

Fig.11. WFC sensitivity to detect a fractional change in sonore intensity.

345

SPACECRAFT. ORBIT AND ORBITAL OPERATION. GROUND SEGMENT

The essential information is given in Tables VIa,b,c.

Table VIa - SAX Spacecraft characteristics

- Total Mass/Power	1200 Kg / 800 W
- Payload Mass/Power	385 Kg / 290 W
- Three axis stabilized	
Abs. Pointing error	3 arcmin
Abs. Pointing stability	1 arcmin
Pointing reconstruction	1 arcmin
- Observation period	10^4 - 10^5 seconds (typically)
- Telemetry	
Mass memmory capacity	450 Mbits per orbit
Effective rate	68 KBPS around orbit
For science data	100 KBPS maximum
Dump rate	1 MBPS

Table VIb - Orbit

- Type	Circular at 2° inclination to the equator
- Height	600 km (BOL)/450 km (EOL)
- Lifetime	2 years extendable to 4 years
- Period	97 min (Sun eclipse 37 min)
- Ground contact	11 min per orbit
- Launch Date/Vehicle	End of 1993 / Atlas G-Centaur

Table VIc - Ground Segment Characteristics

- Ground station at San Marco base, Malindi, Kenya
  Visible for 11 minutes per orbit

- Operations control centre (OCC) in Italy connected by relay satellite perforrms:
  Mission control
  Initial archiving/Data formating
  Scientific data quick look

- SAX data centre (CDS) linked to OCC
  CDS services:
  Observation program coordination
  Final archiving
  Data distribuition via Network to Institutes
  Interface for guest observer program

- Institute centres for each payload instrument
  services:
  Specialist instrument calibration
  maintenance

SAX will be launched by an Atlas G-Centaurus directly into a 600 km orbit at 2 degrees inclination. The payload will thus nearly avoid the South Atlantic Anomaly, and take full advantage of the screening effect of the Earth's magnetic field in reducing the cosmic ray induced background proton fluxes (typically a factor of 20 lower than in the Exosat orbit above the radiation belts), while undergoing the minimum change in magnetic cut-off. This choice of orbit is particularly important in order to minimise the background and systematic effects caused by spallation and changes in incident charged particle fluxes around the orbit to achieve the necessary sensitivity of the PDS for weak source observations, estimated at 1% of the background in this orbit, essential if the broad band sensitivity of SAX is to be ensured. SAX has an official design lifetime of two years but is expected to remain in operation for upto four years. The satellite will achieve one arcminute pointing stability continuosly for a typical maximum single observation time of 10^5 seconds, with a postfacto pointing reconstruction accuracy of 1 arcmin. The chief attitude constraint derives from the need to retain the normal to the solar arrays within 30 degrees of the Sun (with occasional excursion to 45 degrees will be made for the WFC) to ensure proper battery maintenance and allow observations to be continued throughout the Sun eclypse periods with all the instruments. Thus while the whole sky will be available during one year, during a single orbit, and subject to eclipse by the Earth (which has a diameter of about 130 degrees at 600 km), the narrov field instruments will have a band in the sky 60 degrees wide available for observations which includes about 50% of the sky, and theWFC a slightly larged band commensurate with their field of view.

In the mission plan the narrow field instruments will normally have the first choice of observation direction while the WFC will observe the regions of the sky available to them. Periodically the WFC will be given priority to perform long term studies of certain sky regions e.g. the galactic centre and along the galactic plane. During each orbit upto 450 Mbits of information will be stored on board and relayed to the ground during station passage. The average data rate available to the instruments will be approximately 70 kpbs, but peak rates of upto 100 kbps will be catered for. The ground station for telecommand uplink and telemetry retrieval will be situated near the equator (at the San Marco Base, Malindi, Kenya), while the operations control centre connected by a communications relay satellite link will be in Italy.

SAX SCIENTIFIC OBJECTIVES

The scientific objectives of SAX have been described in detail in (2,10). In the mission life of at least two years SAX will perform between 2000 and 3000 separate observations. These will be based on a core program chiefly devoted to systematic studies of various classes of objects, and a guest observer program allocated about 20% of the time. A selection of the areas in which SAX is expected to make its most significant contribution is given below:
- Compact galactic sources: the shape and variabilityu of their continuum and temporal studies of features such as iron fluorescence liens, cyclotron lines and absorption effects as a function of orbital phase and rotation, transient detection and light curve studies
- Supernova remnants: spatially resolved spectra of extended (>>1') galactic SNR's, and the spectra of the Magellanic Clouds remnants
- Stars: coronal emission spectra with a sensitivity comparable to that of the Einstein Observatory upto 10 KeV
- Active galactic nuclei: spectral and temporal variability studies of their continuum upto 200 KeV; the spectra upto 10 KeV of very distant sources (z=3.2 for sources equivalent to 3C273) and soft X-ray excess, photoelectric absorption and iron fluorescence line studies
- Clusters of galaxies: spatially resolved spectra upto 10 KeV of the nearby clusters with iron fluorescence line and temperature gradient studies and high energy spectra for z<0.1 and temperature measurements out to z-1
- Normal galaxies: spectral studies of their extended emission.

The SAX payload has been chosen to cover the energy range 0.1-200 KeV with a series of instruments in a coordinated fashion to notably extend the spectroscopic and time variability studies performed to date. Its launch date at the end of 1993 will give SAX a first opportunity to take advantage in a systematic way of the many new results that will become available from the all-sky imaging survey of ROSAT, and it will precede the large observatory type missions due for the second half of the 1990's which will concentrate on X-ray astronomy upto 10 KeV.

APPENDIX I

The SAX mission development is supported by a consortium of Institutes in Italy together with Institutes in The Netherlands and the Space Science Department of the ESA. A collaboration with the Max Planck Institut for Extraterrestrial Physics also exists for prototype X-ray mirror testing and the calibration of the concentrator mirrors. The composition of the consortium is given below:

- Istituto per le Tecnologie e lo studio delle radiazioni extraterrestri, ITESRE/CNR, Bologna, Italy
- Istituto di Astrofisica Spaziale, IAS/CNR, Frascati, Italy
- Istituto di Fisica Cosmica e Tecnologie Relative, IFCTR/CNR and Unità GIFCO, Milano, Italy
- Istituto di Fisica Cosmica ed Applicazioni dell'Informatica, IFCAI/CNR and Unità GIFCO, Palermo, Italy
- Istituto dell'Osservatorio Astronomico, Università di Roma "La Sapienza", Roma, Italy
- Space Research Utrecht, SRON,The Netherlands
- Space Science Department, SSD, of ESA, Noordwijk, The Netherlands

ACKNOWLEDGEMENTS

The data presented here is the result of the work of many people who are participating in SAX. We wish to acknowledge the contribution of G.C.Perola, G.Boella and G.Di Cocco, and in particular O.Citterio, G.Conti, and B.Sacco for the concentrator mirrors of the concentrator/Spectrometer, A.Smith for its low energy focal plane detector, G.Manzo and S.Re for its medium energy focal plane detector and the HPGSPC, F.Frontera and E.Costa for the PDS, nd J.Bleeker, B.Brinkman, R.Jager and P.Ubertini for the WFC. Finally we wish yto thank from the SAX team at ASI, G.Manarini, M.Casciola and B.Negri.

REFERENCES

1. Spada G., "Proc.of Workshop on Now-thermal and very High Temperature Phenomena in X-Ray Astronomy", eds.G.C.Perola, M.Salvati - Roma 19-20 December 1983, (1983)

2. Perola G.C. i.b.d.

3. Butler R.C. and Scarsi L., Proc.of IAU Symp. 123, G.S.F.C. 1990. To be published

4. Citterio O. et al. Applied Optics 27, 1470, (1988)

5. Citterio O. et al. Proc.of SPIE, 1343 (1990)

6. Favata F. and Smith A., SPIE, 1159, 488 (1989)

7. Giarrusso S. et al. SPIE 1159, 514 (1989)

8. Jager R. et al., SPIE, 2 1159 (1989)

9. Frontera F. et al. Cospar XXVII. The Hague (1990)

10. Perola G.C., Adv. Space Res. 10, N.2, 287 (1990)

SAX MISSION GROUND SUPPORT SYSTEM

M. Manca [1], F. D'Alessandro [1], B. Negri [2], M. Trifoglio [3]

[1] Telespazio S.p.A., Via Tiburtina 965, Rome, Italy
[2] Italian Space Agency (ASI), Via Regina Margherita 202, Rome, Italy
[3] ITSRE-CNR, Via de' Castagnoli 1, Bologna, Italy

ABSTRACT

A description of the SAX Ground Segment is given from the point of view of its architecture and activities, together with a brief overview of the SAX mission.

The principal components of the SAX Ground Segment are the Operational Control Center, the Telemetry Tracking and Command Station and the Data Relay System. The core component of the system is the Control Center, where a powerful hardware and software configuration permits a real time response to the classical needs of telemetry, telecommand, ranging and tracking, attitude and orbit determination functions, and also serves to provide some interesting additional features. Among these, some singular scientific functionalities, such as observation scheduling, raw data archiving, scientific quick look, have to be performed in 'almost real time'. The high performance level required strongly influences the architectural design.

INTRODUCTION

The Satellite for Astronomy in X-rays (SAX), a joint program of the Italian Space Agency (ASI) and the Netherlands Agency for Space Programs (NIVR), is scheduled to be launched at the end of 1993. The expendable launcher Atlas/G-Centaur will directly place SAX, a three axes stabilized satellite, into a circular orbit of 600 Km with an inclination of 2 degrees. SAX has an official design lifetime of two years but is expected to remain in operation for up to four years. In April '89 the SAX Programme entered its C/D Phase (a continous phase).

The SAX satellite is devoted to systematic, integrated and comprehensive studies of galactic and extra-galactic sources in the energy band 0.1-200 KeV. The basic scientific objectives can be summarized as follows:

Data Analysis in Astronomy IV, Edited by V. Di Gesù *et al.*
Plenum Press, New York, 1992

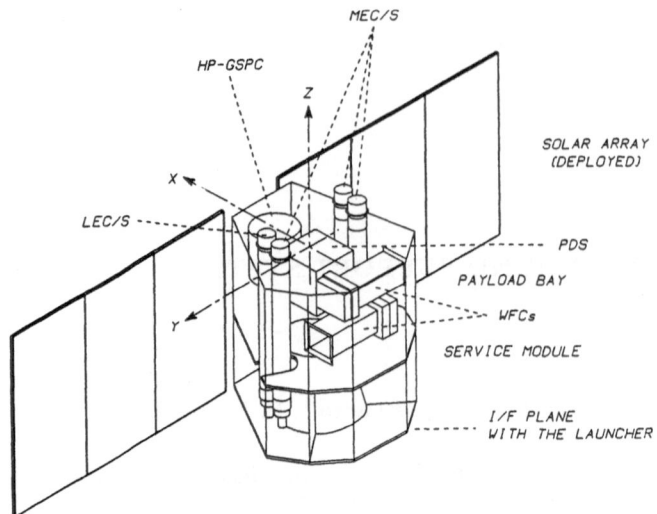

Figure 1. SAX Overall Satellite Configuration

- broad band spectroscopy from 0.1-10 KeV with imaging resolution of 1 arcmin;

- continuum and cyclotron line spectroscopy in the wide energy range of 3-200 KeV;

- variability studies of bright source energy spectra on a timescale from milliseconds to days and months;

- systematic long term variability studies over the entire sky and for source intensities of 1 mCrab or more.

The payload of SAX is composed of the following set of instruments:

- four Concentrator/Spectrometer units (C/S) made up of three Medium Energy C/S (MEC/S) units working in the energy range 1-10 KeV and one Low Energy C/S (LEC/S) unit in the energy range 0.1-10 KeV;

- a High Pressure Gas Scintillation Proportional Counter (HP-GSPC) covering the energy range 3-120 KeV;

- a Phoswich Detector System (PDS) working in the energy range 15-200 KeV;

- two Wide Field Cameras (WFCs) working in the energy range 2-30 KeV.

The four C/S, the HP-GSPC, and the PDS constitute the set of coaligned Narrow Field Instruments (NFI) having a FOV of approximately one degree. The two WFC optical axes lie in a perpendicular plane to the NFI axes, with a degree of freedom of 30 degrees around a direction ortogonal to the sun vector. The two WFCs point in opposite directions with a FOV of 20x20 degrees for each camera.

The SAX satellite is shown in Fig.1. The satellite will achieve 1 arcmin pointing stability continously for a maximum single observation time of 10**5 s, with the postfacto pointing reconstruction accuracy of 1 arcmin. A detailed description of the SAX payload is given in ref.1.

The low orbit (about 600 Km) and the consequent short contact time between the station and the satellite (about 9 minutes) suggest the use of an on-board tape

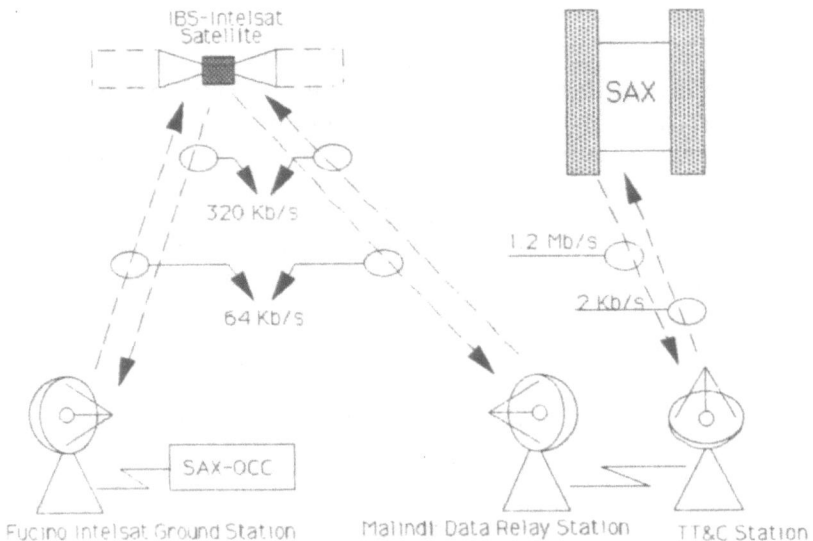

Figure 2. SAX Ground Segment Configuration

recorder of a maximum capacity of 500 Mbit to be reversed to ground during each station passage. The average data rate available for the instruments will be approximately 70 Kbps with peak rates of 100 Kbps.

The SAX Control Center is required to satisfy several particular scientific requirements in addition to the classical functionalities of a control center, determined by the payload complexity and the high volume and rate of data transfer.

The SAX mission development is supported by a consortium of Institutes in Italy together with Institutes in The Netherlands and the Space Science Department of ESA. Telespazio S.p.A. is responsible of the development of the SAX ground segment and the management of spacecraft and payload during the operative life.

THE SAX GROUND SEGMENT ARCHITECTURE

The SAX Ground Support System will manage the mission both during the pre-operational phase and the operational scientific phase. It is characterized by the following components:

- a satellite Operational Control Center (SAX-OCC), to be located in Italy;

- a Telemetry Tracking and Command (TT&C) station, to be located in Kenya.

- a Data Relay System, consisting of a Data Relay ground station, connects the previous components. It will be co-located with the TT&C station and will serve to implement the bidirectional link, via INTELSAT, between the TT&C station and the OCC.

The SAX Ground Segment is shown in Fig.2.

351

The Operational Control Center Architecture

The present baseline is to locate the SAX Control Center at Fucino (Italy). The SAX-OCC will support both the customary functionalities foreseen for a control center and more specific ones strictly related to the fulfillment of the scientific requirements.

The OCC has to be able to continously support the SAX mission over a maximum period of four years. Its hardware configuration will allow for a quick back-up in case of critical component failure. Its architectural design is dictated by consideration of the following features:

- high system reliability: the failure of a single device will not affect system functionality;

- resource sharing: all the processors share the same files with full read/write capability;

- flexible configuration: a quick system reconfiguration is allowed by powerful local communications and variable logical connections;

- potential growth of data storage capacity and of the number of diverse types of resources, such as workstations, processors and mass storage controllers;

- high speed external communications: a remote station X25 link will grant reception of telemetry and telecommands.

The aforementioned requirements lead to the choice of hardware configuration shown in Fig.3. All the elements are connected by a Local Network (LAN). The CPU B in the Vax Cluster will assure the back-up on CPU A or C, respectively performing the following functions:

- telemetry, telecommand, ranging and tracking, station monitoring and control, attitude reconstruction and flight dynamics,

- scientific data processing, observation scheduling, data archiving, recording of data on optical disks and on magnetic tapes.

The seven interchangeable workstations are devoted to:

- spacecraft and station monitoring & control,

- scientific processing and data presentation,

- satellite and station mimics.

For graphic treatment or image processing of WFC data, suitable hardware is foreseen. The choice is determined by the balance between the use of RISC workstations and an array processor attached to the CPU C. The array processor could greatly improve the processing speed of some applications, but it could overload the local network. On the other hand, the alternative solution allows for the displacement of a part of the scientific processing in local workstations resulting in a slower computation but offering a high operational flexibility.

An Optical Disk unit will permit the definitive storage of all the data collected and to be transferred to the SAX Scientific Data Center.

Other hardware components of the SAX-OCC are:

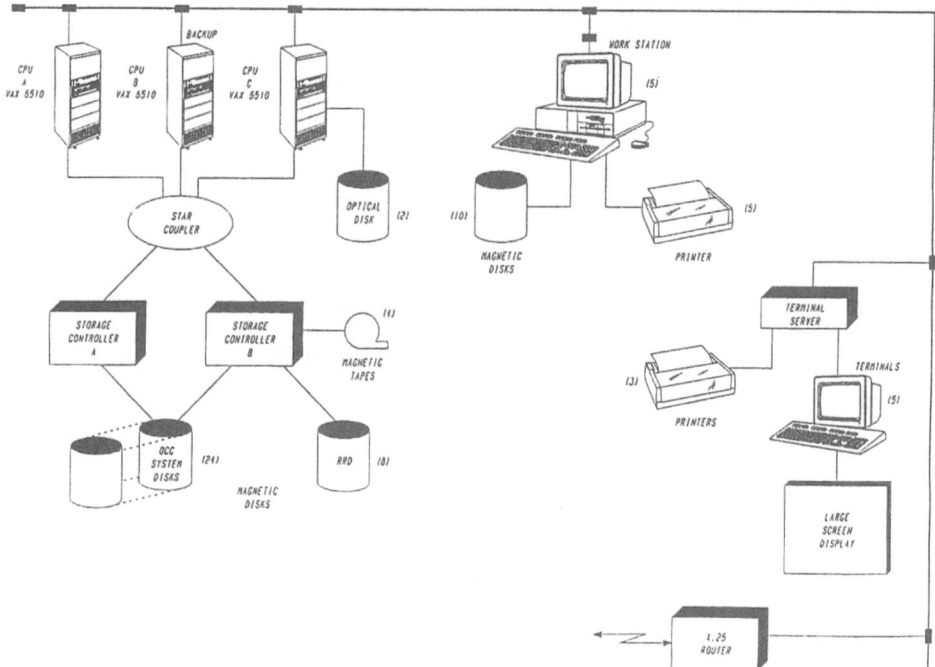

Figure 3. OCC Architecture

- three DEC X25 Routers and two X25 switch nodes, which can handle 12 lines,

- two storage controllers allowing for the management of disks and tapes in the VAX Cluster and the shadowing, in a transparent way of application software.

The TT&C Station

The radio link between the satellite and the ground is assured by the Telemetry Tracking and Command station (TT&C) in the S-band (2 GHz) with several system operating modes, following the Packet Telemetry Standard (ref.4). Fig.4 shows a functional block diagram of the TT&C Station. It performs the following functions at every satellite passage:

- A telemetry function, consisting in the acquisition of all the original source packets. After the acquisition the down-linked data will be relayed to the OCC for further processing. During the operational phase, housekeeping and scientific data are transmitted in High Bit Rate mode over a double physical channel, where the first one supports real time data transmission at a rate of 131 Kbps and the second one is dedicated to the stored data dump at a 917 Kbps rate.

- A telecommand function serves to load all the scheduled instrument operative plans and the related configurations, as well as any command useful for the in orbit spacecraft control. Before the station passage, the commands are generated by the OCC and relayed to the remote Station Computer, where they are formatted according to the ESA Standard, and then stored until the up-linking is requested.

- The tracking function regards satellite acquisition from the beginning of each passage, when the spacecraft rises on the horizon, and its keeping during all the

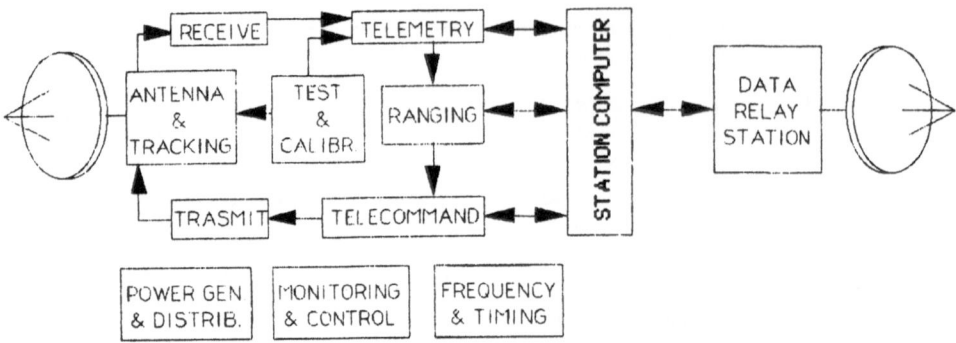

Figure 4. TT&C Station Functional Block Diagram

visibility period. In the operational phase, the tracking function is carried out in programmed mode under the guidance of an antenna control unit which receives the antenna pointing angles directly from the OCC.

THE OCC SOFTWARE SYSTEMS

The need for a high flexibility in the OCC configuration leads to an open software architecture with a number of principal modules. These are connected by a suitable intertask communication system which implements both standard and scientific functions. The following modules are the more important among the standard ones:

- Telemetry subsystem: devoted to the acquisition of the satellite telemetry according to the ESA Packet Telemetry Standard. The communication link contains 8 telemetry virtual channels with a maximum packet length of 8872 bits. The acquisition task must acquire telemetry source packets at a maximum rate of 256 Kbps from a long distance X25 link through the INTELSAT communication system. Each acquired packet will be distributed to the related buffer, on an application identifier basis. After a pre-processing phase, the telemetry is archived on disk for future uses.

- Telecommand subsystem: in charge of the basic function of satellite commanding according to ESA standard. As a first step the specific command is requested, either for immediate issue or for automatic, timed forwarding; just before coding, the commands are checked for compatibility with current or estimated satellite status. As a second step, the commands are coded properly as a string of bits and sent to the TT&C equipment for uplink to the satellite. Uplinked and verified commands are filed on disks.

- Ranging subsystem: dedicated to forwarding ranging requests to the proper station equipment, either on user request or on a time schedule basis.

- Database Maintenance: in order to guarantee maximum flexibility, two kinds of databases are present in the classical part of the SAX-OCC. The first, the satellite database, contains information about satellite features and telemetry data presentation pages. The other, the system database, is used to optimize system performance and user interface.

- Man Machine Interface subsystem: to be developed with an extensive use of windows, icons, menus and logical keys. Standard XWindow environment has been adopted to guarantee software portability.

- Flight Determination subsystem: SAX orbit determination is achieved making use of a determined orbit and historical data, the orbit propagation is also performed to schedule the observations and the station passes; a trajectory optimization is finally calculated to evaluate requirements for satellite manoeuvres.

- Attitude Reconstruction and Control subsystem: this dedicated task serves to evaluate fine spacecraft attitude for scientific purposes. This software computes also the feasibility, the satellite attitude and the related manoeuvres required to reach, given the pointing requests for the scientific instruments from the observation scheduling function.

SCIENTIFIC FUNCTIONALITIES AT THE SAX CONTROL CENTER

Due to the number of instruments in the SAX payload, mentioned in the introduction, and the technical constraints imposed by a maximum data generation capability of 100 Kbps on board, different operative modes for instrument data transmission are foreseen.

Every SAX Narrow Field Instrument has its operative modes characterized by a further distinction in submodes which leads to a variable field size for an information. The main common operative modes for the NFI are listed below:

- direct modes, in which event data is directly transmitted,

- indirect modes, the event data is transmitted after an on-board accumulation,

- stand-by mode, in which the instrument is switched-on but is configured to produce no scientific data.

The operative modes for each Wide Field Camera are:

- Normal and High Time Resolution mode, used during normal scientific observation,

- Diagnostic and Background Analysis mode, used for diagnostic reasons.

The SAX experiments are defined in terms of Observing Periods, during which satellite operations are devoted to a unique pointing, in order to perform routine calibrations, electronic in-flight tests or other non-scientific operations. An Observing Period can last a maximum of 10**5 s, and has an average length of 10**4 s. A given Observing Period consists of consecutive Observations determined by a sequence of changes either in operative mode, in attitude or in the experiment configuration, including relevant changes in the On-Board Data Handling processor. The average length of an Observation is assessed to be about 3000 s.

The concepts of Observing Period and Observation defined above are considered at the SAX-OCC in order to perform the Observation Scheduling task, using as input the requests of the SAX Scientific Data Center (SDC) and the Star Catalogue, generating observing plans and then the Payload Operation Plan (POP). Moreover these concepts are so used in selecting data for a given instrument so that the following functions are accomplished:

- achieve the scientific Quick Look devoted to analyzing, monitoring and evaluating the functional performance of the NFI and the scientific results of the WFCs;

- archive data in almost real time in the Reformatted Raw Data (RRD) archive, a data structure made up of an on-line part on magnetic disks and of a collection of optical disks;

- produce the Final Observation Tapes (FOT) containing the instrument data and all the information useful for the scientific analysis related to a given Observing Period.

Observation Scheduling

Observation Scheduling is the subsystem responsible for preparation of an observation plan, residing at the OCC. This consists of a sequence of operations to be performed in order to execute the experiments, as requested by the Scientific Committee of the SAX-SDC.

Spacecraft and payload activities have to be planned far in advance of the execution, and the complexity of the observation scheduling process has suggested a decomposition of the problem across different time scales. Subsequently, two main processes are envisaged, taking all possible constraints into account:

- Long Term Scheduling, which analyses the feasibility of an observation within a relatively long period of time and which orders observations in an optimal manner. The main input is the Mission Plan prepared by the SAX-SDC Scientific Committee, while the output is the Long Term Plan, which specifies a list of verified activities and their sequential relationships. This plan could be reprocessed at any moment in order to consider all the possible contingences.

- Short Term Scheduling, is a refinement of the previous phase, and serves to prepare detailed schedules for few days, specifing the exact start and end of an Observation. The main input is the Long Term Plan mentioned above, while the output is the Nominal Pointing file contained in the POP; these files contain a list of computer commands to be loaded on board concerning, respectively, the satellite attitude and the payload setting and operation.

In addition, to help the user, the Observation Scheduling System provides an interactive support tool. This produces maps of observability for NFI and WFC instruments, computes the period of visibility for a selected target and its observation efficiency.

The Scientific Quick Look

A scientific Quick Look at the SAX-OCC is required in order to allow personnel responsible for the scientific payload to take decisions regarding the progress of the experiments. This scientific processing is made up of two procedures: Class3 and Class4.

In a first instance, Class3 processing will perform the monitoring, analysis and evaluation of the instrumental health and scientific results in almost real time, within 2 or 3 orbits. During this phase, no more than the 20-30% of the total amount of data is to be processed. Class3 Processing is characterized by some general functions, common to all the instruments, and others which are instrument specific.

A first class of functions, named Primary Functions, will perform primitive accumulation and spectra integrations using the scientific data stored in the RRD structure as input. The output consists of both monodimensional (i.e. histograms, time profiles, energy and position spectra, etc.) and bidimensional arrays (i.e. images, pseudo-images). These output structures will be saved in a scratch work area to be used as input by other classes of functions or displayed on user request.

The successive class of functions, the Manipulation Functions, performs data rearrangements (e.g. smoothing, operations on calibration lines, rebinnings, background extraction) using as input the output files stored in the aforementioned scratch work area or data in memory. Their output could either lie in a file for further manipulation, or be immediately displayed.

The last class of functions, the Display and Monitoring Functions, deals with data presentation, plus other specialized tasks as joint analysis of scientific and HK data and will allows plot or display graphs and images.

Collections of functions contained in the above classes will automatically run in procedures using a set of default parameters in order to make the Class3 Quick Look faster.

Another kind of scientific Quick Look is the Class4 Processing. Its main task is to complete a similar processing of Class3 Quick Look over a long time period (up to one month). It is also concerned with trend analysis performed both on HK and on calibration data. The input data will be either the RRD archive, both on line and dumped on magnetic disk from optical disks, or files produced by the Class3 and saved in the scratch area.

The Reformatted Raw Data Archive Structure

As mentioned before, 8 virtual channels are allotted to each of the two physical channels, as allowed by the Packet Telemetry ESA Standard. Such virtual channels are assigned to each of the following data types: On Board Time (OBT), technical Housekeepings (HK), WFC1 instrument, WFC2 instrument, PDS instrument, LEC/S instrument, MEC/S instruments, and HP-GSPC instrument.

The expected data volume can reach the value of about 60 MBytes for each orbit lasting an average time of 96 minutes.

The formats of the scientific source packets will be instrument specific and will depend on its operating conditions. The packets generated under direct operating modes contain information related to a single photon (i.e. time, energy, xy position and burst length) in a series of event data fields. The packets generated under indirect operating modes contain a series of histograms representing the result of some on-board accumulations.

A field in the source packet header should allow the unique and direct identification of each telemetry packet layout. No additional on-line processing is foreseen for the scientific packets before they are stored in the RRD archive. This archive consists essentially of three parts:

- a RRD database where the information regarding each Observation is stored and used for a fast retrieval of the requested data,

- an on-line file system addressed by the RRD database, where the files concerning the Observations are stored for no more than 1 week,

- an off-line part of the archive residing on optical disks, where the data, organized according to Observing Period, are copied. A copy of these optical disks will stay at the SAX-OCC for a month and will then be sent to the SAX-SDC.

Some collateral files and tables are present too.

The scientific archive design is aimed at a fast, flexible storage and retrieval, keeping the data formats of the telemetry source packets, and the capability of copying the on-line part of the RRD onto the optical disk storage with semplicity in an automatic way. Because of the different nature of the data coming from 6 separate instruments, separate data sets are designed.

The sum of the scientific needs leads to a net time of 20 minutes to be required (during each orbit) for insertion of all the field packets in the raw data files and for a net time of 15 minutes for the transfer of older data to optical disks. It is necessary also to keep track of the Observations within each Observing Period in order to organize both optical disk and FOT layouts.

The packet storage and retrieval are mainly based on the specification of time limits and/or a list of observations. The Data Base Management System is in charge of such operations, at a file level, while suitable application program will perform further selection criteria for data within each packet data field (e.g. position limits for images, energy limits for spectra). At any time the following information at least will be available:

- the list of all the Observing Period,

- the list of all the Observations pertaining to each observing period for a given instrument,

- whether data pertaining to a given Observing Period or Observation resides on magnetic or on optical disk.

The RRD Database component of the archive is made up of:

- the Observing Period Table, referring to information about the pointing, instruments configuration and the storage of data, using a temporal primary key;

- the 6 'Instrument' Observation Data, carrying information about WFC1, WFC2, HP-GSPC, PDS, LEC/S, MEC/S observations such as operative mode, Observation start and end time, on/off line flag, or instrument unit.

The use of the primary temporal key will permit the storage and retrieval information with temporal relationship among the different tables of the Database. Further

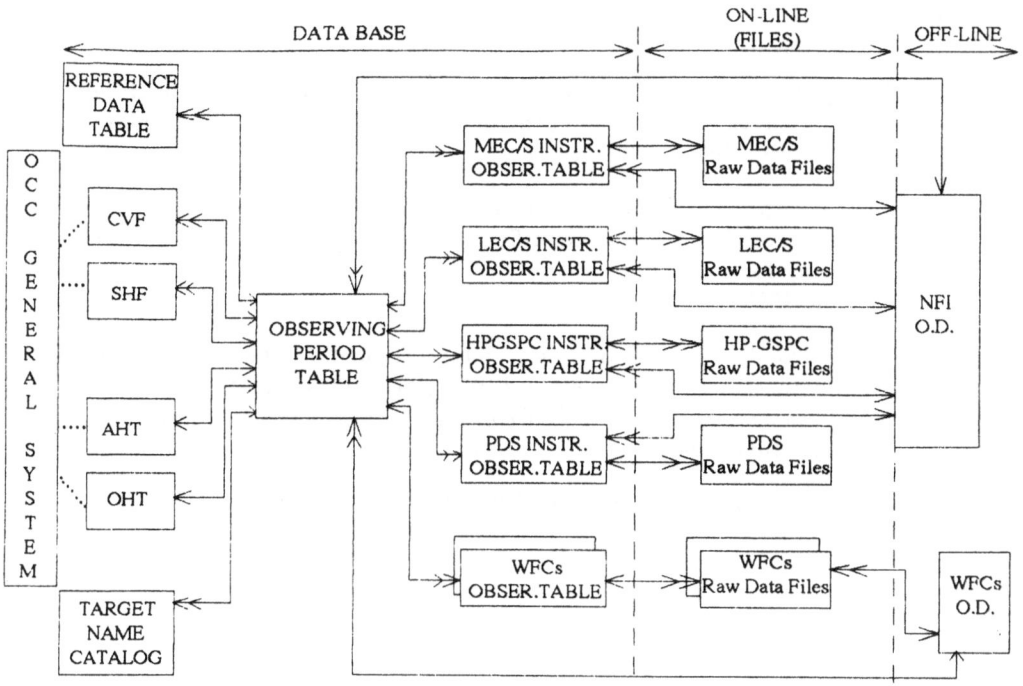

Figure 5. SAX-RRD Logical Structure

details of these RRD Database Tables are given in ref.3.

The On-line part of the RRD archive will lie on magnetic disk. It is made up of 6 data sets, one for each instrument, arranged in files containing the data fields of the raw data packets of a single Observation.

The Off-line part of the RRD archive is made up of optical disks where each is organized so that data pertaining to a single Observing Period is not splitted across different disks.

Other collateral files and tables are related to the database: the Target Name Catalogue, the Attitude (AHT) and Orbital (OHT) Tables related to reconstructed attitude and orbital information relevant to the scientific analysis, the Short History File (SHF) and the Command Verified File (CVF) respectively containing the HK from Virtual Channel 1 and a set of parameters which report the actual execution of commands, and the Reference Data Table (RDT) with some values of scientific variables and/or HKs for comparative use.

The choice of a relational Data Base Management System allows for the SQL standard query language to be used and the dynamical redefinition of relationships to be established without introducing changes in the archive architecture.

The performance of such a structure is evaluated in a series of trials, either on storing and on retrieval capability employing the target hardware and software systems. The results are encouraging because storage times are far below the requirements: about 9 minutes of CPU time and about 20 minutes of elapsed time for orbital data of 44 MBytes, while the retrieval time varies from a few hundreths of a second in the case of a single table query to tenths of second in the case of a complete database query, and to a few seconds when the data extraction from on-line files is also required.

REFERENCES

1. C. Butler and L. Scarsi, The SAX mission, in: Proc. COSPAR: 'Recent Results and Perspective Instrument Development in X and Gamma Ray Astronomy', The Hague, 2-6 July 1990, (in the press)

2. M. Manca, A. Apolloni, B. Negri and M. Trifoglio, The architecture of the SAX-OCC: a Control Centre for a scientific satellite, in: Proc. ESA Symp.: 'Ground data systems for spacecraft control' (ESA Sp-308, Oct.'90), Darmstadt, FRG, (26-29 June 1990)

3. F. D'Alessandro, C. Coletta, B. Negri and M. Trifoglio, The design of the SAX scientific data structure for a quick look at the OCC, in: Proc. ESA Symp.: 'Ground data systems for spacecraft control' (ESA Sp-308, Oct.'90), Darmstadt, FRG, (26-29 June 1990)

4. 'Packet Telemetry Standard' (ESA PSS-04-106) Issue 1, (1988)

CONTRIBUTORS

INDEX

Neural network, 103, 271, 276

Operating Systems
  MS-DOS, 10
  ULTRIX, 142
  UNIX, 10, 18, 62, 142, 305, 329
  VMS, 10, 59, 142, 286, 301

Point spread function, 90, 91, 173
Power density spectrum, 57
PSPC, 117
Pulsar analysis, 55, 225, 229, 290, 296,
    320

Query languages
  EQUEL, 131
  ESQL, 131
  QBE, 6
  QBF, 131
  QUEL, 131
  SQL, 6, 131, 303

ROSAT, 22, 115, 121, 131, 141, 145, 171,
    336

SAX, 330, 336, 341, 355
Software packages
  AGL-ASTRONET, 19
  COMPASS, 217, 229
  EXSAS, 141, 142
  IGORE, 286, 298
  PROS, 145, 150
  SASS, 153
  STSDAS, 145, 327
  XRONOS, 59
Solar system, 79
Source searcing, 175
Spectral analysis, 144, 189, 198, 251,
    290, 293, 299, 327, 330, 331, 342
Spectral classification, 66, 68, 144, 300
Spectral fitting, 66, 151
Spectrometer
  γ-ray, 291
  GSPC, 337, 346
SPECTRUM-XΓ, 329
STARLINK, 164, 166
Stochastic inversion, 84
Superresolution, 92

Telemetry, 77, 121, 282, 305,
    311, 333, 351
Telescopes
  Compton, 225
  extreme ultraviolet, 321
  γ-ray, 51, 303, 311
  optical, 76
  X-ray, 117, 329, 330
Texture analysis, 41, 42
Time analysis, 51, 144, 150

UHURU, 116

Undersampling, 92

Wide field camera, 161, 171, 341,
    337, 355
Window systems
  NSF, 16
  X11, 16, 147

XMM, 117

X-ray, 24, 51, 121, 167, 185,
    335, 341, 355